Pathogenesis of Systemic Lupus Erythematosus

Alberta Hoi
Editor

Pathogenesis of Systemic Lupus Erythematosus

Insights from Translational Research

Springer

Editor
Alberta Hoi
Department of Medicine
Monash University
Clayton, VIC, Australia

ISBN 978-3-030-85160-6   ISBN 978-3-030-85161-3 (eBook)
https://doi.org/10.1007/978-3-030-85161-3

This Springer imprint is published by the registered company Springer Nature Switzerland AG
The registered company address is: Gewerbestrasse 11, 6330 Cham, Switzerland

Contents

Abbreviations

Ab	Antibody
ACR	American College of Rheumatology
ADA	Anti drug antibody
ADAb	Anti-drug antibodies
ADAM10	A disintegrin and metalloproteinase 10
ADCC	Antibody-dependent cellular cytotoxicity
AE	Adverse event
AICDA	Activation-induced cytidine deaminase
AID	Activation induced deaminase
AIRE	Autoimmune regulator
Akt	Protein kinase B
ALP	Alkaline phosphatase
AMPAR	α-amino-3-hydroxy-5-methyl-4-isoxazolepropionic acid receptor
ANA	Anti-nuclear antibody
ANOVA	Analysis of variance
AP-1	Activator Protein 1
APC	Antigen presenting cells
APL	Antiphospholipid
APLC	Asia Pacific Lupus Collaboration
APRIL	A proliferation-inducing ligand
ATAC	Assay for transposase-accessible chromatin
autoAbs	Autoantibodies
β2GP1	Beta-2 glycoprotein 1
BAFF	B-cell activating factor from the tumor necrosis factor family
BAFF-R	BAFF receptor
BATF	Basic leucine zipper ATF-like transcription factor
BBB	Blood brain barrier
Bcl	B-cell lymphoma
BCMA	B-cell maturation antigen
BCR	B-cell receptor
BDCA	Blood Dendritic Cell Antigen
BICLA	BILAG-based Combined Lupus Assessment
BILAG	The British Isles Lupus Assessment Group
BLIMP1	B-lymphocyte-induced maturation protein 1
Blk	B-cell lymphocyte kinase

BLyS	B lymphocyte stimulator
BM	Bone marrow
BMD	Bone mineral density
BMDC	Bone marrow derived dendritic cell
BMP6	Bone morphogenetic protein 6
BRAVE	Study of Baricitinib in Participants With Systemic Lupus Erythematosus
C/EBPδ	CCAAT/enhancer-binding protein delta
C3	complement component 3
C4	complement component 4
CANDLE	Chronic atypical neutrophilic dermatosis with lipodystrophy and elevated temperatures
CAR	Chimeric antigen receptors
CARD9	Caspase recruitment domain-containing protein
CAR-T/CAR-tregs	Chimeric antigen receptors regulatory T-cells
CB	Cannabinoid receptor
CCL	C-C motif chemokine ligands
CD	Cluster of differentiation
CDC	Complement-dependent cytotoxicity
c-fms	Colony-stimulating factor 1 receptor (Feline McDonough Sarcoma)
cGAS	Cytosolic GAMP synthase
CI	Confidence interval
CLASI	Cutaneous Lupus Disease Area and Severity Index
CLE	Cutaneous lupus erythematosus
CNS	Central nervous system
CNV	Copy number variations
COX-2	Cyclo-oxygenase-2
CpG	Cytosine phosphate guanine
CS	Class switched
CSF	Colony-stimulating factor
CTLA	Cytotoxic T-lymphocyte-associated protein
CTX-1	C-terminal telopeptide
CVID	Common variable immunodeficiency
CXCL	Chemokine (C-X-C motif) ligand
CXCR	C-X-C chemokine receptor
D/d	Day
DAI	DNA-dependent activator of IFN-regulatory factor
DB	Double blind
DC	Dendritic cell
DED	Dry eye disease
DN	Double negative
DNA	Deoxyribonucleic acid
DNBS	Dinitrobenzene sulfonic acid
DRESS	Drug reaction with eosinophilia and systemic symptoms
dsDNA	Double stranded DNA

ECLAM	European consensus lupus activity measurement
EMP6	Epithelial membrane protein 3
ENA	Extractable nuclear antigens
ERK	Extracellular signal-regulated kinase
ETS	E-twenty-six (E26)
EULAR	European League Against Rheumatism
FACIT-fatigue	Functional Assessment of Chronic Illness Therapy-Fatigue
FcR	Fc Receptor
FCRL	Fc receptor like
FCγR	Fc gamma receptor
FDA	Food and Drug Administration
FDC	Follicular dendritic cells
FDR	False-discovery rate
Flt3	FMS-like tyrosine kinase 3
FOXP3	Forkhead box P3
FVIII	Factor VIII
Fyk	A Src family tyrosine-protein kinase, encoded by the FYN gene
GAD	Glutamic acid decarboxylase
GAMP	Guanosine monophosphate–adenosine monophosphate
GBM	Glomerular basement membrane
GC	Germinal centres
GC	Glucocorticoid
G-GSF	Granulocyte colony-stimulating facto
GILZ	Glucocorticoid-induced leucine zipper
GILZ-Tg	Glucocorticoid-induced leucin zipper transgenic
GIOP	Glucocorticoid-induced osteoporosis
GM-CSF	Granulocyte-macrophage colony-stimulating factor
GR	Glucocorticoid receptor
GRB2	Growth factor receptor-bound protein 2
GRE	Glucocorticoid response elements
GWAS	Genome-wide association studies
HC	Heathy controls
HCQ	Hydroxychloroquine
hFOB	Human foetal osteoblasts
HRQOL	Health Related Quality of Life (and Well Being)
IC	Immune complex
ICOS	Inducible cell costimulatory
IDO	Indoleamine 2,3-dioxygenase
IFN	Interferon
IFNAR	IFN alpha receptor
IFNGR	IFN gamma receptor
IFNLR	IFN lambda receptor
Ig	Immunoglobulin
IgG1κ	Immunoglobulin G1 kinase

IgG1	Immunoglobin type G1
IKK_	IκB kinase __
Ikzf2	IKAROS Family Zinc Finger 2
IL	Interleukin
IL-1R	Interleukin 1 receptor
ILT	Ig like transcript
IP	Interferon gamma-induced protein 10
IPEX	Immune dysregulation polyendocrinopathy enteropathy X-linked
IRAK	Interleukin-1 receptor associated kinase
IRF	IFN regulatory factor
ISG	IFN-stimulated genes
ISGF	IFN-stimulated gene factor
ISRE	IFN-stimulated response elements
ITAM	Immunoreceptor tyrosine-based activation motif
ITIM	Immunoreceptor tyrosine-based inhibitory motifs
ITT	Intention-to-treat
IV	Intravenous
JAK	Janus kinase
LAG-3	Lymphocyte activating 3
Lck	Lymphocyte-specific protein tyrosine kinase
LINE	Long interspersed nuclear element type
LLDAS	Lupus Low Disease Activity State
LN	Lupus nephritis
LPS	Lipopolysaccharide
LTE	Long term extension
Lyn	Lck/Yes novel tyrosine kinase
mAb	Monoclonal antibody
MAPK	Mitogen-activated protein kinase
MAVS	Mitochondrial antiviral signalling protein
MCP-1	Monocyte chemoattractant protein 1
MDA	Melanoma differentiation-associated gene
MHC	Major histocompatibility class
MIG	Monokine induced by gamma
MIP-1α	Macrophage inflammatory protein 1 alpha
miRNA	MicroRNA
Mod	Moderate
mRNA	Messenger RNA
MSC	Mesenchymal stem cells
mtecs	Medullary thymic epithelial cells
mTOR2	Mammalian target of rapamycin 2
MyD88	Myeloid differentiation primary response 88
NCS	Non-CS
NEJM	New England Journal of Medicine
NET	Neutrophil extracellular trap
NFAT	Nuclear factor of an activated T-cell

NF-κB	Nuclear factor kappa-light-chain-enhancer of activated B-cells
nGRE	Negative glucocorticoid response elements
NK cell	Natural killer (cells)
NMDAR	N-methyl-D-aspartate receptor
NPSLE	Neuro psychiatric SLE
NR2	NR2 subunit of N-methyl D-aspartate receptor
NSPA	Neuronal surface P antigen
O1	Primary outcome
O2	Key secondary outcomes
ODN-CpG/CpG-ODN	CpG oligodeoxynucleotides
OPG	Osteoprotegerin
P1NP	Propeptide of type 1 N-terminal procollagen
PAMP	Pathogen-associated molecular patterns
PAX5	Paired box 5
PB	plasmablast
PBMC	Peripheral blood mononuclear cells
PC	Plasma cell
PD	Pharmacodynamic
pDC	Plasmacytoid dendritic cells
PGA	Physician Global Assessment
PI3K	phosphoinositide 3-kinase
PK	Pharmacokinetic
PLC	Phospholipase C
PNL	Prednisolone
Popn	Population
PPARγ2	Peroxisome proliferator-activated receptor gamma
PRDM1	Positive regulatory domain zinc finger protein 1
PRR	Pattern recognition receptors
pSS	Primary Sjogren's syndrome
Pt	Patient
PTEN	Phosphatase and tensin homolog
PTPB	Protein-tyrosine phosphatase 1B
pTregs	Peripheral Tregs
PY	Patient years
q4w	4 weekly
RANK	Receptor activator of nuclear kappa B
RANKL	Receptor activator of nuclear kappa B ligand
RANTES	Regulated on Activation, Normal T Expressed and Secreted
RAR	Retinoic acid receptor
RBP	RNA binding protein
RCT	Randomised control trial
RibP	Ribosomal P
RIG	Retinoic acid-inducible gene
RNA	Ribonucleic acid
RNase	Ribonuclease

RNA-seq	RNA sequencing
RNP	Ribonucleoprotein
ROR	Related orphan receptor
RORγt	RAR-related orphan receptor gamma
rtPCR	Real-time polymerase chain reaction
S1PR4	Sphingosine-1-phosphate-receptor-4
SAE	Serious adverse events
SC	Subcuntaneous
scRNA-seq	Single-cell RNA-seq
SD	Standard deviation
SELENA-SLEDAI	Safety of Estrogens in Lupus Erythematosus National Assessment version of the SLEDAI
SFK	Src family kinases
Ship-1	SH2-containing inositol 5'-phosphatase 1
SLAM	Systemic Lupus Activity Measure
SLE	Systemic lupus erythematosus
SLEDAI-2K	SLE Disease Activity Index 2000
Sm	Smith
SNP	Single-nucleotide polymorphism
SRI	SLE Responder Index
SSc	Systemic sclerosis
Ssdna	Single stranded DNA
ssRNA	Single stranded RNA
sTACI	Soluble TACI
STAT	Signal transducer and activator of transcription
STING	Stimulator of IFN genes
SYK	Spleen tyrosine kinase
T2T	Treat-to-target
TACI	Transmembrane activator and cyclophilin ligand interactor
T-bet	T-box expressed in T-cells
TBK	TANK-binding kinase
TCR	T-Cell Receptor
TCR-T/TCR-Tregs	T-cell receptor t regulatory cells
T_{FH} cells	T follicular helper cells
Tgfbr	TGF beta receptor
TGFβ	Transforming growth factor beta
Th	T helper
TLR	Toll Like receptor
TNF	Tumour necrosis factor
TNFR	Tumour necrosis factor receptors
TNFRSF	Tumour necrosis factor receptor superfamily
TNFSF	Tumour necrosis factor ligand superfamily
Tr1	Type 1 regulatory
TRA	Tissue restricted antigen
TRACP	Tartrate-resistant acid phosphatase
TRAF	TNF receptor-associated factor

Treg(s)	Regulatory T-cells
TRIF	TIR-domain-containing adaptor-inducing IFNβ
TSC	Transforming growth factor beta-stimulated clone
tTreg	Thymic Tregs/Natural Tregs
TULIP	Treatment of Uncontrolled Lupus via the Interferon Pathway
TYK	Tyrosine kinase
URTI	Upper respiratory tract infection
USP	Ubiquitin-specific protease
UV	Ultraviolet
XBP1	X-box binding protein 1

List of Figures

List of Tables

Advances in Translational Science to Identify New Therapies for Systemic Lupus Erythematosus

1

Alberta Hoi, Fabien Vincent,
and Margaret L. Hibbs

Abstract

The scope of this textbook is to provide an overview of the latest translational research in the field of lupus pathogenesis, with particular emphasis on how these discoveries progress in parallel with therapeutic drug development. Systemic lupus erythematosus (SLE) is a multifaceted disease with a number of well-defined immune pathways that are dysregulated, resulting in an immune-mediated chronic inflammatory injury at target organs. As knowledge of these pathways evolves to provide opportunities for targeted drug therapy and lays the foundation for personalized medicine, clinicians and researchers need to keep up with the ever-expanding medical and scientific literature.

Clinically SLE is an incredibly heterogenous disease, characterised by chronic inflammation in one or more organs, such as kidney, skin, joints and the nervous system as well as serous membranes that line organs, particularly the pericardium and pleura, and less commonly, the peritoneum [1]. In addition to the direct immune mediated effects on organs and tissues, SLE can be associated with a number of morbidities such as accelerated atherosclerosis, with cardiovascular disease among the top causes of mortality in SLE patients [2].

Up to 70% of SLE patients develop some form of kidney disease, known as lupus nephritis (LN), which is a major cause of morbidity and mortality. LN results when ICs, formed between anti-dsDNA and DNA, deposit in the glomerular basement membrane of the kidney. This incites an inflammatory response, involving activation of the complement and Fc receptor pathways and recruitment of inflammatory cells to the kidney. This leads to their activation and amplification of the response with further cellular influx, the induction of mesangial cell proliferation and fibrotic processes, and the secretion of inflammatory mediators by both endogenous and infiltrating cells. The renal tissue becomes damaged from the effects of complement, reactive oxygen species and proteases released from neutrophils and

A. Hoi (✉) · F. Vincent
Rheumatology Research Group, School of Clinical Sciences At Monash Health, Centre for Inflammatory Diseases, Monash University, Clayton, VIC 3168, Australia
e-mail: alberta.hoi@monash.edu

F. Vincent
e-mail: fabien.vincent@monash.edu

A. Hoi
Department of Rheumatology, Monash Health, Clayton, VIC 3168, Australia

A. Hoi
Department of Rheumatology, Austin Health, Heidelberg, VIC 3084, Australia

M. L. Hibbs
Department of Immunology and Pathology, Central Clinical School, Alfred Research Alliance, Monash University, Melbourne, VIC 3004, Australia
e-mail: margaret.hibbs@monash.edu

A. Hoi (ed.), *Pathogenesis of Systemic Lupus Erythematosus*,
https://doi.org/10.1007/978-3-030-85161-3_1

macrophages [3] (Fig. 1.1). Many genes have been implicated in susceptibility to LN, with a large majority in immune and inflammatory markers [4]. Among many, these include genes that encode proteins in the MHC-II complex involved in antigen presentation [5–7], Fcγ receptors (FcγR) that clear immune complexes (ICs) [8, 9], proteins involved in leukocyte migration such as CD11b [10, 11], which is also a receptor for complement, transcription factors such as STAT4 that regulate T cell signalling [12, 13], and proteins that regulate the innate immune response such as toll-like receptors (TLRs) [14]. A number of urine biomarkers have recently emerged that are specific for renal involvement in SLE and many of these are proinflammatory molecules, cytokines and chemokines [15].

A recent study has utilised single-cell transcriptomics to investigate the immune cell environment in the kidneys of patients with LN comparing to healthy subjects [16]. The authors identified a diversity of immune cell populations including multiple subsets of B, T, NK and myeloid cells that demonstrated both proinflammatory and inflammation-resolving responses. The B cell landscape included naïve B cells, activated B cells, B cells with a high interferon (IFN)-stimulated gene signature and plasma cells, while the myeloid cells found included inflammatory, phagocytic and inflammation-resolving M2-like CD16+ macrophages together with tissue-resident macrophages and conventional and plasmacytoid DCs. A clear IFN response was found in most cells as well as upregulation of chemokines involved in cell trafficking. While neutrophils were not observed, this may be reflective of the sample preparation used. Another recent study has utilised a similar approach to compare the transcriptome of skin and renal biopsies from SLE patients, specifically looking at kidney tubular cells and keratinocytes respectively, showing a high type I IFN response signature in both cell types [17]. In addition, high IFN and fibrosis signatures in tubular cells were associated with a failure to respond to treatment. Collectively, these studies highlight the complexity of LN, providing some answers to why this disease is so difficult to treat; however, these types of methods may lead to a more personalised approach to treatment.

Many patients with SLE develop neuropsychiatric symptoms within the first year of diagnosis, commonly referred to as neuropsychiatric SLE (NPSLE) or central nervous system (CNS) SLE [18]. The symptoms can include headaches, cerebrovascular accidents, seizures, cranial nerve disorders, peripheral nervous system disorders (e.g. mononeuropathy, polyneuropathy, plexopathy), cognitive impairment, and psychiatric manifestations (e.g. anxiety, depression, psychosis). While highly prevalent in patients, it is probably the least understood of all of the clinical manifestations of the disease, and some symptoms, for example anxiety and depression, may not reflect true CNS SLE but be driven by the debilitating nature of the disease itself. Nonetheless, it is associated with a poorer quality of life in affected patients.

Neuropsychiatric events in SLE patients likely arise as a consequence of a breach of the blood–brain barrier, allowing access of autoreactive antibodies and inflammatory mediators, which results in neuroinflammation and vascular damage [19]. Histopathology studies of cerebral autopsy material to assess brain injury have revealed diffuse vasculopathy, vasculitis, microthrombi and infarcts in tissue from patients diagnosed with CNS SLE more frequently than in the brain tissue of SLE patients without this disease manifestation [20]. Interestingly, complement deposition was also observed in the cerebral vessels of all SLE patients regardless of CNS SLE diagnosis suggesting the presence of autoantibodies or ICs and providing support that the complement pathway may require a certain threshold to be reached before brain injury occurs. It is probably not surprising that numerous inflammatory cytokines and chemokines have been found to be significantly increased in the cerebrospinal fluid of patients with CNS SLE, including IL-6, IL-8 (CXCL8), IP-10 (CXCL10), RANTES (CCL5), MCP-1 (CCL2), MIG (CXCL9) and G-CSF [21–27].

One key cytokine that is commonly elevated in SLE is IFN-α, and interestingly, IFN-α therapy as a disease management strategy for patients with viral infections can induce neuropsychiatric side-effects [28]. It is known that levels of IFN-α

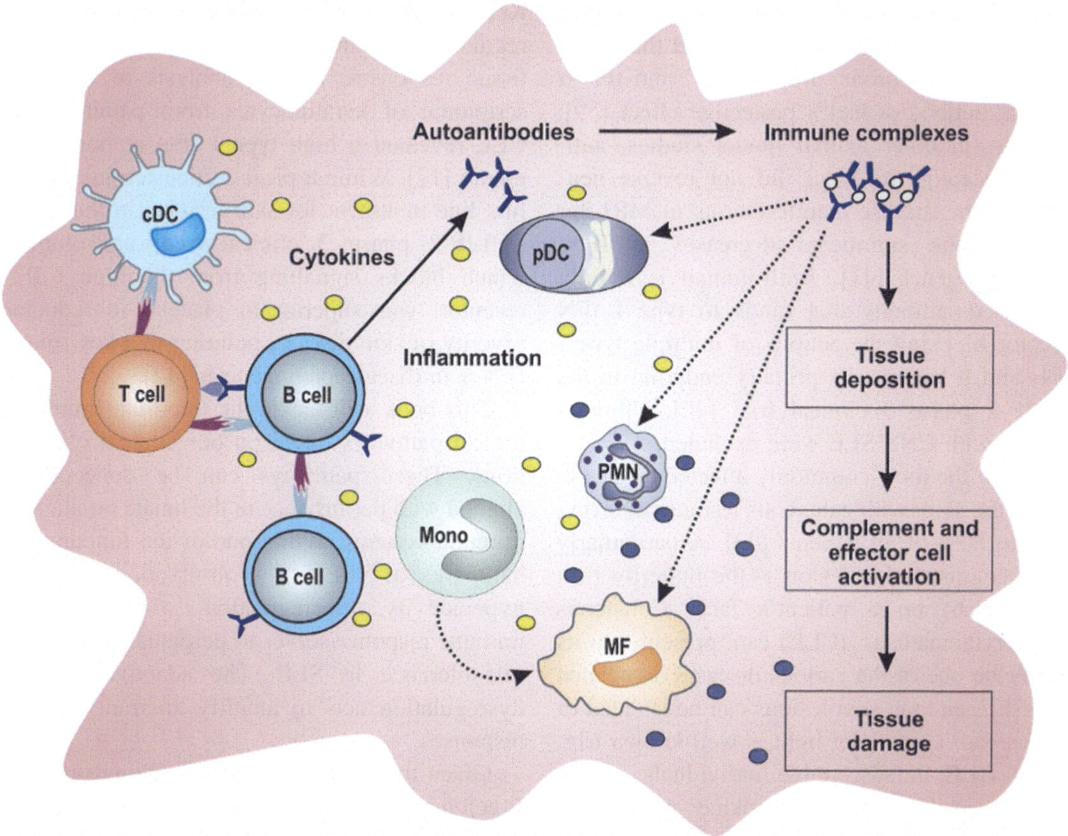

Fig. 1.1 Critical role of inflammation in organ damage in SLE. In susceptible individuals, there is a break down in immune tolerance leading to the production of autoreactive antibodies. This leads to the formation of soluble circulating ICs between DNA and anti-dsDNA, which are deposited in tissues, such as the glomerular basement membrane of the kidney, the basement membrane of skin or a blood vessel wall. IC deposition stimulates an inflammatory cascade involving chemokine and cytokine release and inflammatory cell recruitment and activation. This incites complement activation as well as the production of inflammatory mediators and proteases, which ultimately causes tissue damage. These processes underlie the development of LN and cutaneous lupus erythematosus but the exact mechanisms at play in the development of neuropsychiatric SLE are still to be defined. (Courtesy of Dr Fabien Vincent)

are elevated in the cerebrospinal fluid of patients with CNS SLE [29–31], providing further evidence that it may play a role in the neuropsychiatric disease manifestations. In addition, autoantibodies in the cerebrospinal fluid of CNS SLE patients can form ICs that potently stimulate production of IFN-α [32]. Neuronal and astrocyte damage is a feature of CNS SLE [33], providing a source of autoantigen and there are autoantibodies that are characteristic of CNS SLE such as those that cross-react with the NR2 glutamate receptor [34] and the neuronal surface P antigen [35, 36], as well anti-ribosomal P protein antibodies that are associated with SLE psychoses [37, 38]. A recent study has discovered that, in lupus-prone mice, IFN-α can cross the blood brain barrier to activate brain resident immune cells known as microglia, which can trigger destruction of neuronal synapses [39]. Activated microglia may therefore be a therapeutic target in CNS SLE, and a c-fms tyrosine kinase inhibitor that targets CSF-1 receptor signalling in microglia and macrophages led to an improvement in depression-like behaviour of MRL/lpr mice in addition to attenuating kidney disease [40]. These important findings have revealed how inflammation may underpin CNS SLE. Additional studies showed that treatment of lupus-prone mice with anti-IFN-α receptor antibodies had a protective effect [39]; however, in other animal model studies, anti-IFN-α receptor treatment did not reverse neuropsychiatric disease manifestations in MRL/lpr mice, despite significant decreases in IFN-stimulated genes [41]. Anifrolumab is a fully humanised antibody that binds to type I IFN receptor blocking the actions of multiple type I IFNs and it has met its primary endpoint in the TULIP-2 phase 3 clinical trial [42], although patients with CNS SLE were excluded.

One of the most commonly affected tissues in SLE is the skin, with cutaneous lesions occurring in up to 80% of all patients [43]. A particularly well-recognised skin lesion is the butterfly rash (malar rash) on a patient's face. Cutaneous lupus erythematosus (CLE) can present on its own or be one of the various diseases associated with SLE, and the skin lesions can be isolated or widespread. Ultraviolet light is well-known trigger of CLE in susceptible individuals, which induces cytokine and chemokine production, tissue inflammation and keratinocyte apoptosis providing a source of nuclear autoantigens in the skin. The role of B cells in the pathophysiology of CLE is not well understood and indeed, B cell depleting therapies have shown no efficacy [44].

It has been known for many years that long-term IFN-α treatment induces autoantibodies against the epidermis [45]. Plasmacytoid dendritic cells (pDC), which are the major producers of IFN-α, are normally located in primary lymphoid tissues, however they have been found to accumulate in CLE lesions [46], as well as in non-inflammatory skin of SLE patients [47]. Enhanced IFN signalling in the skin of SLE patients has been shown to induce a Th1-biased immune response and the recruitment of activated T cells to the skin lesion [48], and this can be amplified following ultraviolet light-induced injury [49]. In lupus-prone mice, autoimmune inflammation in the skin was found to be dependent on the chronic activation of pDCs by nucleic acids binding to TLR7 and 9 and their sustained production of IFN-α [50]. Keratinocytes can also produce IFN and IFN-regulated cytokines and chemokines to amplify tissue responses, and an analysis of the transcriptome of keratinocytes from patients with CLE revealed a high type I IFN response signature [17]. While topical corticosteroids are the first line treatment for skin lesions, in the recent TULIP-2 phase 3 clinical trial, anifrolumab, which blocks signalling from the type I IFN receptor, was superior to placebo in reducing severity of skin disease, pointing to a key role of IFN-α in disease pathogenesis [42].

This book will tackle the important immunological pathways based on our clinical observations. These pathways can be conceptually aligned with perturbance in the innate or adaptive immune systems. While one of the fundamental hallmarks of SLE is the observation of B cell hyperactivity, dysregulation of the innate immune response serves to perpetuate the loss of self-tolerance in SLE. The adaptive immune dysregulation acts to amplify aberrant immune responses.

Given the hallmark of SLE is the presence of B cell hyperactivity, we start by a comprehensive discussion on the pivotal roles B cells play in SLE pathogenesis. The pathogenic role of autoantibodies and their clinical relevance will be discussed. The findings of antinuclear antibodies and autoreactive B cells in SLE have led to broad efforts to study how B cells act as antibody-producing or as antigen-presenting cells that can prime autoreactive T cell activation. The maintenance and survival of B cells is dependent on B-lymphocyte stimulator (BLyS), which is otherwise known as B-cell activating factor of the TNF family (BAFF). The interplay between B and T cells, the dysregulation of signaling and cytokine network from the perspective of the B cell will be explained in Chap. 2.

In Chap. 3 we will review the current status of B cell-targeted therapies in SLE, including targeting strategy achieved via B cell depletion (CD20), through negative regulation of B cell antigen receptor (BCR) signalling (CD22, Fc gamma receptor (FcγRIIb), or via neutralizing one or several components of the BAFF/a proliferation-inducing ligand (APRIL) system.

Type I interferons are a large group of pleiotropic cytokines with potent anti-viral and immunoregulatory functions, produced by many cellular sources but most dominantly by plasmacytoid dendritic cells. In Chap. 4, the biology of type I interferons will be presented, and the signaling pathway of type I interferon is explained, highlighting its potential for therapeutic intervention. The downstream effects of type I interferons mediate most of the self-amplification processes, bridging both the innate and adaptive immune responses. The regulation of interferon production is a key pathogenic pathway recognized in SLE microarray studies.

While an interferon gene signature is present in the majority of SLE patients, its utility is still to be determined at the individual patient level. In Chap. 5 we summarize the potential avenues to target the IFN pathway therapeutically with or without using the interferon gene signature. Strategies such as neutralizing IFNs themselves or interfering with the production of IFN or its downstream effect involving the Janus Kinase/Signal Transducer and Activator of Transcription (JAK/STAT) signaling pathway. Further discussion on intracellular targets will also be covered in Chap. 8.

The activation of T cells and their subsequent orchestration of a range of immunological functions is an important mechanism in lupus pathogenesis. In Chap. 6, we explored a number of co-stimulatory and co-inhibitory molecules which serve as cell surface receptors and ligands that can positively or negatively regulate T cell function. These constitute important targets for immune modulation in potential therapeutic development in SLE.

A number of key T cell cytokines have also been shown to play important roles in amplifying lupus pathogenesis. In Chap. 7, we discussed how cytokine dysregulation contributes to each phase of disease development in SLE, from failure of restoration of self-tolerance, aberrant production of pathogenic autoantibodies, perpetuation of systemic or end organ inflammation, or promotion of local tissue damage. Some of these key cytokines that have emerged as important players in our current understanding of lupus pathogenesis could bring us closer to potential therapeutics in SLE.

Further to the development of monoclonal antibodies that directly inhibit cytokine actions, potential inhibition of intracellular targets has gained significant interest due to the much lower cost of production. The effects of cytokines can be ameliorated by small molecules that interfere with the downstream effects of cytokines. Among these, the Janus Kinase inhibitors have been best-studied. In Chap. 8, we discuss a number of intracellular targets, such as the calcium-calcineurin-Nuclear Factor of an Activated T cell (NFAT) and Mitogen-Activated Protein Kinase (MAPK) pathways, and B cell receptor-mediated pathways such as Bruton's tyrosine kinase (BTK), or spleen tyrosine kinase (SYK), and the Janus Kinase (JAK), ubiquitin-proteasome system (UPS) and the Mechanistic Target of Rapamycin (mTOR) pathways.

Finally, there are two anti-inflammatory pathways that are particularly relevant in immune homeostasis. T regulatory cells (Tregs) can be utilized to help restore immune tolerance and are amongst the latest focus of therapeutic development in SLE. They play an important role in preventing the induction of autoimmunity. Recent studies have revealed a paradigm shift in the utility of antigen-specific Tregs in bolstering the naturally occurring anti-inflammatory mechanisms, and suppressing proinflammatory autoreactive T cells thereby providing a strategy of protection against autoimmune disease.

Glucocorticoid (GC), as a potent natural anti-inflammatory hormone, can mediate its effects by recruiting histone deacetylase that serve to repress gene transcription. The harmful metabolic effects of GC are the major limitation of utilising GC as therapeutics in SLE, as it is often a double-edged sword in the arsenal of lupus treatment. Glucocorticoid-induced leucine zipper (GILZ) is a gene upregulated by glucocorticoid that can be a potential target for development of anti-inflammatory strategies. Preliminary evidence suggests that GILZ may be as protective as GC in its anti-inflammatory effects but may not evoke the metabolic pathways.

This book will provide an overview of the latest translational research in the field of lupus pathogenesis. Despite a lag in the development of therapeutics for SLE, the understanding of lupus pathogenesis has progressed, and the therapeutic horizon is looking promising for the next few years and coming decades. There are ongoing unmet clinical needs and an unwavering commitment of researchers and clinicians in the lupus community to develop novel targeted drug therapies. This book will serve as an aid for clinicians and researchers to keep up with the ever-expanding medical and scientific literature in these areas.

References

1. Kaul A, Gordon C, Crow MK, Touma Z, Urowitz MB, van Vollenhoven R et al (2016) Systemic lupus erythematosus. Nat Rev Dis Primers 2:16039
2. Liu Y, Kaplan MJ (2018) Cardiovascular disease in systemic lupus erythematosus: an update. Curr Opin Rheumatol 30(5):441–448
3. Davidson A (2016) What is damaging the kidney in lupus nephritis? Nat Rev Rheumatol 12(3):143–153
4. Song K, Liu L, Zhang X, Chen X (2020) An update on genetic susceptibility in lupus nephritis. Clin Immunol 210:108272. https://doi.org/10.1016/j.clim.2019.108272
5. Chung SA, Brown EE, Williams AH, Ramos PS, Berthier CC, Bhangale T et al (2014) Lupus nephritis susceptibility loci in women with systemic lupus erythematosus. J Am Soc Nephrol 25(12):2859–2870
6. Niu Z, Zhang P, Tong Y (2015) Value of HLA-DR genotype in systemic lupus erythematosus and lupus nephritis: a meta-analysis. Int J Rheum Dis 18(1):17–28
7. Xu R, Li Q, Liu R, Shen J, Li M, Zhao M et al (2017) Association analysis of the MHC in lupus nephritis. J Am Soc Nephrol 28(11):3383–3394
8. Salmon JE, Millard S, Schachter LA, Arnett FC, Ginzler EM, Gourley MF et al (1996) Fc gamma RIIA alleles are heritable risk factors for lupus nephritis in African Americans. J Clin Invest 97(5):1348–1354
9. Tsang ASMW, Nagelkerke SQ, Bultink IE, Geissler J, Tanck MW, Tacke CE et al (2016) Fc-gamma receptor polymorphisms differentially influence susceptibility to systemic lupus erythematosus and lupus nephritis. Rheumatology (Oxford) 55(5):939–948
10. Sanchez E, Nadig A, Richardson BC, Freedman BI, Kaufman KM, Kelly JA et al (2011) Phenotypic associations of genetic susceptibility loci in systemic lupus erythematosus. Ann Rheum Dis 70(10):1752–1757
11. Khan SQ, Khan I, Gupta V (2018) CD11b activity modulates pathogenesis of lupus nephritis. Front Med (Lausanne). 5:52
12. Taylor KE, Remmers EF, Lee AT, Ortmann WA, Plenge RM, Tian C, et al (2008) Specificity of the STAT4 genetic association for severe disease manifestations of systemic lupus erythematosus. PLoS Genet 4(5):e1000084. https://doi.org/10.1371/journal.pgen.1000084
13. Bolin K, Sandling JK, Zickert A, Jonsen A, Sjowall C, Svenungsson E, et al (2013) Association of STAT4 polymorphism with severe renal insufficiency in lupus nephritis. PLoS One 8(12):e84450. https://doi.org/10.1371/journal.pone.0084450
14. Conti F, Spinelli FR, Alessandri C, Valesini G (2011) Toll-like receptors and lupus nephritis. Clin Rev Allergy Immunol 40(3):192–198
15. Soliman S, Mohan C (2017) Lupus nephritis biomarkers. Clin Immunol 185:10–20
16. Arazi A, Rao DA, Berthier CC, Davidson A, Liu Y, Hoover PJ et al (2019) The immune cell landscape in kidneys of patients with lupus nephritis. Nat Immunol 20(7):902–914
17. Der E, Suryawanshi H, Morozov P, Kustagi M, Goilav B, Ranabothu S et al (2019) Tubular cell and keratinocyte single-cell transcriptomics applied to lupus nephritis reveal type I IFN and fibrosis relevant pathways. Nat Immunol 20(7):915–927
18. Hanly JG (2014) Diagnosis and management of neuropsychiatric SLE. Nat Rev Rheumatol 10(6):338–347
19. Schwartz N, Stock AD, Putterman C (2019) Neuropsychiatric lupus: new mechanistic insights and future treatment directions. Nat Rev Rheumatol 15(3):137–152
20. Cohen D, Rijnink EC, Nabuurs RJ, Steup-Beekman GM, Versluis MJ, Emmer BJ et al (2017) Brain histopathology in patients with systemic lupus erythematosus: identification of lesions associated with clinical neuropsychiatric lupus syndromes and the role of complement. Rheumatology (Oxford) 56(1):77–86
21. Hirohata S, Miyamoto T (1990) Elevated levels of interleukin-6 in cerebrospinal fluid from patients with systemic lupus erythematosus and central nervous system involvement. Arthritis Rheum 33(5):644–649
22. Trysberg E, Carlsten H, Tarkowski A (2000) Intrathecal cytokines in systemic lupus erythematosus with central nervous system involvement. Lupus 9(7):498–503
23. Okamoto H, Katsumata Y, Nishimura K, Kamatani N (2004) Interferon-inducible protein 10/CXCL10 is increased in the cerebrospinal fluid of patients with central nervous system lupus. Arthritis Rheum 50(11):3731–3732
24. Iikuni N, Okamoto H, Yoshio T, Sato E, Kamitsuji S, Iwamoto T et al (2006) Raised monocyte chemotactic protein-1 (MCP-1)/CCL2 in cerebrospinal fluid of

patients with neuropsychiatric lupus. Ann Rheum Dis 65(2):253–256
25. Fragoso-Loyo H, Richaud-Patin Y, Orozco-Narvaez A, Davila-Maldonado L, Atisha-Fregoso Y, Llorente L et al (2007) Interleukin-6 and chemokines in the neuropsychiatric manifestations of systemic lupus erythematosus. Arthritis Rheum 56(4):1242–1250
26. Fragoso-Loyo H, Atisha-Fregoso Y, Llorente L, Sanchez-Guerrero J (2013) Inflammatory profile in cerebrospinal fluid of patients with headache as a manifestation of neuropsychiatric systemic lupus erythematosus. Rheumatology (Oxford) 52 (12):2218–2222
27. Yoshio T, Okamoto H, Kurasawa K, Dei Y, Hirohata S, Minota S (2016) IL-6, IL-8, IP-10, MCP-1 and G-CSF are significantly increased in cerebrospinal fluid but not in sera of patients with central neuropsychiatric lupus erythematosus. Lupus 25 (9):997–1003
28. Raison CL, Demetrashvili M, Capuron L, Miller AH (2005) Neuropsychiatric adverse effects of interferon-alpha: recognition and management. CNS Drugs 19(2):105–123
29. Winfield JB, Shaw M, Silverman LM, Eisenberg RA, Wilson HA 3rd, Koffler D (1983) Intrathecal IgG synthesis and blood-brain barrier impairment in patients with systemic lupus erythematosus and central nervous system dysfunction. Am J Med 74 (5):837–844
30. Shiozawa S, Kuroki Y, Kim M, Hirohata S, Ogino T (1992) Interferon-alpha in lupus psychosis. Arthritis Rheum 35(4):417–422
31. Fragoso-Loyo H, Atisha-Fregoso Y, Nunez-Alvarez CA, Llorente L, Sanchez-Guerrero J (2012) Utility of interferon-alpha as a biomarker in central neuropsychiatric involvement in systemic lupus erythematosus. J Rheumatol 39(3):504–509
32. Santer DM, Yoshio T, Minota S, Moller T, Elkon KB (2009) Potent induction of IFN-alpha and chemokines by autoantibodies in the cerebrospinal fluid of patients with neuropsychiatric lupus. J Immunol 182 (2):1192–1201
33. Trysberg E, Nylen K, Rosengren LE, Tarkowski A (2003) Neuronal and astrocytic damage in systemic lupus erythematosus patients with central nervous system involvement. Arthritis Rheum 48(10):2881–2887
34. DeGiorgio LA, Konstantinov KN, Lee SC, Hardin JA, Volpe BT, Diamond B (2001) A subset of lupus anti-DNA antibodies cross-reacts with the NR2 glutamate receptor in systemic lupus erythematosus. Nat Med 7(11):1189–1193
35. Matus S, Burgos PV, Bravo-Zehnder M, Kraft R, Porras OH, Farias P et al (2007) Antiribosomal-P autoantibodies from psychiatric lupus target a novel neuronal surface protein causing calcium influx and apoptosis. J Exp Med 204(13):3221–3234
36. Hanly JG, Urowitz MB, Su L, Bae SC, Gordon C, Clarke A et al (2011) Autoantibodies as biomarkers for the prediction of neuropsychiatric events in systemic lupus erythematosus. Ann Rheum Dis 70 (10):1726–1732
37. Bonfa E, Golombek SJ, Kaufman LD, Skelly S, Weissbach H, Brot N et al (1987) Association between lupus psychosis and anti-ribosomal P protein antibodies. N Engl J Med 317(5):265–271
38. Yoshio T, Hirata D, Onda K, Nara H, Minota S (2005) Antiribosomal P protein antibodies in cerebrospinal fluid are associated with neuropsychiatric systemic lupus erythematosus. J Rheumatol 32 (1):34–39
39. Bialas AR, Presumey J, Das A, van der Poel CE, Lapchak PH, Mesin L et al (2017) Microglia-dependent synapse loss in type I interferon-mediated lupus. Nature 546(7659):539–543
40. Chalmers SA, Wen J, Shum J, Doerner J, Herlitz L, Putterman C (2017) CSF-1R inhibition attenuates renal and neuropsychiatric disease in murine lupus. Clin Immunol 185:100–108
41. Huang MW, Stock AD, Mike EV, Herlitz L, Kolbeck R, Putterman C (2019) Anti-IFNAR treatment does not reverse neuropsychiatric disease in MRL/lpr lupus mice. Lupus 28(13):1510–1523
42. Morand EF, Furie R, Tanaka Y, Bruce IN, Askanase AD, Richez C et al (2020) Trial of anifrolumab in active systemic lupus erythematosus. N Engl J Med 382(3):211–221
43. Wenzel J (2019) Cutaneous lupus erythematosus: new insights into pathogenesis and therapeutic strategies. Nat Rev Rheumatol 15(9):519–532
44. Vital EM, Wittmann M, Edward S, Md Yusof MY, MacIver H, Pease CT et al (2015) Brief report: responses to rituximab suggest B cell-independent inflammation in cutaneous systemic lupus erythematosus. Arthritis Rheumatol 67(6):1586–1591
45. Fleischmann M, Celerier P, Bernard P, Litoux P, Dreno B (1996) Long term interferon alpha therapy induces autoantibodies against epidermis. Dermatology 192(1):50–55
46. Farkas L, Beiske K, Lund-Johansen F, Brandtzaeg P, Jahnsen FL (2001) Plasmacytoid dendritic cells (natural interferon- alpha/beta-producing cells) accumulate in cutaneous lupus erythematosus lesions. Am J Pathol 159(1):237–243
47. Blomberg S, Eloranta ML, Cederblad B, Nordlin K, Alm GV, Ronnblom L (2001) Presence of cutaneous interferon-alpha producing cells in patients with systemic lupus erythematosus. Lupus 10(7):484–490
48. Wenzel J, Worenkamper E, Freutel S, Henze S, Haller O, Bieber T et al (2005) Enhanced type I interferon signalling promotes Th1-biased

inflammation in cutaneous lupus erythematosus. J Pathol 205(4):435–442
49. Meller S, Winterberg F, Gilliet M, Muller A, Lauceviciute I, Rieker J et al (2005) Ultraviolet radiation-induced injury, chemokines, and leukocyte recruitment: an amplification cycle triggering cutaneous lupus erythematosus. Arthritis Rheum 52(5):1504–1516
50. Guiducci C, Tripodo C, Gong M, Sangaletti S, Colombo MP, Coffman RL et al (2010) Autoimmune skin inflammation is dependent on plasmacytoid dendritic cell activation by nucleic acids via TLR7 and TLR9. J Exp Med 207(13):2931–2942

Hallmark of Systemic Lupus Erythematosus: Role of B Cell Hyperactivity

2

Fabien B. Vincent, William A. Figgett, and Margaret L. Hibbs

Abstract

B cells have been the focus of systemic lupus erythematosus (SLE) research for the past two decades since they are acknowledged to play a central role in disease pathogenesis. This has been somewhat fruitful, yielding the approval in 2011 by the Food and Drug Administration of the only (as of today) biological therapy in SLE, belimumab, a humanised monoclonal antibody (mAb) targeting the key B cell survival factor B cell-activating factor of the tumour necrosis factor (TNF) family (BAFF). However, given the modest clinical benefit of this biological agent, this condition is still burdened by a lack of new targeted therapies, without which many patients will continue to suffer from severe manifestations of the disease leading to irreversible organ damage, affecting their quality of life while increasing morbidity and mortality. This is in stark contrast to other autoimmune diseases, such as rheumatoid arthritis, where a revolution of targeted treatments, which started more than a decade ago, led to a new era of therapeutic management. Facing this unacceptable situation, a comprehensive understanding of the perturbed B cell biology operative in SLE is crucial in order to fine-tune and develop new targeted therapy alternatives in a precision medicine approach. In this chapter, we will discuss the pivotal roles of B cells in SLE pathogenesis, particularly changes that are observed in B cell subsets, the presence of B cell hyperactivity and loss of tolerance, the clinical relevance and pathogenic roles of autoantibodies, the operative signalling pathways, the interplay between B and T cells, the B cell cytokine network, and B cell-associated genetic factors in SLE.

F. B. Vincent (✉)
Rheumatology Research Group, School of Clinical Sciences at Monash Health, Centre for Inflammatory Diseases, Monash University, Clayton, VIC 3168, Australia
e-mail: fabien.vincent@monash.edu

W. A. Figgett
Department of Microbiology and Immunology at the Peter Doherty Institute for Infection and Immunity, University of Melbourne, Melbourne, VIC 3000, Australia
e-mail: w.figgett@garvan.org.au

W. A. Figgett
Kinghorn Centre for Clinical Genomics, Garvan Institute of Medical Research, Darlinghurst, NSW 2010, Australia

M. L. Hibbs
Department of Immunology and Pathology, Central Clinical School, Alfred Research Alliance, Monash University, Melbourne, VIC 3004, Australia
e-mail: margaret.hibbs@monash.edu

A. Hoi (ed.), *Pathogenesis of Systemic Lupus Erythematosus*,
https://doi.org/10.1007/978-3-030-85161-3_2

Introduction

Systemic lupus erythematosus, an autoimmune connective tissue disease, is a multifactorial condition influenced by many genetic, environmental and hormonal factors [1]. The disease predominantly affects women of childbearing age, with a female:male ratio of 9:1. The condition affects approximately 1:1000 people in Australia, and more than 5 million people worldwide. Prevalence of the disease is higher in some populations, including Indigenous Australians, African Americans, African Caribbeans, Hispanics, Asians, and North American Indians [2]. The disease is highly heterogenous, both at the clinical and biological levels. Indeed, virtually any organ can be variably affected, such as the lungs, heart, kidneys, brain, skin and joints, with the central nervous system and renal manifestations being the most devastating. The disease remains challenging to diagnose and manage. The mainstay of the therapeutic arsenal encompasses antimalarials, corticosteroids, and immunosuppressants which are characterised by limited effectiveness and significant toxicity. In the past 60 years, only one new drug has been approved for SLE, belimumab, the first and currently sole biological therapy in use in SLE. This drastically contrasts with some other autoimmune diseases, such as rheumatoid arthritis, where a new era of successful targeted therapy began in the mid-2000's, which now sees a proper arsenal of biologics for disease management. As a consequence of this lack of new targeted therapies, some SLE patients still suffer from severe manifestations, along with inexorable accrual of organ damage, a major cause of morbidity and mortality.

B cells are key players in SLE pathogenesis, and the successful phase 3 clinical trials of belimumab in SLE [3, 4] further confirm the pathogenic role of a crucial B cell maturation, differentiation and survival factor, known as B cell-activating factor of the tumour necrosis factor (TNF) family (BAFF) [5]. However, the clinical benefit of this agent, albeit significant, was modest in these trials. There is a pressing need for new targeted therapies for SLE patients, and considering the highly heterogenic profile of SLE, a precision medicine approach based on patient stratification for selection of the appropriate treatment founded on the dominant pathway at play seems attractive. In this chapter, we will comprehensively review the key B cell biology aspects that are perturbed in this condition, which is essential knowledge for the successful development of new targeted therapies, fine-tuning the use of existing ones, and development of patient stratification methods, without which a precision medicine approach could not be achieved.

B Cell Hyperactivity/Loss of Tolerance in SLE

It is widely accepted that B cells, as the source of autoantibodies (autoAbs), play an essential role in the development of SLE, and experimental model studies have demonstrated that genetic deletion of B cells in SLE-susceptible strains of mice such as MRL lpr [6, 7] and Lyn-deficient mice [8, 9] is protective against disease development. While tissue pathology is mediated by autoAbs deposition and the ensuing inflammatory cascade, this often results from enhanced B cell function or activity, although it is clear that augmented function of other cells such as T lymphocytes or dendritic cells (DCs) can also drive autoAbs production from B cells.

B cell hyperactivity, which is their excessive response to stimulation, is the cardinal feature of SLE. The origins of B cell hyperactivation are still not fully elucidated but likely involve many intersecting pathways. B cell hyperactivity can be demonstrated in several ways including enhanced survival, increased expression of antigen-presenting major histocompatibility class II (MHC-II) molecules as well as molecules involved in T cell co-stimulation such as CD40, CD80 and CD86, which can drive T cell activation. B cell hyperactivation can also be demonstrated by enhanced plasma cell (PC) maturation, augmented immunoglobulin

(Ig) class-switching and increased production of antibodies. While many cytokines can influence B cell responses, activated B cells can also produce cytokines to regulate immune responses, and this feature is augmented in autoimmune settings.

B cells can contribute to the development of SLE via numerous mechanisms (Fig. 2.1). The production of excess circulating B cell survival factors such as BAFF (*TNFSF13b*) and a proliferation-inducing ligand (APRIL, *TNFS13*) by innate cells, can induce autoreactive B cell survival, enhance their differentiation into PC and promote immunoglobulin class-switch recombination to generate pathogenic autoAbs isotypes (Fig. 2.1A). The production of autoAbs by autoreactive B cells can lead to the formation of pro-inflammatory immune complexes (ICs) which can then deposit in tissues and stimulate effector cells and processes that can lead to tissue damage (Fig. 2.1B). B cells can also function as antigen-presenting cells activating autoreactive T cells through antigen presentation and co-stimulation which then leads to augmented T cell effector activity and promotes the expression of T cell cytokines such as interleukin (IL)-17, interferon (IFN)-γ and GM-CSF (Fig. 2.1C). B cells can furthermore produce factors such as the pro-inflammatory cytokine IL-6 which can have effects on numerous cell types to drive extramedullary hematopoiesis, stimulate acute phase response proteins, promote inflammation and inhibit the generation of immune suppressive regulatory T cells (Treg) (Fig. 2.1D).

Over-expression of BAFF is closely linked to SLE pathogenesis [10–13]. BAFF gene expression is induced by many pathogens including viruses, bacteria, fungi and parasites [14], highlighting a connection between infection and the development of SLE. BAFF is primarily expressed by innate immune cells including monocytes, macrophages, neutrophils and DCs [15–17], and several cytokines such as IFN-α, IFN-γ and TNF-α can also promote its expression (18). Over-expression of BAFF in mice promotes B cell survival and the differentiation of B cells, leading to the development of a SLE-like disease [19]. BAFF-transgenic mice not only have augmented numbers of immature, follicular and marginal zone B cells and PC, but their B cells exhibit activation as evidenced by augmented MHC-II expression [19]. While numbers of effector CD4 and CD8 T cells are also increased in BAFF-transgenic mice, BAFF-transgenic mice lacking T cells develop SLE comparable to their T cell sufficient counterparts [20] indicating that the alterations in the B cell compartment driven by BAFF are sufficient to induce the disease.

APRIL is another B cell survival factor that can be produced by myeloid cells and it shares significant homology with BAFF [21]. APRIL likely acts at a later stage of B cell development supporting the survival of antigen-experienced B cells and bone marrow PC [22]. Elevated levels of APRIL have been observed in SLE patients with levels correlating with disease activity and lupus nephritis (LN) [23–25]. While loss of APRIL in mice does not affect B cell development or most humoral immune responses [26], it seems likely that it contributes to autoimmune disease by supporting the survival of autoreactive B cells and PC, and APRIL signalling is needed for antibody class switching to IgA which is important for mucosal immunity but is implicated in IgA nephropathy in SLE [27].

SLE is frequently associated with increased production of type I IFN, and it is clear that type I IFN can influence B cell survival and function in numerous ways. Type I IFN induces macrophages and DCs to produce BAFF and APRIL and promote CD40-independent Ig class-switching [21]. IFN-α is also highly implicated in autoimmune responses as it promotes B cell survival, induces B cell activation including upregulation of CD86, CD69 and CD25 and enhances their responses to B cell antigen receptor cross-linking [28]. Thus, via prolonged effects on B cell survival, activation and function, type I IFN are highly implicated in B cell-dependent autoimmune responses.

The type II IFN, IFN-γ, is also highly implicated in B cell hyperactivation and SLE pathogenesis. It is produced by NK cells, NKT cells, effector CD4 and CD8 T cells, and ILC1 cells,

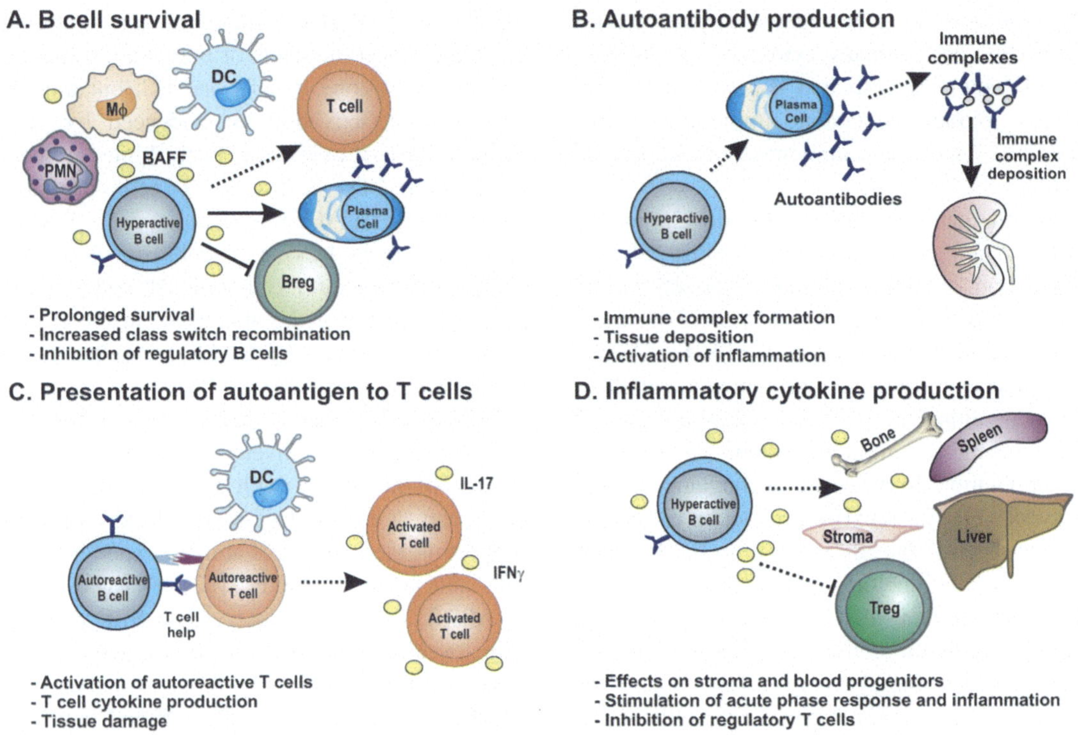

Fig. 2.1 Overview of how B cells contribute to SLE. B cells can have pathogenic roles in SLE via (**A**) augmented survival and activation of self-reactive B cells; (**B**) autoantibody production causing tissue pathology; (**C**) antigen presentation and activation of autoreactive T cells; and, (**D**) inflammatory cytokine production

and while it has effects on numerous immune cell types, it plays a key role in the upregulation of MHC-I and MHC-II expression, augmenting antigen presentation on B cells and DCs. IFN-γ is also involved in B cell isotype switching, promoting the production of pathogenic IgG2a and IgG3 isotypes from PC.

IL-6 is another cytokine that is linked with SLE [29–31]. It is produced in response to pathogens by non-haematopoietic cells as well as most immune cells including B cells. While it has pleiotropic functions on numerous cell types, on B cells it induces their maturation into antibody-producing cells, promotes the survival of PC and enhances antibody production [32]. IL-6 also inhibits the generation of inducible regulatory T cells, thereby preventing the control of autoimmune responses [33].

IL-21 is a pleiotropic T cell cytokine that is strongly implicated in SLE pathogenesis. It is secreted by activated T cells as well as T follicular helper (Tfh) cells in B cell follicles where it acts as a potent inducer of B cell proliferation and differentiation into PC [34] and serves to regulate germinal centre (GC) responses [35]. Increased numbers of IL-21-producing T cells are found in SLE patients [36], elevated levels of IL-21 are associated with T and B cell alterations in SLE [37], and increased IL-21 expression correlates with increased numbers of PC in patients with active SLE [38]. Furthermore, both the genes encoding IL-21 and the IL-21 receptors are risk factors for SLE [39, 40]. The optimal formation of Tfh cells requires both IL-6 and IL-21 [41] so these two cytokines almost certainly cooperate in SLE pathogenesis. Collectively, these data suggest that overexpression of IL-21 contributes to SLE via the production of pathogenic autoAbs.

IL-17 is a proinflammatory cytokine produced by both innate and adaptive T cells. It acts on many cells including fibroblasts, keratinocytes, epithelium and endothelium to induce the production of chemokines that promote the trafficking of immune cells to sites of infection and inflammation. Enhanced levels of IL-17A have been observed in SLE patients [42, 43], and numbers of IL-17-producing T cells are increased in peripheral blood [44, 45]. In addition to its role in regulating the recruitment of inflammatory cells, IL-17-producing T cells have been shown to have strong effects on B cells, functioning as effective B cell helpers in stimulating B cell proliferation and a strong class-switched antibody response [46]. In lupus-prone BXD2 mice, IL-17 has been shown to underlie their spontaneous autoreactive GC development that leads to production of pathogenic autoAbs [47]. Thus IL-17 plays a multifaceted role in SLE.

Aside from genes encoding soluble factors, numerous other genes that are expressed in B cells have been implicated in SLE disease susceptibility and animal model studies have demonstrated that knockout mutations in many genes can enhance B cell reactivity, promote autoreactive PC and autoAbs production. Many of these genes are involved in controlling B cell signalling pathways, with their disruption leading to increased B cell responsiveness (Lyn, SHIP-1, SHP, FcγRs, BLK, NFkB1, CD40, etc.).

B Cell Subsets in SLE

Several B cell subsets have been implicated in SLE pathogenesis, particularly memory B cells and PC, and more recently regulatory B cells (Breg), which are the focus of this chapter section.

B cells begin life in the bone marrow (BM), where they develop from haematopoietic stem cells in an antigen-independent manner into common lymphoid precursors, then through pro- and pre-B developmental stages into immature B cells. Immature B cells leave the BM equipped with a B cell receptor (BCR) to enter the blood stream and access secondary lymphoid tissues such as the spleen and lymph nodes. These immature B cells are termed transitional B cells, which may start an antigen-dependent maturation process to differentiate into memory B cells or terminally differentiate into Ab-secreting PC [48]. GC host the antigen-dependent stage of naïve B cell activation and differentiation, which includes somatic hypermutation and Ab class switching. Importantly, class-switched (CS) memory B cells can differentiate into PC to greatly improve immune responses when the same antigen is encountered again. Of note, most long-lived PC will return to the BM (where the PC mostly have an IgG isotype); PC also migrate to gut-associated lymphoid tissues (with increased proportion of IgA isotype PC).

In SLE, patients are characterised by an absolute and relative peripheral B lymphocytopenia [49, 50]. This perturbation of peripheral B cell homeostasis in SLE is further characterised by unbalanced proportions of specific B cell subsets. This is acknowledged to reflect a shift toward an increase in transitional, increased CS memory B cells as well as increased Ab-secreting plasmablasts (PB) and PC subsets, while non-CS (NCS) memory B cells and naïve B cells are usually decreased. Naïve B cell lymphopenia plays a role in the relative expansion of memory B cells, and drug-induced depletion may further disrupt B cell populations. However, as detailed below in this section, there are differences between studies, where discrepancies may be explained by variation in study population, including age, gender, ethnicity, as well as disease activity state and treatment used.

Naïve B Cells

Naïve B cells are reported to be decreased in proportion in SLE patients compared to healthy controls (HC) [51]. However, other studies have described higher proportions of naïve B cells in SLE patients compared to HC [52], as well as in active SLE (SLEDAI > 4) [53], while further studies have shown no difference [49, 50, 53, 54]. Patients with inactive LN characterised by recent relapses harboured lower proportions of naive B cells as compared to those with no relapse [55]. Of note, no association with anti-dsDNA or C3 was observed with this B cell subset in both patients groups [55].

Memory B Cells

Memory B cells have been reported to be increased in percentage in patients with new-onset SLE compared to HC [56], while other studies have reported lower proportions and absolute numbers in SLE patients [52, 57]. While another study has observed no significant difference in percentage of memory B cells between the two groups, inactive LN patients characterised by recent relapses harboured a higher memory B cell to naïve B cell ratio as compared to those with no relapse [55]. No association with anti-dsDNA or C3 was observed with this B cell subset in both patients groups [55]. Of note, memory B cells have been reported lower in active SLE (SLEDAI > 4) [53].

Some studies have further investigated subsets of memory B cells in SLE, particularly NCS and CS memory B cells. While $CD27^+$ is the "gold standard" cell surface marker for identifying memory B cells, CS memory B cells are characterised by loss of expression of IgM and IgD, since they have undergone CS recombination toward a different Ig isotype (IgG, IgA or IgE). Several studies have demonstrated that absolute and relative numbers of NCS memory B cells are lower in SLE versus HC [49, 52–54], while no differences were found in other studies [50]. CS memory B cells were reported higher in recently diagnosed SLE patients that had not received corticosteroid and immunosuppressant treatments compared to HC [52], lower in active SLE (SLEDAI > 4) [53], while no differences between SLE versus HC were found in other studies [49, 50, 53, 54]. A moderate negative correlation was observed with markers of disease activity, as assessed by SLEDAI and C3, and CS memory B cells in recently diagnosed SLE patients naïve from corticosteroid and immunosuppressant treatments [52].

Several groups have identified and investigated a potentially distinct subset of CS memory B cells termed "double negative" (DN) that lacks both CD27 and IgD cell surface markers [58]; these appear to be the same subset that have been described as "atypical" memory B cells in the literature. This DN subset has been reported higher in SLE versus HC [53, 54, 58], while no differences were found in other studies [50]. In one of these studies, SLE patients harboured a higher percentage of DN memory B cells compared to HC, however results should be interpreted with caution due to apparent incorrect statistical analysis methods used (*t*-test instead of ANOVA), and also since DN form a minor proportion of cells (normally < 10% of B cells) their measurement may be easily confounded by other cell populations if only considering proportion (e.g. if another cell subset decreases, DN % may appear to increase) and not absolute numbers as well [58]. Proportion of activated ($CD95^+$) DN memory B cells was also reported higher in percentage and absolute number in active (SLEDAI > 4) versus inactive SLE patients [59]. SLE patients with a higher proportion of DN memory B cells harboured augmented LN and high anti-dsDNA and anti-RNP/Sm autoAbs titres [58].

In line with the aforementioned characteristics of DN B cells in SLE, proportions of "atypical" memory B cells have also been reported to be increased in patients with SLE compared to HC [49, 60–62]. They are particularly increased in LN [60–62] and active mucocutaneous manifestations [62], and positively associated with autoAbs production (including ssDNA, dsDNA, Sm, nucleosome, chromatin, La, histone, C1q and RNP) [60–62], positively correlated with proportion of PC [62], as well as disease activity in some studies [60, 62] but not in another [49]. It is noteworthy that this B cell subset has also been identified in kidney B cell infiltrates of active LN patients [62]. Bearing in mind a small sample size, one study suggested that the presence of this subset may be a predictive biomarker of poor long-term outcome in SLE [63]. Amongst all the aforementioned studies focused on DN and "atypical" memory B cells, an activated phenotype with expression of the following surface markers can be delineated: $CD10^{-}CD19^{hi}CD20$-$^{hiCD21lo}CD21R^{hi}CD23^{-}CD24^{-}CD27^{-/lo}CD32^{hi}$C-$D38^{lo}CD39^{-}CD40^{lo}IgD^{-/lo}CD11c^{+}CD75^{hi}CD77$ $^{-}CD86^{int}CD95^{hi}$, with expression of the transcription factor T-bet. Interestingly, there cells are also reported BAFF-$R^{hi}TACI^{int}BCMA^{lo}$ [62].

Of particular interest, Nicholas et al. have shown that, upon stimulation, this "atypical" memory B cell subset can differentiate into autoAb-secreting PC, particularly autoreactive anti-Sm IgG-secreting PC in patients with SLE known to be seropositive for anti-Sm autoAbs [63]. RNA-seq analysis provided further characterisation of this B cell subset in SLE, harbouring a transcriptomic signature close to PB, with upregulation of *PRDM1* (BLIMP1), *AICDA* (AID) and *XBP1*, and some more PC-specific genes including, *BMP6*, *EMP3*, and *S100A4* [62]. The authors suggested that, together with their phenotypically low expression of CD38 and CD27, this subset may represent precursors of PB [62]. Importantly, similar to memory B cells, this B cell subset terminally differentiated into IgG-secreting PC upon triggering via activated T cells, which was IL-21 dependent [62]. They, however, harboured a significantly higher propensity to autoAbs production as compared to memory B cell-derived PC, particularly including those targeting histone, nucleosome, RNP and Sm [62]. Of note, this B cell subset did not spontaneously secrete IgG Abs [62]. Furthermore, Wang et al. demonstrated that IL-21 was a key driver for the differentiation of naïve into this "atypical" memory B cell subset [62].

Close in time to the paper by Wang et al. [62], Jenks et al. also published a pivotal study focused on this DN B cell subset in SLE [54]. The authors demonstrated that the DN memory B cell subset can be further subdivided; DN memory B cells indeed encompass the so-called DN2 B cell subset which is $CXCR5^{-}FCRL4^{-}FCRL5^{+}$, and constituted the major part of the DN memory pool in SLE [54]. Importantly, similar to the findings of Wang et al. [62], evidence suggested that this DN2 B cell subset was a precursor of PC, sharing transcriptional similarity with PC, including upregulation of *PRDM1* (BLIMP-1) and downregulation of *ETS1* [54]. Moreover, this DN2 subset originated from naïve B cells in a TLR-7 and IL-21-dependent manner, and not from memory B cells [54]. Interestingly, this DN2 subset was positively correlated with SLE overall and renal disease activity as well as with anti-Sm and anti-RNA autoAbs [54].

Plasmablasts and Plasma Cells

PB are reported to be expanded in percentage in patients with SLE [49, 52, 57] and active disease (SLEDAI > 4) [57] compared to HC, as well as in LN patients compared to patients without LN [57]. This PB expansion is moderately positively correlated with disease activity, as measured by complement levels [49] or SLEDAI in recently diagnosed SLE patients that are naïve from corticosteroids and immunosuppressants [52].

Similar to PB, PC are reportedly increased in percentage in patients with SLE, including those with new-onset SLE, as compared to HC [51, 56]. PC (proportion and/or absolute numbers) are moderately positively correlated with measures of disease activity, including SLEDAI and ECLAM [51, 64]. As expected, SLE patients seropositive for autoAbs such as anti-dsDNA, anti-Sm, anti-histone, anti-Ro, anti-La are characterised by increased PC numbers [64].

B Regulatory Cells

Breg are a regulatory B cell subset whose function mirrors that of Treg, where they are involved in peripheral immune tolerance [48]. They have a pivotal role in negative control of $CD4^{+}$ T cell activation and production of proinflammatory cytokines IFNγ and TNF [65], and of pDC-produced IFNα [66], both via IL-10 secretion [65, 66]. Importantly, in SLE, the Breg subset can become functionally impaired, mainly due to signalling defects in CD40 [65] and Type I IFN

[66], both involving the STAT3 pathway [65, 66]. This B cell subset is characterised by the production of several cytokines with inhibitory functions including IL-10 and TGF-β, and the more recently discovered IL-35 [67].

The percentage of IL-35-producing Breg and plasma levels of IL-35 have both been reported decreased and positively correlated in patients with new-onset SLE compared to HC [56]. IL-10-producing Breg and $CD5^+$ Breg have also both been reported decreased in percentage in patients with new-onset SLE [56]. Of note, percentage of IL-10-producing Breg were also positively correlated with both percentage of IL-35-producing Breg and plasma levels of IL-35 in patients with new-onset SLE [56]. Conversely, another study showed increased proportion, but not absolute number, of $CD5^+CD1d^{hi}$ Breg in active SLE versus both inactive SLE and HC [68], as well as positive correlation between proportion and absolute number of $CD5^+CD1d^{hi}$ Breg and disease activity (SLEDAI) [68]. Similarly, proportion, but not absolute number, of $CD19^+IL\text{-}10^+$ B cells and $CD19^+CD38^{hi}CD24^{hi}$ B cells were higher in SLE patients, including active SLE, versus HC [68].

Recent work has uncovered a crucial role for $CD4^+$ Tfh and plasmacytoid dendritic cells (pDC) in the regulation of Breg, with impaired mechanisms identified in SLE. In an elegant study, Yoshizaki et al. demonstrated that the maturation, expansion and the IL-10 production of functional IL-10-producing Breg depended on several key signals including IL-21, CD40 and MHC-II derived from T cells [69], suggested to be $CD4^+$ Tfh cells [68]. Conversely, TGF-β and IFNγ may negatively regulate IL-10-producing Breg [69]. Of particular interest, TLR-9-stimulated-pDCs have also been demonstrated to induce Breg expansion and their IL-10 production, to a higher extent compared to $CD4^+$ T cells [66]. The authors discovered that these IL-10-producing $CD38^{hi}CD24^+$ Breg originated from immature B cells upon pDC stimulation involving IFNα and CD40 signalling [66]. Importantly, in SLE, impairment in pDCs and B cells were observed [66]. Ultimately, the authors demonstrated that pDC-produced IFNα is negatively controlled by Breg via IL-10, a mechanism defective in SLE [66]. Interestingly, IL-10-producing Breg are upregulated by in vivo treatment with myeloid-derived suppressor cells in the lupus-prone Sanroque mice, which exhibited improvement in disease manifestations including lower anti-dsDNA AutoAbs levels and proteinuria, and improved kidney histopathological features [70]. Of note, glatiramer acetate has also been shown to induce differentiation of memory B cells into functional Breg cells in SLE [71].

In summary, B cells in the immune system follow a complex path of development and differentiation into specialised subsets of cells responsible for protective antibody production and regulation. Rather than a single B cell defect in SLE, a variety of defects may arise in patients, often leading to studies with opposite conclusions. Nevertheless, B cell subsets remain an essential part of the immune system and a key player in the pathogenesis of SLE.

Autoantibodies: The Immunological Hallmark of SLE

SLE and many other autoimmune diseases such as type 1 diabetes, primary Sjogren's syndrome (pSS), rheumatoid arthritis, Hashimoto thyroiditis, and systemic sclerosis (SSc), are all characterised by the presence of circulating autoAbs. One of the hallmarks of SLE is the presence of autoAbs against nuclear antigens. These antigens can be nucleic acid such as DNA (single stranded (ss) or double stranded (ds) DNA) and RNA, protein such as histone, or DNA/histone complexes (nucleosome). The presence of anti-nuclear Ab (ANA) is, however, not an immunological characteristic specific to SLE; other autoimmune diseases, such as pSS or SSc, are also characterised by their presence. Some autoAbs such as anti-Ro, anti-La or anti-RNP Abs can be detected in patients with SLE, but are not specific to SLE. For instance, anti-Ro and anti-La Abs are more specific to pSS.

Furthermore, circulating autoAbs, including anti-dsDNA and anti-Smith (Sm) autoAbs, can also be detected in healthy individuals [72, 73].

Patients diagnosed with SLE are typically positive for ANA, with most also positive for anti-dsDNA autoAbs which are specific for this condition. Anti-Sm Abs are less prevalent in SLE, but are also highly specific [74]. While also detected in many patients diagnosed with idiopathic SLE, anti-histone autoAbs are considered more specific to drug-induced SLE [75]. In clinical practice, an ANA test will return positive if the patient has autoAbs against nuclear antigens. In this situation, knowing which nuclear antigen(s) is/are targeted by autoAbs is important for diagnostic purposes. ELISA or multiplex assays are commonly used to identify anti-dsDNA autoAbs and autoAbs to extractable nuclear antigens (ENA). ENA autoAb panels differ according to accredited laboratories, but usually will test for the presence of anti-Sm, anti-SSA/Ro, anti-SSB/La, anti-Scl-70, anti-RNP, and anti-Jo1.

According to the 2019 European League Against Rheumatism (EULAR)/American College of Rheumatology (ACR) criteria for SLE classification, ANA positivity is now a mandatory entry criterion for an individual to be classified as having SLE [76]. AutoAbs listed in the immunological domain of this scoring system includes APL Abs (anti-cardiolipin, β2GP1 or lupus anticoagulant), anti-dsDNA and anti-Sm Abs [76].

AutoAb Production in SLE

AutoAbs, as any Abs, are secreted by B cells that reached their terminal differentiation stage. These are PB and PC, key subsets of B cells involved in SLE pathogenesis (see Sect. "B cell Subsets in SLE"). Anti-RNA binding protein (RBP) autoAbs, which particularly include anti-Sm, anti-RNP, anti-Ro and anti-La, appear to be produced by GC-derived long-lived PC [77], while short-lived PB may be the source of anti-dsDNA autoAbs [78]. The difference in life span between those two Ab-secreting B cell subsets could explain the observed difference in longitudinal assessment of anti-Sm and anti-dsDNA titres over time; while anti-Sm Abs are generally stable over time, anti-dsDNA Ab are more characterised by fluctuation of their titre [78], and reported to be predictive of severe disease flare [79].

A pivotal study has investigated the development of circulating autoAbs in individuals prior to SLE diagnosis and disease onset, contributing to a deeper understanding of their kinetics in SLE pathogenesis [72]. Leveraging the unique US Department of Defense Serum Repository, authors retrospectively assessed biobanked samples from 130 military personnel diagnosed with SLE, collected on average 4.4 years and up to 9.4 years before diagnosis. ANA and a panel of six autoAbs most commonly found in SLE were tested. AutoAbs were detected on average 3.3 and up to 9.4 years before SLE diagnosis in 115/130 patients. Furthermore, most patients were seropositive before the first recorded non-immunological disease manifestation that occurred on average 1.5 years before diagnosis. Importantly, autoAbs were identified in 90/130 of these patients in their first ever collected sample, suggesting an underestimation of the seropositive preclinical period, as underlined by the authors [72]. Prior to SLE diagnosis, ANA, APL, anti-Ro and anti-La were detected significantly earlier than anti-Sm and anti-nRNP (mean time interval 3.4 vs 1.2 years). Anti-dsDNA was detected significantly earlier (mean time interval 2.2 years) than anti-nRNP and non-significantly later than ANA. Development of anti-Sm and anti-nRNP autoAbs were very close to non-immunological clinical onset (mean time interval 0.47 and 0.2 years, respectively). The authors also showed that diversification of antigenic specificity of autoAbs gradually accrued over time to peak at and plateau from disease diagnosis.

In contrast to the few autoAbs tested in routine clinical settings when ordering an anti-ENA Ab panel, autoAbs against over 180 different targets have been reported in SLE [80]. In SLE, not only are autoAbs directed against nuclear autoantigens (ssDNA, dsDNA, nucleosome,

RBP), but also against cytoplasmic components, such as ribosome (anti-ribosomal P (RibP), highly specific for SLE) [81], as well as mitochondrial DNA and RNA [82], cytokines including Type I, II and III IFNs [83] and BAFF [84], and complement components (C), such as C1q [85], and C3b [86]. It is worth noting that some anti-cytokine autoAbs, such as those targeting Type I IFN are neutralising autoAbs [83]. The clinical relevance of neutralising autoAbs against cytokines that are targeted by biological therapy in clinical trial design, patient stratification and use of approved biological drugs warrants further investigations [84]. The pressing unmet need in SLE for a precision medicine approach where patients could be stratified according to their dominant biological pathway at play to select the appropriate therapy may need to consider the potential presence of these anti-cytokine autoAbs.

Pathogenic Roles of AutoAbs in SLE

Pathogenicity of autoAbs in SLE does not only rely on one unique mechanism; rather, multiple mechanisms have been described, including IC-mediated glomerular damage, synaptic transmission defects and neuronal damage, cytokine up-regulation, and neutrophil extracellular trap (NET) formation. A crucial point needs to be underlined: the pathogenicity of an autoAb is independent of its disease-specificity [78].

IC-Mediated Glomerular Damage

When complexed with autoantigens, autoAbs can be pathogenic in SLE. This mechanism has been well described in LN pathogenesis, particularly with IgG anti-dsDNA autoAbs, although other autoAbs such as those targeting C1q, Sm, Ro, La, chromatin can also deposit in the glomerular basement membrane (GBM) [87]. Interestingly, anti-DNA autoAb-transgenic mice developed nephritis, but to a lower extent compared to lupus-prone mice which spontaneously generate various autoAbs [88]. The formation of the so-called IC in the circulation can then deposit in the GBM, or form in situ when circulating autoAbs recognise a cognate self-nuclear antigen "planted" in the GBM, inducing tissue inflammation and glomerular damage via complement activation. While promising assays assessing the presence of circulating IC have been developed [89], there is currently no validated test used in routine clinical settings. However, indirect estimation is achieved by the concomitant measurement of serum levels of anti-dsDNA autoAbs and C3 and C4. Those blood tests are essential in monitoring patient disease activity, and are part of validated composite disease activity scoring systems, such as the SLE Disease Activity Index (SLEDAI), forming its immunological domain.

Synaptic Transmission Defect and Neuronal Damage

Two autoAbs appear pathogenic in neuropsychiatric SLE (NPSLE) via potential synaptic transmission defects and neuronal damage: anti-*N*-methyl-D-aspartate receptor (NMDAR) and anti-RibP autoAbs, both linked to cognitive impairment in SLE [90]. Human anti-DNA autoAbs cross-reacting with the extra-cellular domain of the neuronal NMDAR, ubiquitously expressed in the central nervous system (CNS), has been shown to induce murine hippocampal neuronal damage-associated cognitive dysfunction (flexible memory task defect) following blood brain barrier (BBB) disruption by systemic LPS treatment [91]. While the presence of autoAbs against NMDAR was reported in SLE patients' serum and cerebrospinal fluid [92], Kowal et al. demonstrated not only their presence in the brain of patients with neuro-psychiatric SLE manifestations, but also their ability to induce murine hippocampal neuronal damage once injected in situ [91].

Anti-RibP autoAbs are associated with NPSLE, particularly so with the psychotic manifestation of diffuse NPSLE [81, 93]. Their link with neuropsychiatric pathogenicity may be explained by cross-reactivity with neuronal surface P antigen (NSPA), and involve calcium influx-dependent neuronal death as well as perturbation of glutamatergic transmission via both NMDAR and α-amino-3-hydroxy-5-methyl-4-

isoxazolepropionic acid receptor (AMPAR) [94]. Importantly, as for anti-NMDAR autoAbs, BBB disruption appeared mandatory for anti-RibP autoAbs to exert their neuronal pathogenic effect [94].

Cytokine Up-Regulation

Anti-dsDNA and anti-RBP autoAbs upregulate cytokine expression in SLE. IgG autoAbs to RNA or DNA autoantigens, forming IC, can bind their cell-surface cognate Fc gamma receptor (FcγR) on pDCs, particularly FcγRIIa. After internalisation, DNA/RNA is transported to the endosome to signal though Toll-like receptor (TLR)-7 and -9 [95]. This signalling leads to IFNα secretion by pDCs, one of the key pathogenic pathways involved in SLE pathogenesis. Anti-dsDNA autoAbs can upregulate glomerular expression of pro-inflammatory cytokines, chemokines and pro-fibrotic factors. IC formed by nucleosome and anti-dsDNA Abs, as well as free nucleosome or anti-dsDNA Abs, can induce chemokine up-regulation by murine primary mesangial cells [96]. One of these IC-induced chemokines is C–C motif chemokine 2 (CCL2) (also known as monocyte chemoattractant protein 1 (MCP-1)), acknowledged as one of the best urinary biomarkers for LN [97]. Anti-DNA autoAbs were demonstrated to induce pro-inflammatory cytokines such as IL-6 by rat glomerular mesangial cells [98], as well as pro-fibrotic factors such as TGF-β1 and fibronectin in cultured human mesangial cells [99].

NETosis Formation Induced by Autoantibodies

NETosis is a defence mechanism used by neutrophils to trap and kill pathogens, by deploying neutrophil extracellular traps or NETs, which are an extracellular web-like fibre network made of DNA and globular proteins that can bind to microorganisms including bacteria, fungi, and viruses. In this process the neutrophil may die (suicidal NETosis) or remain as an intact cell but without a nucleus (vital NETosis). NETosis is acknowledged to play a role in SLE pathogenesis, exposing NET-component self-antigens extra-cellularly, such as dsDNA, chromatin, and histone [100]. Patients with SLE are characterised by spontaneous NETosis formation in vitro, suggesting a high propensity for extracellular nuclear autoantigen exposure [100]. NET formation is associated with ANA and anti-dsDNA levels in SLE [101]. Furthermore, some autoAbs, targeting NET, LL37, and ribonuclear particles, can play a pathogenic role in SLE by inducing NETosis formation [100].

Signalling Pathways Operative in SLE

B cell function is controlled primarily via signalling through the B cell receptor (BCR) and thus, it is probably not surprising that mutations in this pathway can lead to altered B cell behaviour and cause either immunodeficiency or result in autoimmunity. The BCR, which has tremendous diversity in both sequence and specificity, comprises membrane-bound immunoglobulin (Ig) non-covalently associated with Igα/Igβ signalling heterodimers (Fig. 2.2). Igα/Igβ are the key molecules responsible for transmitting information from the BCR, operating via Src family non-receptor protein tyrosine kinases (SFKs) [102]. Antigenic engagement of the BCR results in activation of SFKs, which then coordinate the activation of a number of signalling pathways including calcium, phosphoinositide 3-kinase (PI3K) and mitogen-activated protein kinase (MAPK) to regulate B cell development, activation and survival (Fig. 2.2). While expression of the BCR is essential for B cell development and survival, one of the three cognate receptors for BAFF (BAFF receptor, BAFF-R), also plays an indispensable role in these processes via binding to BAFF and activating the non-canonical nuclear factor for activated B cells (NF-κB) signalling pathway [103] (Fig. 2.2). CD19 is another key receptor on B cells, which represents a membrane recruitment site for PI3K, and its role is to enhance signals delivered through the BCR by reducing the threshold for B cell activation [104] (Fig. 2.2). There is now good evidence that signalling from the BCR, BAFF-R and CD19 are closely linked [104]. BAFF (targeted by

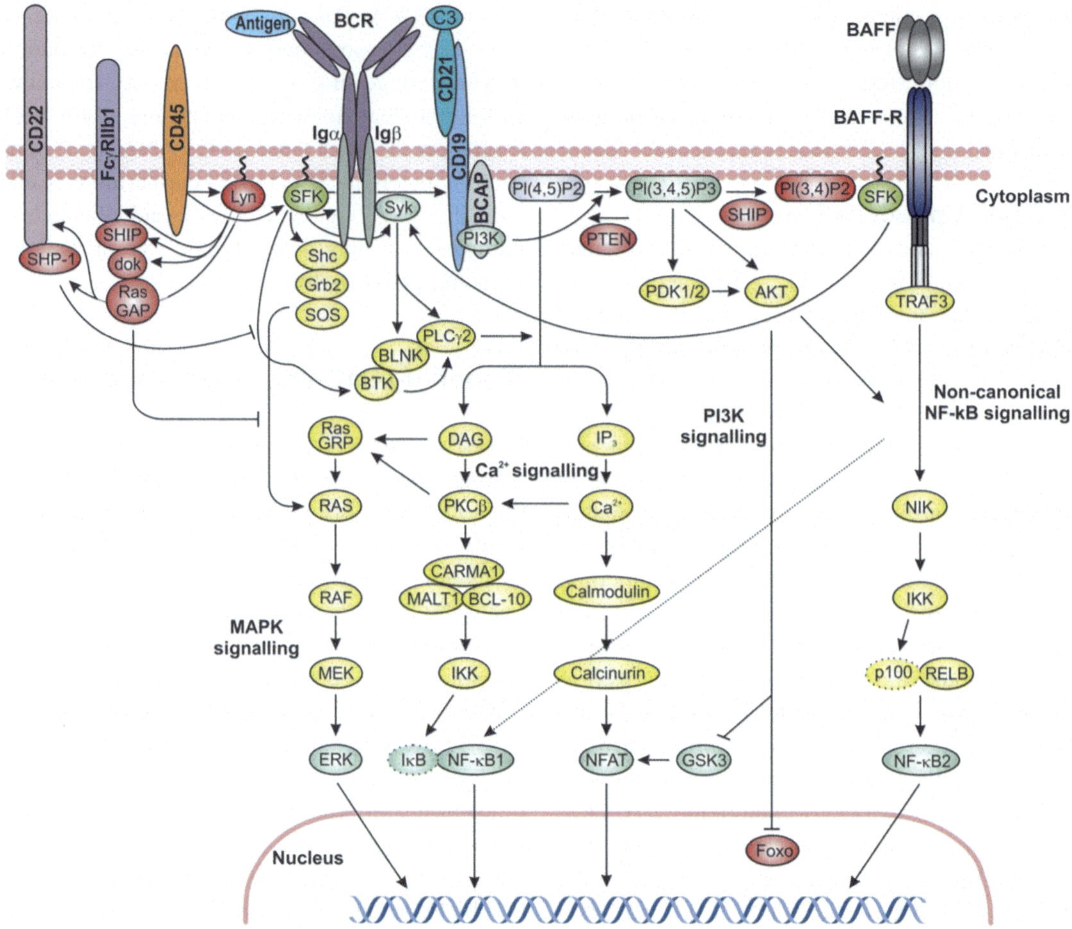

Fig. 2.2 **B cell signalling pathways operative in SLE.** Crosslinking of the BCR activates Src family non-receptor protein tyrosine kinases (SFK) which coordinate pathways that regulate B cell proliferation, activation, cytokine production and survival. Signalling through the BAFF-R promotes B cell development and survival, while CD19 enhances signals from the BCR. Signalling through the inhibitory receptors FcγRIIb and CD22 is initiated by the SFK Lyn to negatively regulate the B cell response

belimumab, tabalumab, blisibimod, atacicept, telitacicept) and BAFF-R (ianalumab) are two of the many targets that have been investigated in SLE clinical trials (*see Chap.* Chap. 3).

In addition to receptors that promote B cell survival and activation, there are a number of other receptors on B cells that participate in negative regulation of the B cell response and these include CD22 and FcγRIIb, which transmit their inhibitory signals via the SFK Lyn [105] (Fig. 2.2). Lyn activity is indispensable for tyrosine phosphorylation of inhibitory motifs (ITIMs) in the cytoplasmic tails of CD22 and FcγRIIb1 [106–109], which generates membrane recruitment sites for negative regulatory phosphatases such as the protein tyrosine phosphatase SHP-1 and the lipid phosphatases SHIP-1. Once recruited to ITIMs, SHP-1 and SHIP-1 are juxtaposed to their substrates where they are able to terminate signalling by dephosphorylation of receptors, enzymes and adaptor molecules in the case of SHP-1, while SHIP-1 hydrolyses the lipid second messenger phosphatidylinositol [3, 4, 5] trisphosphate, otherwise known as PIP3. Thus, Lyn acts via inhibitory receptors and phosphatases to switch off B cell activation thereby negatively regulating the B cell response [105]. CD22 (epratuzumab) and FcγRIIb (obexelimab) have been investigated in SLE clinical trials (*see Chap.* Chap. 3).

Many experimental model studies have shown that over-expression or loss of certain proteins, including receptors, signalling molecules and factors, can promote SLE-like autoimmune disease. Transgenic over-expression of BAFF in mice generates excess B cell survival signals, resulting in an expanded B cell compartment and the development of autoimmunity resembling SLE [19]. Over-expression of CD19, which is a major recruiter of PI3K, leads to autoimmunity, presumably through augmented PI3K activity and enhanced B cell survival [110]. Loss of Lyn, the essential inhibitory enzyme in B cells, leads to SLE-like autoimmune disease in mice [111–113]. Loss-of-function mutations in FcγRIIb1 or CD22 incites immune system defects and SLE [114, 115], while the motheaten mouse, which has a naturally occurring deleterious mutation in the SHP-1 gene [116, 117], rapidly develops systemic inflammation, particularly affecting the skin and lungs, produces autoAbs and exhibits immune-mediated glomerulonephritis [118]. B cell-specific deletion of SHP-1 leads to increases in innate B1a cells, and with age, the mutant mice develop ANA and IC-mediated glomerulonephritis, indicating the key role of SHP-1 in restraining B cell activation and autoimmunity [119]. In many instances, B cells from these inhibitory signalling mutants show heightened activation of calcium signalling, MAPK and PI3K pathways.

The importance of regulation of PI3K activity in B cells is exemplified in mice harbouring mutations in negative controllers of this pathway. Monoallelic deletion of PTEN causes hyper-activation of Akt and expansion of activated B and T cells, which are protected from apoptosis mediated via Fas [120]. PTEN$^{\pm}$ mice develop lethal autoimmune disease characterised by high titre ANA, autoimmune glomerulonephropathy and systemic inflammation [120], and simultaneous haploinsufficiency of SHIP-1 exacerbates this autoimmune phenotype [121]. Loss of PTEN specifically in B cells causes an expansion of innate B cell subsets, B cell hyper-proliferation, elevated serum IgM and anti-DNA autoAbs but reduced serum IgG and Ig class switching [122], suggesting that it acts as both a positive and negative regulator of B cell functions. Germline SHIP-1-deficiency leads to systemic inflammation, and infiltration of leukocytes into several organs including the lungs, which reduces lifespan [123, 124]. This is due to the inability of SHIP-1-deficient immune cells to limit signalling [125]. Conditional deletion of SHIP-1 in B cells results in B cell hyperresponsiveness, plasmacytosis, pathogenic autoAb production and antibody-mediated autoimmune disease characterised by IC deposition and glomerular pathology resembling SLE [126, 127], indicating that SHIP-1 is a general negative regulator of B cell functions. SHIP-1 recruits the adaptor Dok, and mice simultaneously deficient in Dok-1 and Dok-2 in all cells develop class-switched autoAbs with age and SLE-like glomerulonephritis, although this phenotype may be driven by uncontrolled T cell receptor signalling that augments B cell class switching [128].

The Importance of Lyn in Regulating B Cell Signalling and Averting SLE

As discussed above, germline deletion of Lyn leads to the development of autoimmune-mediated glomerulonephritis analogous to human LN [111–113]. Lyn plays a key role in B cells, and mice deficient in Lyn exhibit defects in their B cell compartment such as B cell lymphopenia, the loss of marginal zone B cells, an expansion of PC and elevations in serum titres of IgM, IgA and IgE. Lyn-deficient B cells have an activated phenotype characterised by BCR down-regulation and enhanced expression of MHC-II, CD80 and CD86 [9, 107], and signalling studies show that Lyn-deficient B cells are hyperactive [106–109, 113]. Older Lyn-deficient mice have increased numbers of activated myeloid cells, splenomegaly and inflammatory cytokine imbalances which contribute to disease pathogenesis. Although Lyn is not expressed in conventional T cells, a key feature of older Lyn-deficient mice is the presence of activated T cells brought about by the augmented inflammatory environment [9, 129], which induces B cell class switching and the production of pathogenic antinuclear IgG and IgA autoAbs [9, 130]. Lyn-deficient mice lacking T cells have greatly reduced levels of class-switched autoAbs, indicating the role of T cells and the GC response in disease pathogenesis in this model [131]. Collectively, this immune system dysregulation leads to IC deposition in glomeruli, with activation of innate inflammatory mechanisms causing glomerular destruction.

An important study in 2007 identified the essential role of TLR signalling in autoimmune disease in Lyn-deficient mice [132]. The authors showed that deletion of the MyD88 signalling intermediate from Lyn-deficient mice resulted in an attenuation of ANA and protection from glomerulonephritis, which highlighted a major role for innate immune mechanisms underpinning their development of autoimmune disease.

Subsequent studies have shown that inflammation plays a key role in promoting autoimmune disease development in Lyn-deficient mice, which was demonstrated by deletion of the pro-inflammatory cytokine IL-6; these mice remained autoimmune-prone but did not develop disease [9, 133]. Lyn-deficient mice lacking IL-6 exhibited B cell developmental defects and B cell hyper-responsiveness, indicating that these phenotypes are intrinsic to Lyn deficiency, however age-dependent myeloid cell expansion and T cell activation were absent. Thus, pathogenic autoAb production, activation of the innate immune system and glomerular pathology were ameliorated. These studies have contributed to an improved understanding of the contribution of IL-6-mediated inflammation to the development of SLE.

The PI3K pathway is hyperactivated in Lyn-deficient mice, which is thought to contribute to their autoimmune disease development. Studies have demonstrated that monoallelic disruption of the PI3K isoform p110δ markedly decreased autoimmune-mediated kidney pathology in Lyn-deficient mice [134]. Interestingly, haploinsufficiency of p110δ did not alter B cell hyper-responsiveness in Lyn-deficient mice but it significantly attenuated inflammation, restrained T cell signalling and activation and reduced pathogenic autoAb production. This highlights the involvement of inflammation and T cell activation in driving the disease into a pathogenic state [134].

Additional studies have also emphasised the importance of inflammation in autoimmune disease development in Lyn-deficient mice by knockout of the pro-inflammatory cytokine IFN-γ, which was shown to accumulate in the serum of aged mice, together with an elevation in BAFF, presumably from myeloid cells [129]. Lyn-deficient mice lacking IFN-γ showed a significant improvement in glomerulonephritis and their treatment with anti-BAFF antibody ameliorated glomerulonephritis. The role of IL-10 has also been explored in Lyn-deficient mice with the inflammatory environment shown to induce an expansion of IL-10-producing B cells [135]. Deletion of IL-10 from Lyn-deficient mice resulted in more severe disease characterised by splenomegaly and lymphadenopathy, dramatic increases in proinflammatory cytokines and severe tissue inflammation, supporting the key role of IL-10 in inhibiting inflammation [135].

More recent studies have dissected the role of immune cell subtypes in the development of SLE in Lyn-deficient mice. Conditional deletion of Lyn from B cells was sufficient to induce autoimmunity in mice, with disease reversed by MyD88 deficiency [136]. DC-specific deletion of Lyn also induced MyD88-dependent autoimmune disease, although disease was more rapid and more severe than that observed in the global Lyn-knockout strain [137]. DC deletion of caspase recruitment domain-containing protein 9 (CARD9), an adaptor protein downstream of TLR2 and TLR4 in DCs, reversed the development of the severe autoimmune phenotype that develops in mice lacking Lyn in DCs [138], providing further evidence for the role of TLR signalling in the development of autoimmune disease in this SLE-susceptible strain. While it is unclear why DC deletion leads to exacerbated disease, it is possible that Lyn deficiency in other cell compartments may restrain disease in mice lacking Lyn globally.

A recent study has shown that Lyn binds to IFN regulatory factor-5 (IRF5), a transcription factor that acts downstream of TLR signalling, where it inhibits the production of IFN-α [139]. Monoallelic or biallelic deletion of IRF5 ameliorated splenomegaly, autoAb production and glomerulonephritis in Lyn-deficient mice, suggesting that Lyn is a specific suppressor of this pathway, acting to inhibit TLR signalling and IFN-α production to prevent SLE development. pDCs are the major producers of IFN-α, though their numbers are not altered in Lyn-deficient mice and IFN-α is undetectable in their serum [137]. However, Lyn-deficient mice are hyper-responsive to TLR9 agonists, over-producing IFN-α, and Lyn-deficient DCs derived from the bone marrow in the presence of Flt3 ligand, which generates a high proportion of pDCs, are hyper-responsive to CpG producing more IFN-α compared to those derived from control mice [137], providing further evidence that Lyn is a negative regulator of this pathway.

B Cell/ T Cell Interactions in SLE

The generation of productive, high affinity antibody responses involves the interaction between B cells and T cells, which must be tightly regulated, and thus it is not surprising that in autoimmune diseases, a disruption of this collaboration occurs [140]. This section will describe the role of B/T lymphocyte interaction in normal immune responses and autoimmunity, with attention on the molecules facilitating this interaction, among which many are from the TNF family [48], which are potential therapeutic targets (*see Chap.* 3).

In healthy individuals, B cells and T cells cooperate in the initiation and refinement of antigen-specific antibody responses, leading to protective antibody production from mature PC and the formation of immunological memory cells to assist in subsequent immune responses [141]. Antibody responses are improved by affinity maturation in GC, which are structures formed in lymphoid tissues facilitating multiple rounds of mutation and selection of better BCR [142]. GC B cells rapidly proliferate, upregulate Fas (*TNFRSF6*; an apoptosis receptor) and mutate their BCR genes, giving rise to BCR variants that then compete for selection by Tfh cells and follicular dendritic cells (FDC), which are responsible for selecting higher-affinity BCR variants to be clonally expanded [143–145]. Disadvantageous lower-affinity or non-specific B cells, including self-reactive B cells, fail to compete for BCR engagement (which blocks *Fas* signalling) and are removed by apoptosis, maintaining immunological tolerance (discussed in section "B cell hyperactivity/loss of tolerance in SLE"). Lack of functional *Fas* or its ligand *FasL* results in severe autoimmunity and lymphoproliferation [143, 146], including if *Fas* is deleted specifically on GC B cells [147]. However, mice with a *Fas* mutation stopping palmitoylation had reduced autoimmunity despite being inefficient at apoptosis, suggesting that there are additional *Fas*-dependent tolerance mechanisms apart from apoptosis [148].

B and T cells depend on each other in this setting. $CD4^+$ Tfh precursor cells require cognate B cells to become mature Tfh cells, and then Tfh cells may licence B cells for antibody production [145, 149]. Antigen presenting cells (APC), which can be innate immune cells such as DCs, or B cells, can provide T cells with activation signalling from an antigen in the context of MHC molecules. This signal alone is not enough to activate T cells; additional stimulation such as that from CD80 and CD86 is needed, acting via CD28 expressed on the T cell [150]. CTLA-4 expression on T cells is important for negatively regulating T-cell activation, which arises when CTLA-4 outcompetes the binding of CD28 to CD80 and CD86 on APC, to replace activation signalling from CD28 with inhibitory signalling from CTLA-4 [150]. Genetic association studies suggest genetic variations in CTLA-4 co-stimulation pathway genes may be linked with SLE, and impaired suppression of pathogenic T cell activation may be a risk factor for SLE [151]. CTLA-4 antagonists aiming to increase immune activation (in settings such as cancer, to help anti-tumour responses) and CTLA-4 agonists directed at suppressing immune hyperactivity have been developed. SLE clinical trials with CTLA-4 agonists include CTLA-4 immunoglobulin Fc (CTLA4-Ig) fusion proteins abatacept and belatacept [152] (see Chap. 6).

CD40 (*TNFRSF5*) is a costimulatory receptor that is expressed on B cells and its binding partner CD40L (*TNFSF5*) on activated T cells are critical for GC responses against T-dependent antigens [153]. Important effects of CD40 activation include the upregulation of Fas and MHC molecules on B cells [154], promoting proliferation and differentiation to class-switched antibody secreting cells [155]. Lack of CD40/CD40L signalling in humans impairs antibody class switching and is linked to immunodeficiency syndromes such as hyper IgM syndrome, although CD40/CD40L signalling is considered as a potential therapeutic target in suppressing immune activation in SLE [156] (see Chap. 6).

OX40 (*TNFRSF4*) and OX40L (*TNFSF4*) are another TNF receptor/ligand pair involved in costimulation of T and B lymphocytes. In patients with LN, abundant OX40L was found in renal sections, accumulating in the glomeruli and possibly contributing to proinflammatory IC [157]. Agonistic anti-OX40 antibodies exacerbated renal disease in a murine model of SLE, whereas an OX40 antagonist fusion protein was protective, further suggesting a pathogenic role for OX40 and the potential for therapeutic targeting with antagonists [158].

Losing components of regulatory signalling or abnormal interactions between B and T cells can lead to sub-optimal antibody production or accumulation of autoAb-producing cells (Fig. 2.3). Although the immune system employs various safeguards to reduce autoimmunity, a single nucleotide mutation in a critical regulator can be enough to cause severe autoimmunity. For example, sanroque mice have a point mutation in the *Roqin* gene, which impairs the repression of inducible cell costimulatory (ICOS) expression on T cells, leading in turn to accumulation of spontaneous Tfh cells and GC driving a severe SLE-like autoimmunity phenotype [159].

B Cell Cytokine/Chemokine Network in SLE

Soluble cytokines and chemokine gradients control many activities and homing of immune cells throughout the body. Deregulation of their signalling can lead to hyperactivation of immune cells, infiltration of effector cells causing damage to tissues, and a proinflammatory environment conducive to autoimmune disease development. Several key cytokines are under extensive investigation as therapeutic targets for SLE, the foremost being BAFF and Type I interferon (IFN). The role of, and the therapeutic strategy targeting Type I IFN in SLE will be discussed in Chap. 4. "Type I interferons and the perpetuation of a loss of tolerance" and Chap. 5 "Therapeutic Modulation of the Interferon Pathway in Systemic Lupus Erythematosus".

A substantial proportion of patients with SLE have high circulating levels of BAFF (also known as *TNFSF13B* and B lymphocyte stimulator (BLyS)) [160], which is an essential B cell survival and maturation cytokine [161]. While a limited normal amount of BAFF is advantageous for maintaining B cell immune tolerance, sustained high BAFF levels impair immune tolerance by allowing the escape of low-affinity self-reactive B cells [162]. Belimumab, a humanised mAb against BAFF, showed a significant improvement in a proportion of patients with SLE, and was Food and Drug Administration (FDA)-approved for SLE (discussed further in Chap. 3), highlighting the danger of high BAFF levels in SLE.

Genetically-modified mice with high circulating levels of BAFF exhibit SLE-like disease, which does not require T cells but does require $CD19^+$ B cells and MyD88 signalling [20, 163]. MyD88 is an intracellular signalling adapter involved in innate activation by most TLR, which can broadly detect conserved foreign microbial structures termed pathogen-associated molecular patterns (PAMPs) such as bacterial cell-wall components or viral nucleic acids. MyD88 is mobilised downstream of several TLR and one of the three cognate receptors of BAFF,

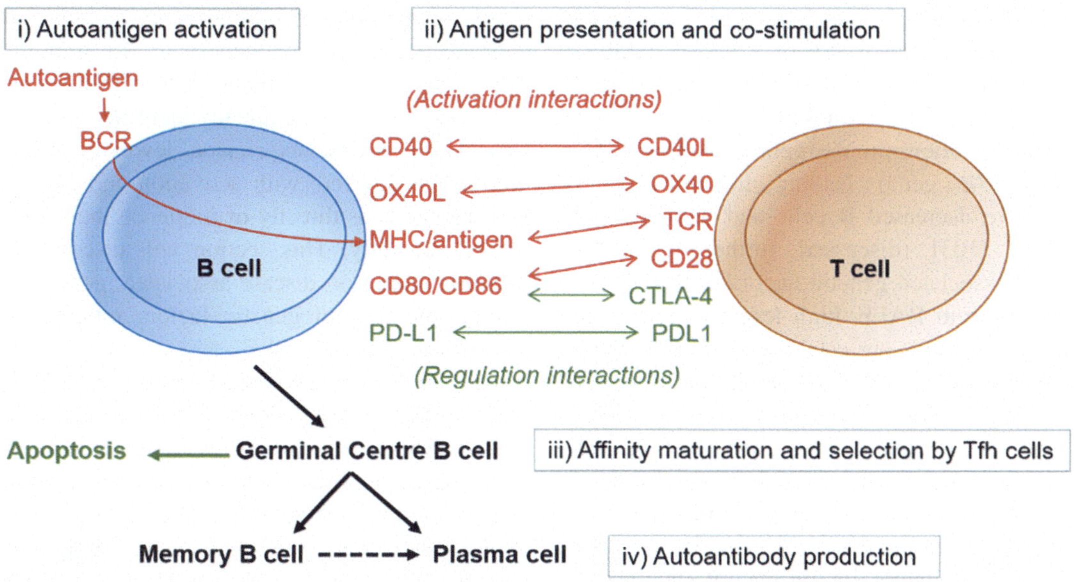

Fig. 2.3 Interactions between B cells and T cells in SLE. (i) Self-reactive B cells expressing a B cell receptor specific for a self-antigen, can internalise the BCR with antigen in a lysosome and process the antigen for presentation. (ii) Processed antigen in the context of MHC is presented to T cells, and co-stimulation is mediated by receptors and ligands expressed on B cells and T cells, which are activating (red) or regulatory (green). Defects in regulatory interactions and/or excessive activating interactions can lead to B cell hyperactivity. (iii) Germinal centre cycling repeatedly modifies the BCR and higher-affinity clones are selected by Tfh cells, or they undergo apoptosis. iv) B cells that differentiate into memory B cells persist and augment subsequent autoimmune activation, and self-reactive B cells that terminally differentiate into plasma cells can secret pathogenic proinflammatory autoantibodies

called transmembrane activator and cyclophilin ligand interactor (TACI, also known as *TNFRSF13B*), and MyD88 can transduce activation signalling to pathways such as NF-κB [164]. Such activation in subsets of innate-like B cells can trigger antibody class-switching and promote proliferation, in an antigen-non-specific manner, to provide a rapid defence to pathogens while adaptive immune responses are being generated [164]. Non-specific B cell activation by TLR ligands and BAFF can be controlled by mechanisms such as TACI-dependent upregulation of Fas, which can trigger apoptosis in TLR-activated B cells to limit their responses [165].

TACI is expressed on mature B cells and can bind to BAFF or a homologous TNF family cytokine called a proliferation-inducing ligand (APRIL, also known as *TNFSF13*). Deleting TACI protects from BAFF-mediated autoimmunity in mouse models [166–168]. In humans, lack of TACI expression is linked to impaired antibody class switching, and mutations in the gene encoding TACI (*TNFRSF13B* gene) are associated with common variable immunodeficiency (CVID) [169]. *TNFRSF13B* mutations such as C104R and A181E impair ligand binding or downstream signalling, leading to weaker memory B cell responses [170]. Membrane-bound TACI can be cleaved by a disintegrin and metalloproteinase 10 (ADAM10), producing soluble TACI (sTACI), which is found at elevated levels in SLE compared to healthy controls and may be a potential biomarker of disease [160, 171].

APRIL and BAFF bind to another receptor called B cell maturation antigen (BCMA, also known as *TNFRSF17*), which is expressed on PC and is required for long-lived PC survival [172]. BAFF transmits survival signalling through BAFF Receptor (BAFF-R, also known as

TNFRSF13), which is expressed on immature B cells. Mice lacking BAFF or BAFF-R exhibit a development block in early transitional stage B cells [173–176]. Similarly, humans with genetic variations in the gene encoding BAFF (*TNFSF13B* gene) leading to low expression have decreased B cells and antibody production [103] (discussed further in section "B cell-associated genetic factors in SLE").

Apart from BAFF, high levels of proinflammatory cytokines are established in SLE [177]. Among those, elevated IL-18, IL-17, IL-6, and IL-8 may be correlated with aspects of disease activity in SLE and/or organ damage [43], with IL-18 associated with renal involvement [178, 179]. A proportion of patients with SLE have higher levels of circulating IL-6 [180], or higher IL-6 receptor expression on their B cells [181].

Abnormalities in chemokine gradients or the receptors for sensing them can affect B cell migration and can lead to the infiltration of inflammatory immune cells into various tissues and organs, causing life-threatening damage. Of note, CXCL13 is a B cell-attracting chemokine which can be increased in SLE inflammatory lesions, and increased circulating levels are correlated with disease activity [182]. CXCL13 is normally expressed in B cell follicles in lymphoid tissues, but has been found to be aberrantly increased in kidney tissues correlating with LN [183]. The receptor of CXCL13 is CXCR5, which is required for B cell trafficking and GC formation, and circulating SLE B cells have increased CXCR5 expression suggesting a pathogenic role [184–187]. The function of CXCR3 and CXCR4 receptors on SLE B cells may also be reduced, further altering the migratory behaviours of SLE B cells [184].

B Cell-Associated Genetic Factors in SLE

A complex contribution of genetic and environmental factors may predispose individuals to SLE. Female preponderance and differences in SLE prevalence according to racial ancestry suggest genetic factors play an important role [188]. Identifying specific genetic factors and their mechanism of pathogenesis in SLE immune cells is an ongoing challenge, although many B cell-related genes were already identified to have variants or altered expression levels in SLE; these often interfere with activation and regulation mechanisms directly or influence these processes indirectly. This section will discuss the methods used for disease-associated gene discovery and several genetic factors influencing pathogenic B cells in SLE.

Genome-wide association studies (GWAS) are designed to search for variations in genomic DNA sequence which may be found in patients with disease significantly more frequently than in healthy control populations [189]. Genetic variations, including single-nucleotide polymorphisms (SNP) and copy number variations (CNV), may be tested as potential disease-associated features or risk-factors. Since many features are tested simultaneously, statistical procedures need to stringently consider the false-discovery rate (FDR) (i.e. accounting for false-positive hits which are generated by performing many tests). For some diseases with a high-penetrance single-locus cause, GWAS may identify a single high-association locus. In contrast, GWAS have identified hundreds of low-association loci in SLE. In this heterogeneous disease, a variety of genetic factors may contribute to risk of disease, but not necessarily the same factors in each patient. Apart from B cell activation genes for which there is strong evidence [13], most of the identified loci are only correlative, and complementary mechanistic investigations need to be performed to establish a causal link. Nevertheless, the identified loci may provide clues to perturbed signalling pathways or cell subsets affecting risk of autoimmunity.

There are remaining challenges in interpreting GWAS-identified loci which are not yet resolved to specific genes or are non-coding sequences which may exert more complex regulatory functions on other genes. GWAS methods also have a limitation for associating rare alleles, for which other approaches are needed [189]. Genome sequencing and pedigree studies of patient probands and their families are suited for focusing on rare risk-associated alleles such as those reported for *BLK* and *BANK1* [190]. In this case, several allele variants had a strong effect on disease risk, while being present at a very low frequency in the wider population. In the example with *BLK*, complementary mechanistic studies with mice carrying an orthologous *Blk* gene variant revealed the consequences of this allele in B cells, which caused an accumulation of pathogenic lymphocytes, and in human B cell lines, the SLE-associated gene variants impaired suppression of IRF5 and Type I IFN [190].

SLE transcriptomic studies have been performed using RNA microarrays, or RNA-sequencing (RNA-seq), or more recently single-cell RNA-seq (scRNA-seq) [191]. These methods compare the expression level of all genes in cohorts of patients with SLE and healthy controls. Expression analyses are most often targeted to peripheral blood mononuclear cells (PBMC), providing a valuable snapshot of circulating immune system cells. Several studies have investigated stratification of SLE patients based on gene expression patterns, to help resolve patient heterogeneity and to distinguish clusters of patients with different pathogenesis mechanisms or clinical features [192–194]. For instance, well-characterised SLE-associated B cell genes such as *TNFSF13B (BAFF)* were strongly upregulated in some clusters of patients, while the B cell homeostasis gene *PAX5* was downregulated in patients with flares [192].

Several variants in the *TNFSF13B* (BAFF) gene were associated with increased risk of SLE and atherosclerosis disease features, suggesting a role for pathogenic B cells in SLE-related vascular pathologies [195]. A SLE-associated gene variant of *TNFSF13B* (BAFF) was identified with a shorter transcript which escapes miRNA regulation, resulting in pathogenic higher expression of BAFF [13].

Growth factor receptor-bound protein 2 (GRB2), a mediator of RAS signalling [196], plays a role in B cells directing GC formation and promoting lymphotoxin-β expression (197), and GRB2 plays a role regulating B cell memory responses [198]. Genetic variants of *GRB2* have also been associated with SLE by GWAS, and one gene variant was found with possibly protective effects [199], exemplifying how different genetic variations in the same gene can potentially have opposite consequences.

Perturbations in transcription factors and chromatin accessibility can have important effects on B cells, affecting their differentiation and activation. Chromatin accessibility, examined by Assay for Transposase-Accessible Chromatin (ATAC)-seq, revealed SLE B cells have increased accessibility at certain genes including: *NRF1*, *CTCF*, and *STAT5* gene promoter regions, as well as enrichment for B cell activating transcription factors including NFKB, AP-1, and BATF [200]. Impairment of the Ets-1 transcription factor promotes more differentiation to PC, by reducing transcription of Pax5 and B cell identity genes and loss of repression of Blimp-1 (a transcription factor for PC differentiation) [201]. Another example is *STAT4*; variants have been identified in SLE [202–204] which affect its role as a transcriptional regulator mediating IL-12 signalling [205]. Increased IL-12 inhibits differentiation of B cells to GC and promotes differentiation to short-lived PB [206].

In summary, the heterogeneity of SLE involves a multiplicity of possible pathogenesis contributors, although as may be expected, genetic variations affecting B cells are highly relevant in SLE. Genetic studies in the past decades have pointed us towards a large number of disease-associated B cell genes, as summarised in Table 2.1. Complementary mechanistic studies are needed to continue to dissect the specific mechanisms by which altered B cell genes contribute to the risk of SLE and worsen severity, taking us beyond correlations to a deeper understanding of the numerous underlying causes of disease.

This non-exhaustive list suggests a range of possible disease mechanisms including impaired tolerance (removal of autoreactive cells), defective B cell development and differentiation, hyperactivation by receptors (BCR, TLR), inflammation, and complex indirect effects on other genes and signalling pathways. Rather than monogenic causes of disease, factors such as these contribute to disease risk or increase

Table 2.1 Numerous genes modulating B cells that may increase SLE risk and/or severity when they are disrupted, mutated or overexpressed

B cell genes	Pathogenic Abnormalities	References
BANK1	Variant forms of the BANK1 scaffolding may alter BCR/CD40 signalling and increase memory B cell generation	[190, 207]
BLK	Impaired BLK reduces repression of IRF5 and type I interferon	[190]
Carabin	Lack of Carabin removes repression of autoreactive B cells stimulated with CpG-DNA via TLR9	[208]
ETS1	Deficiency of Ets-1 transcription factor weakens B cell identity genes (with Pax5) and increases progression to plasma cells (via Blimp-1)	[201]
Fkbp11	Overexpression impairs tolerance to DNA and promotes plasma cell differentiation via the *Pax5* regulator gene	[209]
GRB2	Impaired GRB2 reduces regulation of B cell maturation and memory responses; gene variant associations increase with complement activation	[196–199]
LYN	Decreased Lyn reduces regulation of BCR signalling, since Lyn normally dampens BCR signals by several mechanisms, such as activating negative receptors and mediating BCR internalisation	[210]
TNFSF13B (*BAFF*)	Overexpression can occur by transcripts lacking mRNA level regulation. Excess BAFF impairs B cell tolerance, allowing more self-reactive B cells to persist. High BAFF levels are pro-inflammatory, correlate with autoAbs and global disease severity, and are associated with cardiovascular involvement	[13, 160, 162, 192, 195]
Trib1	Loss of negative regulation (via CD72 binding partner) of BCR activation and IgG production	[211]

severity in established disease, along with a large number of possible genetic and environmental factors.

References

1. Lisnevskaia L, Murphy G, Isenberg D (2014) Systemic lupus erythematosus. Lancet 384 (9957):1878–1888
2. Vincent FB, Bourke P, Morand EF, Mackay F, Bossingham D (2013) Focus on systemic lupus erythematosus in indigenous Australians: towards a better understanding of autoimmune diseases. Intern Med J 43(3):227–234
3. Navarra SV, Guzman RM, Gallacher AE, Hall S, Levy RA, Jimenez RE et al (2011) Efficacy and safety of belimumab in patients with active systemic lupus erythematosus: a randomised, placebo-controlled, phase 3 trial. Lancet 377(9767):721–731
4. Furie R, Petri M, Zamani O, Cervera R, Wallace DJ, Tegzova D et al (2011) A phase III, randomized, placebo-controlled study of belimumab, a monoclonal antibody that inhibits B lymphocyte stimulator, in patients with systemic lupus erythematosus. Arthritis Rheum 63(12):3918–3930
5. Vincent FB, Morand EF, Schneider P, Mackay F (2014) The BAFF/APRIL system in SLE pathogenesis. Nat Rev Rheumatol 10(6):365–373
6. Shlomchik MJ, Madaio MP, Ni D, Trounstein M, Huszar D (1994) The role of B cells in lpr/lpr-induced autoimmunity. J Exp Med 180(4):1295–1306
7. Chan O, Shlomchik MJ (1998) A new role for B cells in systemic autoimmunity: B cells promote spontaneous T cell activation in MRL-lpr/lpr mice. J Immunol 160(1):51–59
8. Harder KW, Quilici C, Naik E, Inglese M, Kountouri N, Turner A et al (2004) Perturbed myelo/erythropoiesis in Lyn-deficient mice is similar to that in mice lacking the inhibitory phosphatases SHP-1 and SHIP-1. Blood 104(13):3901–3910
9. Tsantikos E, Oracki SA, Quilici C, Anderson GP, Tarlinton DM, Hibbs ML (2010) Autoimmune disease in Lyn-deficient mice is dependent on an inflammatory environment established by IL-6. J Immunol 184(3):1348–1360
10. Zhang J, Roschke V, Baker KP, Wang Z, Alarcon GS, Fessler BJ et al (2001) Cutting edge: a role for B lymphocyte stimulator in systemic lupus erythematosus. J Immunol 166(1):6–10
11. Stohl W, Metyas S, Tan SM, Cheema GS, Oamar B, Xu D et al (2003) B lymphocyte stimulator overexpression in patients with systemic lupus

erythematosus: longitudinal observations. Arthritis Rheum 48(12):3475–3486
12. Pers JO, Daridon C, Devauchelle V, Jousse S, Saraux A, Jamin C et al (2005) BAFF overexpression is associated with autoantibody production in autoimmune diseases. Ann N Y Acad Sci 1050:34–39
13. Steri M, Orru V, Idda ML, Pitzalis M, Pala M, Zara I et al (2017) Overexpression of the cytokine BAFF and autoimmunity risk. N Engl J Med 376 (17):1615–1626
14. Sakai J, Akkoyunlu M (2017) The role of BAFF system molecules in host response to pathogens. Clin Microbiol Rev 30(4):991–1014
15. Nardelli B, Belvedere O, Roschke V, Moore PA, Olsen HS, Migone TS et al (2001) Synthesis and release of B-lymphocyte stimulator from myeloid cells. Blood 97(1):198–204
16. Craxton A, Magaletti D, Ryan EJ, Clark EA (2003) Macrophage- and dendritic cell–dependent regulation of human B-cell proliferation requires the TNF family ligand BAFF. Blood 101(11):4464–4471
17. Scapini P, Nardelli B, Nadali G, Calzetti F, Pizzolo G, Montecucco C et al (2003) G-CSF-stimulated neutrophils are a prominent source of functional BLyS. J Exp Med 197(3):297–302
18. Ittah M, Miceli-Richard C, Eric Gottenberg J, Lavie F, Lazure T, Ba N et al (2006) B cell-activating factor of the tumor necrosis factor family (BAFF) is expressed under stimulation by interferon in salivary gland epithelial cells in primary Sjogren's syndrome. Arthritis Res Ther 8(2):R51
19. Mackay F, Woodcock SA, Lawton P, Ambrose C, Baetscher M, Schneider P et al (1999) Mice transgenic for BAFF develop lymphocytic disorders along with autoimmune manifestations. J Exp Med 190(11):1697–1710
20. Groom JR, Fletcher CA, Walters SN, Grey ST, Watt SV, Sweet MJ et al (2007) BAFF and MyD88 signals promote a lupuslike disease independent of T cells. J Exp Med 204(8):1959–1971
21. Litinskiy MB, Nardelli B, Hilbert DM, He B, Schaffer A, Casali P et al (2002) DCs induce CD40-independent immunoglobulin class switching through BLyS and APRIL. Nat Immunol 3 (9):822–829
22. Belnoue E, Pihlgren M, McGaha TL, Tougne C, Rochat AF, Bossen C et al (2008) APRIL is critical for plasmablast survival in the bone marrow and poorly expressed by early-life bone marrow stromal cells. Blood 111(5):2755–2764
23. Koyama T, Tsukamoto H, Miyagi Y, Himeji D, Otsuka J, Miyagawa H et al (2005) Raised serum APRIL levels in patients with systemic lupus erythematosus. Ann Rheum Dis 64(7):1065–1067
24. Hegazy M, Darwish H, Darweesh H, El-Shehaby A, Emad Y (2010) Raised serum level of APRIL in patients with systemic lupus erythematosus: correlations with disease activity indices. Clin Immunol 135(1):118–124
25. Treamtrakanpon W, Tantivitayakul P, Benjachat T, Somparn P, Kittikowit W, Eiam-ong S et al (2012) APRIL, a proliferation-inducing ligand, as a potential marker of lupus nephritis. Arthritis Res Ther 14 (6):R252
26. Varfolomeev E, Kischkel F, Martin F, Seshasayee D, Wang H, Lawrence D et al (2004) APRIL-deficient mice have normal immune system development. Mol Cell Biol 24(3):997–1006
27. da Silva LS, Almeida BL, de Melo AK, de Brito DC, Braz AS, Freire EA (2016) IgA nephropathy in systemic lupus erythematosus patients: case report and literature review. Rev Bras Reumatol Engl Ed 56(3):270–273
28. Braun D, Caramalho I, Demengeot J (2002) IFN-alpha/beta enhances BCR-dependent B cell responses. Int Immunol 14(4):411–419
29. Linker-Israeli M, Deans RJ, Wallace DJ, Prehn J, Ozeri-Chen T, Klinenberg JR (1991) Elevated levels of endogenous IL-6 in systemic lupus erythematosus. A putative role in pathogenesis. J Immunol 147(1):117–123
30. Abdel Galil SM, Ezzeldin N, El-Boshy ME (2015) The role of serum IL-17 and IL-6 as biomarkers of disease activity and predictors of remission in patients with lupus nephritis. Cytokine 76(2):280–287
31. Lu R, Munroe ME, Guthridge JM, Bean KM, Fife DA, Chen H et al (2016) Dysregulation of innate and adaptive serum mediators precedes systemic lupus erythematosus classification and improves prognostic accuracy of autoantibodies. J Autoimmun 74:182–193
32. Tanaka T, Narazaki M, Kishimoto T (2014) IL-6 in inflammation, immunity, and disease. Cold Spring Harb Perspect Biol 6(10):a016295
33. Fujimoto M, Nakano M, Terabe F, Kawahata H, Ohkawara T, Han Y et al (2011) The influence of excessive IL-6 production in vivo on the development and function of Foxp3+ regulatory T cells. J Immunol 186(1):32–40
34. Ozaki K, Spolski R, Ettinger R, Kim HP, Wang G, Qi CF et al (2004) Regulation of B cell differentiation and plasma cell generation by IL-21, a novel inducer of Blimp-1 and Bcl-6. J Immunol 173 (9):5361–5371
35. Linterman MA, Beaton L, Yu D, Ramiscal RR, Srivastava M, Hogan JJ et al (2010) IL-21 acts directly on B cells to regulate Bcl-6 expression and germinal center responses. J Exp Med 207(2):353–363
36. Dolff S, Abdulahad WH, Westra J, Doornbos-van der Meer B, Limburg PC, Kallenberg CG et al (2011) Increase in IL-21 producing T-cells in patients with systemic lupus erythematosus. Arthritis Res Ther 13(5):R157
37. Terrier B, Costedoat-Chalumeau N, Garrido M, Geri G, Rosenzwajg M, Musset L et al (2012) Interleukin 21 correlates with T cell and B cell

subset alterations in systemic lupus erythematosus. J Rheumatol 39(9):1819–1828
38. Nakou M, Papadimitraki ED, Fanouriakis A, Bertsias GK, Choulaki C, Goulidaki N et al (2013) Interleukin-21 is increased in active systemic lupus erythematosus patients and contributes to the generation of plasma B cells. Clin Exp Rheumatol 31 (2):172–179
39. Sawalha AH, Kaufman KM, Kelly JA, Adler AJ, Aberle T, Kilpatrick J et al (2008) Genetic association of interleukin-21 polymorphisms with systemic lupus erythematosus. Ann Rheum Dis 67 (4):458–461
40. Webb R, Merrill JT, Kelly JA, Sestak A, Kaufman KM, Langefeld CD et al (2009) A polymorphism within IL21R confers risk for systemic lupus erythematosus. Arthritis Rheum 60(8):2402–2407
41. Eto D, Lao C, DiToro D, Barnett B, Escobar TC, Kageyama R et al (2011) IL-21 and IL-6 are critical for different aspects of B cell immunity and redundantly induce optimal follicular helper CD4 T cell (Tfh) differentiation. PLoS One 6(3): e17739
42. Wong CK, Lit LC, Tam LS, Li EK, Wong PT, Lam CW (2008) Hyperproduction of IL-23 and IL-17 in patients with systemic lupus erythematosus: implications for Th17-mediated inflammation in auto-immunity. Clin Immunol 127(3):385–393
43. Vincent FB, Northcott M, Hoi A, Mackay F, Morand EF (2013) Clinical associations of serum interleukin-17 in systemic lupus erythematosus. Arthritis Res Ther 15(4):R97
44. Crispin JC, Oukka M, Bayliss G, Cohen RA, Van Beek CA, Stillman IE et al (2008) Expanded double negative T cells in patients with systemic lupus erythematosus produce IL-17 and infiltrate the kidneys. J Immunol 181(12):8761–8766
45. Shah K, Lee WW, Lee SH, Kim SH, Kang SW, Craft J et al (2010) Dysregulated balance of Th17 and Th1 cells in systemic lupus erythematosus. Arthritis Res Ther 12(2):R53
46. Mitsdoerffer M, Lee Y, Jager A, Kim HJ, Korn T, Kolls JK et al (2010) Proinflammatory T helper type 17 cells are effective B-cell helpers. Proc Natl Acad Sci USA 107(32):14292–14297
47. Hsu HC, Yang P, Wang J, Wu Q, Myers R, Chen J et al (2008) Interleukin 17-producing T helper cells and interleukin 17 orchestrate autoreactive germinal center development in autoimmune BXD2 mice. Nat Immunol 9(2):166–175
48. Figgett WA, Vincent FB, Saulep-Easton D, Mackay F (2014) Roles of ligands from the TNF superfamily in B cell development, function, and regulation. Semin Immunol 26(3):191–202
49. Wehr C, Eibel H, Masilamani M, Illges H, Schlesier M, Peter HH et al (2004) A new CD21low B cell population in the peripheral blood of patients with SLE. Clin Immunol 113(2):161–171
50. Wardowska A, Komorniczak M, Skoniecka A, Bullo-Piontecka B, Lisowska KA, Debska-Slizien MA et al (2020) Alterations in peripheral blood B cells in systemic lupus erythematosus patients with renal insufficiency. Int Immunopharmacol 83:106451
51. Biajoux V, Bignon A, Freitas C, Martinez V, Thelen M, Lima G et al (2012) Expression of CXCL12 receptors in B cells from Mexican Mestizos patients with systemic Lupus erythematosus. J Transl Med 10:251
52. Peng Y, Guo F, Liao S, Liao H, Xiao H, Yang L et al (2020) Altered frequency of peripheral B-cell subsets and their correlation with disease activity in patients with systemic lupus erythematosus: a comprehensive analysis. J Cell Mol Med 24 (20):12044–12053
53. Zhu L, Yin Z, Ju B, Zhang J, Wang Y, Lv X et al (2018) Altered frequencies of memory B cells in new-onset systemic lupus erythematosus patients. Clin Rheumatol 37(1):205–212
54. Jenks SA, Cashman KS, Zumaquero E, Marigorta UM, Patel AV, Wang X et al (2018) Distinct effector B cells induced by unregulated toll-like receptor 7 contribute to pathogenic responses in systemic lupus erythematosus. Immunity 49 (4):725–39 e6
55. Yap DYH, Yung S, Lee P, Yam IYL, Tam C, Tang C et al (2020) B cell subsets and cellular signatures and disease relapse in lupus nephritis. Front Immunol 11:1732
56. Ye Z, Jiang Y, Sun D, Zhong W, Zhao L, Jiang Z (2019) The plasma interleukin (IL)-35 level and frequency of circulating IL-35(+) regulatory B cells are decreased in a cohort of Chinese patients with new-onset systemic lupus erythematosus. Sci Rep 9 (1):13210
57. Fang C, Luo T, Lin L (2017) The correlational research among serum CXCL13 levels, circulating plasmablasts and memory B cells in patients with systemic lupus erythematosus: a STROBE-compliant article. Medicine (Baltimore) 96(48): e8675
58. Wei C, Anolik J, Cappione A, Zheng B, Pugh-Bernard A, Brooks J et al (2007) A new population of cells lacking expression of CD27 represents a notable component of the B cell memory compartment in systemic lupus erythematosus. J Immunol 178(10):6624–6633
59. Jacobi AM, Reiter K, Mackay M, Aranow C, Hiepe F, Radbruch A et al (2008) Activated memory B cell subsets correlate with disease activity in systemic lupus erythematosus: delineation by expression of CD27, IgD, and CD95. Arthritis Rheum 58(6):1762–1773
60. Wu C, Fu Q, Guo Q, Chen S, Goswami S, Sun S et al (2019) Lupus-associated atypical memory B cells are mTORC1-hyperactivated and functionally dysregulated. Ann Rheum Dis 78(8):1090–1100
61. Liu Y, Zhou S, Qian J, Wang Y, Yu X, Dai D et al (2017) T-bet(+)CD11c(+) B cells are critical for

antichromatin immunoglobulin G production in the development of lupus. Arthritis Res Ther 19(1):225
62. Wang S, Wang J, Kumar V, Karnell JL, Naiman B, Gross PS et al (2018) IL-21 drives expansion and plasma cell differentiation of autoreactive CD11c (hi)T-bet(+) B cells in SLE. Nat Commun 9(1):1758
63. Nicholas MW, Dooley MA, Hogan SL, Anolik J, Looney J, Sanz I et al (2008) A novel subset of memory B cells is enriched in autoreactivity and correlates with adverse outcomes in SLE. Clin Immunol 126(2):189–201
64. Jacobi AM, Odendahl M, Reiter K, Bruns A, Burmester GR, Radbruch A et al (2003) Correlation between circulating CD27high plasma cells and disease activity in patients with systemic lupus erythematosus. Arthritis Rheum 48(5):1332–1342
65. Blair PA, Norena LY, Flores-Borja F, Rawlings DJ, Isenberg DA, Ehrenstein MR et al (2010) CD19(+) CD24(hi)CD38(hi) B cells exhibit regulatory capacity in healthy individuals but are functionally impaired in systemic Lupus Erythematosus patients. Immunity 32(1):129–140
66. Menon M, Blair PA, Isenberg DA, Mauri C (2016) A regulatory feedback between plasmacytoid dendritic cells and regulatory B cells is aberrant in systemic lupus erythematosus. Immunity 44 (3):683–697
67. Shen P, Roch T, Lampropoulou V, O'Connor RA, Stervbo U, Hilgenberg E et al (2014) IL-35-producing B cells are critical regulators of immunity during autoimmune and infectious diseases. Nature 507(7492):366–370
68. Yang X, Yang J, Chu Y, Xue Y, Xuan D, Zheng S et al (2014) T follicular helper cells and regulatory B cells dynamics in systemic lupus erythematosus. PLoS One 9(2):e88441
69. Yoshizaki A, Miyagaki T, DiLillo DJ, Matsushita T, Horikawa M, Kountikov EI et al (2012) Regulatory B cells control T-cell autoimmunity through IL-21-dependent cognate interactions. Nature 491 (7423):264–268
70. Park MJ, Lee SH, Kim EK, Lee EJ, Park SH, Kwok SK et al (2016) Myeloid-derived suppressor cells induce the expansion of regulatory B cells and ameliorate autoimmunity in the sanroque mouse model of systemic lupus erythematosus. Arthritis Rheumatol 68(11):2717–2727
71. Amrouche K, Pers JO, Jamin C (2019) Glatiramer acetate stimulates regulatory B cell functions. J Immunol 202(7):1970–1980
72. Arbuckle MR, McClain MT, Rubertone MV, Scofield RH, Dennis GJ, James JA et al (2003) Development of autoantibodies before the clinical onset of systemic lupus erythematosus. N Engl J Med 349(16):1526–1533
73. Slight-Webb S, Lu R, Ritterhouse LL, Munroe ME, Maecker HT, Fathman CG et al (2016) Autoantibody-positive healthy individuals display unique immune profiles that may regulate autoimmunity. Arthritis & rheumatology. 68 (10):2492–2502
74. Flechsig A, Rose T, Barkhudarova F, Strauss R, Klotsche J, Dahnrich C et al (2017) What is the clinical significance of anti-Sm antibodies in systemic lupus erythematosus? A comparison with anti-dsDNA antibodies and C3. Clin Exp Rheumatol 35(4):598–606
75. Vaglio A, Grayson PC, Fenaroli P, Gianfreda D, Boccaletti V, Ghiggeri GM et al (2018) Drug-induced lupus: traditional and new concepts. Autoimmun Rev 17(9):912–918
76. Aringer M, Costenbader K, Daikh D, Brinks R, Mosca M, Ramsey-Goldman R et al (2019) 2019 European League Against Rheumatism/American College of Rheumatology classification criteria for systemic lupus erythematosus. Arthritis & Rheumatol 71(9):1400–1412
77. Han S, Zhuang H, Shumyak S, Yang L, Reeves WH (2015) Mechanisms of autoantibody production in systemic lupus erythematosus. Front Immunol 6:228
78. Pisetsky DS, Lipsky PE (2020) New insights into the role of antinuclear antibodies in systemic lupus erythematosus. Nat Rev Rheumatol 16(10):565–579
79. Pan N, Amigues I, Lyman S, Duculan R, Aziz F, Crow MK et al (2014) A surge in anti-dsDNA titer predicts a severe lupus flare within six months. Lupus 23(3):293–298
80. Yaniv G, Twig G, Shor DB, Furer A, Sherer Y, Mozes O et al (2015) A volcanic explosion of autoantibodies in systemic lupus erythematosus: a diversity of 180 different antibodies found in SLE patients. Autoimmun Rev 14(1):75–79
81. Sciascia S, Bertolaccini ML, Roccatello D, Khamashta MA, Sanna G (2014) Autoantibodies involved in neuropsychiatric manifestations associated with systemic lupus erythematosus: a systematic review. J Neurol 261(9):1706–1714
82. Becker Y, Marcoux G, Allaeys I, Julien AS, Loignon RC, Benk-Fortin H et al (2019) Autoantibodies in systemic lupus erythematosus target mitochondrial RNA. Front Immunol 10:1026
83. Gupta S, Tatouli IP, Rosen LB, Hasni S, Alevizos I, Manna ZG et al (2016) Distinct functions of autoantibodies against interferon in systemic lupus erythematosus: a comprehensive analysis of anticytokine autoantibodies in common rheumatic diseases. Arthritis Rheumatol 68(7):1677–1687
84. Howe HS, Thong BYH, Kong KO, Chng HH, Lian TY, Chia FL et al (2017) Associations of B cell-activating factor (BAFF) and anti-BAFF autoantibodies with disease activity in multi-ethnic Asian systemic lupus erythematosus patients in Singapore. Clin Exp Immunol 189(3):298–303
85. Eggleton P, Ukoumunne OC, Cottrell I, Khan A, Maqsood S, Thornes J et al (2014) Autoantibodies against C1q as a diagnostic measure of lupus nephritis: systematic review and meta-analysis. J Clin Cell Immunol. 5(2):210

86. Birmingham DJ, Bitter JE, Ndukwe EG, Dials S, Gullo TR, Conroy S et al (2016) Relationship of circulating anti-C3b and anti-C1q IgG to lupus nephritis and its flare. Clin J Am Soc Nephrol 11 (1):47–53
87. Mannik M, Merrill CE, Stamps LD, Wener MH (2003) Multiple autoantibodies form the glomerular immune deposits in patients with systemic lupus erythematosus. J Rheumatol 30(7):1495–1504
88. Tsao BP, Ohnishi K, Cheroutre H, Mitchell B, Teitell M, Mixter P et al (1992) Failed self-tolerance and autoimmunity in IgG anti-DNA transgenic mice. J Immunol 149(1):350–358
89. Stopforth RJ, Oldham RJ, Tutt AL, Duriez P, Chan HTC, Binkowski BF et al (2018) Detection of experimental and clinical immune complexes by measuring SHIP-1 recruitment to the inhibitory FcgammaRIIB. J Immunol 200(5):1937–1950
90. Massardo L, Bravo-Zehnder M, Calderon J, Flores P, Padilla O, Aguirre JM et al (2015) Anti-N-methyl-D-aspartate receptor and anti-ribosomal-P autoantibodies contribute to cognitive dysfunction in systemic lupus erythematosus. Lupus 24(6):558–568
91. Kowal C, Degiorgio LA, Lee JY, Edgar MA, Huerta PT, Volpe BT et al (2006) Human lupus autoantibodies against NMDA receptors mediate cognitive impairment. Proc Natl Acad Sci U S A 103(52):19854–19859
92. Yoshio T, Onda K, Nara H, Minota S (2006) Association of IgG anti-NR2 glutamate receptor antibodies in cerebrospinal fluid with neuropsychiatric systemic lupus erythematosus. Arthritis Rheum 54(2):675–678
93. Choi MY, FitzPatrick RD, Buhler K, Mahler M, Fritzler MJ (2020) A review and meta-analysis of anti-ribosomal P autoantibodies in systemic lupus erythematosus. Autoimmun Rev 19(3):102463
94. Gonzalez A, Massardo L (2018) Antibodies and the brain: antiribosomal P protein antibody and the clinical effects in patients with systemic lupus erythematosus. Curr Opin Neurol 31(3):300–305
95. Vincent FB, Morand EF, Mackay F (2012) BAFF and innate immunity: new therapeutic targets for systemic lupus erythematosus. Immunol Cell Biol 90(3):293–303
96. Kanapathippillai P, Hedberg A, Fenton CG, Fenton KA (2013) Nucleosomes contribute to increase mesangial cell chemokine expression during the development of lupus nephritis. Cytokine 62 (2):244–252
97. Vincent FB, Kandane-Rathnayake R, Hoi AY, Slavin L, Godsell JD, Kitching AR et al (2018) Urinary B-cell-activating factor of the tumour necrosis factor family (BAFF) in systemic lupus erythematosus. Lupus 27(13):2029–2040
98. Yu CL, Sun KH, Tsai CY, Hsieh SC, Yu HS (2001) Anti-dsDNA antibody up-regulates interleukin 6, but not cyclo-oxygenase, gene expression in glomerular mesangial cells: a marker of immune-mediated renal damage? Inflamm Res 50(1):12–18
99. Yung S, Zhang Q, Zhang CZ, Chan KW, Lui SL, Chan TM (2009) Anti-DNA antibody induction of protein kinase C phosphorylation and fibronectin synthesis in human and murine lupus and the effect of mycophenolic acid. Arthritis Rheum 60(7):2071–2082
100. Apel F, Zychlinsky A, Kenny EF (2018) The role of neutrophil extracellular traps in rheumatic diseases. Nat Rev Rheumatol 14(8):467–475
101. van der Linden M, van den Hoogen LL, Westerlaken GHA, Fritsch-Stork RDE, van Roon JAG, Radstake T et al (2018) Neutrophil extracellular trap release is associated with antinuclear antibodies in systemic lupus erythematosus and anti-phospholipid syndrome. Rheumatology (Oxford) 57(7):1228–1234
102. Mackay F, Figgett WA, Saulep D, Lepage M, Hibbs ML (2010) B-cell stage and context-dependent requirements for survival signals from BAFF and the B-cell receptor. Immunol Rev 237 (1):205–225
103. Smulski CR, Eibel H (2018) BAFF and BAFF-receptor in B Cell selection and survival. Front Immunol 9:2285
104. Schweighoffer E, Tybulewicz VL (2018) Signalling for B cell survival. Curr Opin Cell Biol 51:8–14
105. Xu Y, Harder KW, Huntington ND, Hibbs ML, Tarlinton DM (2005) Lyn tyrosine kinase; accentuating the positive and the negative. Immunity 22 (1):9–18
106. Smith KG, Tarlinton DM, Doody GM, Hibbs ML, Fearon DT (1998) Inhibition of the B cell by CD22: a requirement for Lyn. J Exp Med 187(5):807–811
107. Cornall RJ, Cyster JG, Hibbs ML, Dunn AR, Otipoby KL, Clark EA et al (1998) Polygenic autoimmune traits: Lyn, CD22, and SHP-1 are limiting elements of a biochemical pathway regulating BCR signaling and selection. Immunity 8 (4):497–508
108. Chan VWF, Lowell CA, DeFranco AL (1998) Defective negative regulation of antigen receptor signaling in lyn-deficient B lymphocytes. Curr Biol 8(10):545–553
109. Nishizumi H, Horikawa K, Mlinaric-Rascan I, Yamamoto T (1998) A double-edged kinase Lyn: a positive and negative regulator for antigen receptor-mediated signals. J Exp Med 187 (8):1343–1348
110. Inaoki M, Sato S, Weintraub BC, Goodnow CC, Tedder TF (1997) CD19-regulated signaling thresholds control peripheral tolerance and autoantibody production in B lymphocytes. J Exp Med 186 (11):1923–1931
111. Hibbs ML, Tarlinton DM, Armes J, Grail D, Hodgson G, Maglitto R et al (1995) Multiple defects in the immune system of lyn-deficient mice, culminating in autoimmune disease. Cell 83:301–311

112. Nishizumi H, Taniuchi I, Yamanashi Y, Kitamura D, Iiic D, Mori S et al (1995) Impaired proliferation of peripheral B cells and indication of autoimmune disease in Lyn-deficient mice. Immunity 3:549–560
113. Chan VWF, Meng F, Soriano P, DeFranco AL, Lowell CA (1997) Characterization of the B lymphocyte populations in lyn deficient mice and the role of lyn in signal initiation and down regulation. Immunity 7:69–81
114. Bolland S, Ravetch JV (2000) Spontaneous autoimmune disease in Fc(gamma)RIIB-deficient mice results from strain-specific epistasis. Immunity 13 (2):277–285
115. O'Keefe TL, Williams GT, Batista FD, Neuberger MS (1999) Deficiency in CD22, a B cell-specific inhibitory receptor, is sufficient to predispose to development of high affinity autoantibodies. J Exp Med 189(8):1307–1313
116. Shultz LD, Schweitzer PA, Rajan TV, Yi T, Ihle JN, Matthews RJ et al (1993) Mutations at the murine motheaten locus are within the hematopoietic cell protein-tyrosine phosphatase (Hcph) gene. Cell 73:1445–1454
117. Tsui HW, Siminovitch KA, de Souza L, Tsui FW (1993) Motheaten and viable motheaten mice have mutations in the hematopoietic cell phosphatase gene. Nat Genet 4:124–129
118. Shultz LD, Coman DR, Bailey CL, Beamer WG, Sidman CL (1984) Viable motheaten," a new allele at the motheaten locus. I Pathology Am J Pathol 116(2):179–192
119. Pao LI, Lam KP, Henderson JM, Kutok JL, Alimzhanov M, Nitschke L et al (2007) B cell-specific deletion of protein-tyrosine phosphatase Shp1 promotes B-1a cell development and causes systemic autoimmunity. Immunity 27(1):35–48
120. Di Cristofano A, Kotsi P, Peng YF, Cordon-Cardo C, Elkon KB, Pandolfi PP (1999) Impaired fas response and autoimmunity in Pten+/- mice. Science 285(5436):2122–2125
121. Moody JL, Pereira CG, Magil A, Fritzler MJ, Jirik FR (2003) Loss of a single allele of SHIP exacerbates the immunopathology of Pten heterozygous mice. Genes Immun 4(1):60–66
122. Suzuki A, Kaisho T, Ohishi M, Tsukio-Yamaguchi M, Tsubata T, Koni PA et al (2003) Critical roles of Pten in B cell homeostasis and immunoglobulin class switch recombination. J Exp Med 197(5):657–667
123. Helgason CD, Damen JE, Rosten P, Grewal R, Sorensen P, Chappel SM et al (1998) Targeted disruption of SHIP leads to hemopoietic perturbations, lung pathology, and a shortened life span. Genes Dev 12:1610–1620
124. Liu Q, Sasaki T, Kozieradzki I, Wakeham A, Itie A, Dumont DJ et al (1999) SHIP is a negative regulator of growth factor receptor-mediated PKB/Akt activation and myeloid cell survival. Genes Dev 13:786–791
125. Hibbs ML, Raftery AL, Tsantikos E (2018) Regulation of hematopoietic cell signaling by SHIP-1 inositol phosphatase: growth factors and beyond. Growth Factors 36:213–231
126. Maxwell MJ, Duan M, Armes JE, Anderson GP, Tarlinton DM, Hibbs ML (2011) Genetic segregation of inflammatory lung disease and autoimmune disease severity in SHIP-1-/- mice. J Immunol 186 (12):7164–7175
127. O'Neill SK, Getahun A, Gauld SB, Merrell KT, Tamir I, Smith MJ et al (2011) Monophosphorylation of CD79a and CD79b ITAM motifs initiates a SHIP-1 phosphatase-mediated inhibitory signaling cascade required for B cell anergy. Immunity 35 (5):746–756
128. Yasuda T, Bundo K, Hino A, Honda K, Inoue A, Shirakata M et al (2007) Dok-1 and Dok-2 are negative regulators of T cell receptor signaling. Int Immunol 19(4):487–495
129. Scapini P, Hu Y, Chu CL, Migone TS, Defranco AL, Cassatella MA et al (2010) Myeloid cells, BAFF, and IFN-gamma establish an inflammatory loop that exacerbates autoimmunity in Lyn-deficient mice. J Exp Med 207(8):1757–1773
130. Oracki SA, Tsantikos E, Quilici C, Light A, Schmidt T, Lew AM et al (2010) CTLA4Ig alters the course of autoimmune disease development in Lyn-/- mice. J Immunol 184(2):757–763
131. Hua Z, Gross AJ, Lamagna C, Ramos-Hernandez N, Scapini P, Ji M et al (2014) Requirement for MyD88 signaling in B cells and dendritic cells for germinal center anti-nuclear antibody production in Lyn-deficient mice. J Immunol 192(3):875–885
132. Silver KL, Crockford TL, Bouriez-Jones T, Milling S, Lambe T, Cornall RJ (2007) MyD88-dependent autoimmune disease in Lyn-deficient mice. Eur J Immunol 37(10):2734–2743
133. Gutierrez T, Halcomb KE, Coughran AJ, Li QZ, Satterthwaite AB (2010) Separate checkpoints regulate splenic plasma cell accumulation and IgG autoantibody production in Lyn-deficient mice. Eur J Immunol 40(7):1897–1905
134. Maxwell MJ, Tsantikos E, Kong AM, Vanhaesebroeck B, Tarlinton DM, Hibbs ML (2012) Attenuation of phosphoinositide 3-kinase delta signaling restrains autoimmune disease. J Autoimmun 38 (4):381–391
135. Scapini P, Lamagna C, Hu Y, Lee K, Tang Q, DeFranco AL et al (2011) B cell-derived IL-10 suppresses inflammatory disease in Lyn-deficient mice. Proc Natl Acad Sci USA 108(41):E823–E832
136. Lamagna C, Hu Y, DeFranco AL, Lowell CA (2014) B cell-specific loss of Lyn kinase leads to autoimmunity. J Immunol 192(3):919–928
137. Lamagna C, Scapini P, van Ziffle JA, DeFranco AL, Lowell CA (2013) Hyperactivated MyD88 signaling in dendritic cells, through specific deletion of Lyn kinase, causes severe autoimmunity and inflammation. Proc Natl Acad Sci USA 110(35): E3311–E3320

138. Ma J, Abram CL, Hu Y, Lowell CA (2019) CARD9 mediates dendritic cell-induced development of Lyn deficiency-associated autoimmune and inflammatory diseases. Sci Signal 12(602)
139. Ban T, Sato GR, Nishiyama A, Akiyama A, Takasuna M, Umehara M et al (2016) Lyn kinase suppresses the transcriptional activity of IRF5 in the TLR-MyD88 pathway to restrain the development of autoimmunity. Immunity 45(2):319–332
140. Petersone L, Edner NM, Ovcinnikovs V, Heuts F, Ross EM, Ntavli E et al (2018) T cell/B cell collaboration and autoimmunity: an intimate relationship. Front Immunol 9:1941
141. Mesin L, Ersching J, Victora GD (2016) Germinal center B cell dynamics. Immunity 45(3):471–482
142. De Silva NS, Klein U (2015) Dynamics of B cells in germinal centres. Nat Rev Immunol 15(3):137–148
143. Takahashi Y, Ohta H, Takemori T (2001) Fas is required for clonal selection in germinal centers and the subsequent establishment of the memory B cell repertoire. Immunity 14(2):181–192
144. Strasser A, Jost PJ, Nagata S (2009) The many roles of FAS receptor signaling in the immune system. Immunity 30(2):180–192
145. Crotty S (2011) Follicular helper CD4 T cells (TFH). Annu Rev Immunol 29:621–663
146. Watanabe-Fukunaga R, Brannan CI, Copeland NG, Jenkins NA, Nagata S (1992) Lymphoproliferation disorder in mice explained by defects in Fas antigen that mediates apoptosis. Nature 356(6367):314–317
147. Hao Z, Duncan GS, Seagal J, Su YW, Hong C, Haight J et al (2008) Fas receptor expression in germinal-center B cells is essential for T and B lymphocyte homeostasis. Immunity 29(4):615–627
148. Cruz AC, Ramaswamy M, Ouyang C, Klebanoff CA, Sengupta P, Yamamoto TN et al (2016) Fas/CD95 prevents autoimmunity independently of lipid raft localization and efficient apoptosis induction. Nat Commun 7:13895
149. Biram A, Davidzohn N, Shulman Z (2019) T cell interactions with B cells during germinal center formation, a three-step model. Immunol Rev 288 (1):37–48
150. Beyersdorf N, Kerkau T, Hunig T (2015) CD28 co-stimulation in T-cell homeostasis: a recent perspective. Immunotargets Ther 4:111–122
151. Julia A, Lopez-Longo FJ, Perez Venegas JJ, Bonas-Guarch S, Olive A, Andreu JL et al (2018) Genome-wide association study meta-analysis identifies five new loci for systemic lupus erythematosus. Arthritis Res Ther 20(1):100
152. Crepeau RL, Ford ML (2017) Challenges and opportunities in targeting the CD28/CTLA-4 pathway in transplantation and autoimmunity. Expert Opin Biol Ther 17(8):1001–1012
153. Kawabe T, Naka T, Yoshida K, Tanaka T, Fujiwara H, Suematsu S et al (1994) The immune responses in CD40-deficient mice: impaired immunoglobulin class switching and germinal center formation. Immunity 1(3):167–178
154. Garrone P, Neidhardt EM, Garcia E, Galibert L, van Kooten C, Banchereau J (1995) Fas ligation induces apoptosis of CD40-activated human B lymphocytes. J Exp Med 182(5):1265–1273
155. Rousset F, Garcia E, Banchereau J (1991) Cytokine-induced proliferation and immunoglobulin production of human B lymphocytes triggered through their CD40 antigen. J Exp Med 173 (3):705–710
156. Karnell JL, Rieder SA, Ettinger R, Kolbeck R (2019) Targeting the CD40-CD40L pathway in autoimmune diseases: humoral immunity and beyond. Adv Drug Deliv Rev 141:92–103
157. Aten J, Roos A, Claessen N, Schilder-Tol EJ, Ten Berge IJ, Weening JJ (2000) Strong and selective glomerular localization of CD134 ligand and TNF receptor-1 in proliferative lupus nephritis. J Am Soc Nephrol 11(8):1426–1438
158. Sitrin J, Suto E, Wuster A, Eastham-Anderson J, Kim JM, Austin CD et al (2017) The Ox40/Ox40 ligand pathway promotes pathogEnic Th cell responses, plasmablast accumulation, and lupus nephritis in NZB/W F1 mice. J Immunol 199 (4):1238–1249
159. Yu D, Tan AH, Hu X, Athanasopoulos V, Simpson N, Silva DG et al (2007) Roquin represses autoimmunity by limiting inducible T-cell co-stimulator messenger RNA. Nature 450 (7167):299–303
160. Vincent FB, Kandane-Rathnayake R, Koelmeyer R, Hoi AY, Harris J, Mackay F et al (2019) Analysis of serum B cell-activating factor from the tumor necrosis factor family (BAFF) and its soluble receptors in systemic lupus erythematosus. Clin Transl Immunol 8(4):e01047
161. Mackay F, Woodcock SA, Lawton P, Ambrose C, Baetscher M, Schneider P et al (1999) Mice transgenic for BAFF develop lymphocytic disorders along with autoimmune manifestations. J Exp Med 190(11):1697–1710
162. Thien M, Phan TG, Gardam S, Amesbury M, Basten A, Mackay F et al (2004) Excess BAFF rescues self-reactive B cells from peripheral deletion and allows them to enter forbidden follicular and marginal zone niches. Immunity 20(6):785–798
163. Fairfax KA, Tsantikos E, Figgett WA, Vincent FB, Quah PS, LePage M et al (2015) BAFF-driven autoimmunity requires CD19 expression. J Autoimmun 62:1–10
164. He B, Santamaria R, Xu W, Cols M, Chen K, Puga I et al (2010) The transmembrane activator TACI triggers immunoglobulin class switching by activating B cells through the adaptor MyD88. Nat Immunol 11(9):836–845
165. Figgett WA, Fairfax K, Vincent FB, Le Page MA, Katik I, Deliyanti D et al (2013) The TACI receptor regulates T-cell-independent marginal zone B cell responses through innate activation-induced cell death. Immunity 39(3):573–583

166. Figgett WA, Deliyanti D, Fairfax KA, Quah PS, Wilkinson-Berka JL, Mackay F (2015) Deleting the BAFF receptor TACI protects against systemic lupus erythematosus without extensive reduction of B cell numbers. J Autoimmun 61:9–16
167. Jacobs HM, Thouvenel CD, Leach S, Arkatkar T, Metzler G, Scharping NE et al (2016) Cutting edge: BAFF promotes autoantibody production via TACI-dependent activation of transitional B cells. J Immunol 196(9):3525–3531
168. Arkatkar T, Jacobs HM, Du SW, Li QZ, Hudkins KL, Alpers CE et al (2018) TACI deletion protects against progressive murine lupus nephritis induced by BAFF overexpression. Kidney Int 94 (4):728–740
169. Salzer U, Bacchelli C, Buckridge S, Pan-Hammarstrom Q, Jennings S, Lougaris V et al (2009) Relevance of biallelic versus monoallelic TNFRSF13B mutations in distinguishing disease-causing from risk-increasing TNFRSF13B variants in antibody deficiency syndromes. Blood 113 (9):1967–1976
170. Romberg N, Virdee M, Chamberlain N, Oe T, Schickel JN, Perkins T et al (2015) TNF receptor superfamily member 13b (TNFRSF13B) hemizygosity reveals transmembrane activator and CAML interactor haploinsufficiency at later stages of B-cell development. J Allergy Clin Immunol 136(5):1315–1325
171. Hoffmann FS, Kuhn PH, Laurent SA, Hauck SM, Berer K, Wendlinger SA et al (2015) The immunoregulator soluble TACI is released by ADAM10 and reflects B cell activation in autoimmunity. J Immunol 194(2):542–552
172. O'Connor BP, Raman VS, Erickson LD, Cook WJ, Weaver LK, Ahonen C et al (2004) BCMA is essential for the survival of long-lived bone marrow plasma cells. J Exp Med 199(1):91–98
173. Sasaki Y, Casola S, Kutok JL, Rajewsky K, Schmidt-Supprian M (2004) TNF family member B cell-activating factor (BAFF) receptor-dependent and -independent roles for BAFF in B cell physiology. J Immunol 173(4):2245–2252
174. Ng LG, Sutherland AP, Newton R, Qian F, Cachero TG, Scott ML et al (2004) B cell-activating factor belonging to the TNF family (BAFF)-R is the principal BAFF receptor facilitating BAFF costimulation of circulating T and B cells. J Immunol 173(2):807–817
175. Schiemann B, Gommerman JL, Vora K, Cachero TG, Shulga-Morskaya S, Dobles M et al (2001) An essential role for BAFF in the normal development of B cells through a BCMA-independent pathway. Science 293(5537):2111–2114
176. Mackay F, Schneider P, Rennert P, Browning J (2003) BAFF AND APRIL: a tutorial on B cell survival. Annu Rev Immunol 21:231–264
177. Gottschalk TA, Tsantikos E, Hibbs ML (2015) Pathogenic inflammation and its therapeutic targeting in systemic lupus erythematosus. Front Immunol 6:550
178. Ruchakorn N, Ngamjanyaporn P, Suangtamai T, Kafaksom T, Polpanumas C, Petpisit V et al (2019) Performance of cytokine models in predicting SLE activity. Arthritis Res Ther 21(1):287
179. Mende R, Vincent FB, Kandane-Rathnayake R, Koelmeyer R, Lin E, Chang J et al (2018) Analysis of serum inteRleukin (IL)-1beta and IL-18 in systemic lupus erythematosus. Front Immunol 9:1250
180. Grondal G, Gunnarsson I, Ronnelid J, Rogberg S, Klareskog L, Lundberg I (2000) Cytokine production, serum levels and disease activity in systemic lupus erythematosus. Clin Exp Rheumatol 18 (5):565–570
181. Nagafuchi H, Suzuki N, Mizushima Y, Sakane T (1993) Constitutive expression of IL-6 receptors and their role in the excessive B cell function in patients with systemic lupus erythematosus. J Immunol 151(11):6525–6534
182. Schiffer L, Kumpers P, Davalos-Misslitz AM, Haubitz M, Haller H, Anders HJ et al (2009) B-cell-attracting chemokine CXCL13 as a marker of disease activity and renal involvement in systemic lupus erythematosus (SLE). Nephrol Dial Transplant 24(12):3708–3712
183. Ishikawa S, Sato T, Abe M, Nagai S, Onai N, Yoneyama H et al (2001) Aberrant high expression of B lymphocyte chemokine (BLC/CXCL13) by C11b+CD11c+ dendritic cells in murine lupus and preferential chemotaxis of B1 cells towards BLC. J Exp Med 193(12):1393–1402
184. Henneken M, Dorner T, Burmester GR, Berek C (2005) Differential expression of chemokine receptors on peripheral blood B cells from patients with rheumatoid arthritis and systemic lupus erythematosus. Arthritis Res Ther 7(5):R1001–R1013
185. Wiener A, Schippers A, Wagner N, Tacke F, Ostendorf T, Honke N et al (2016) CXCR5 is critically involved in progression of lupus through regulation of B cell and double-negative T cell trafficking. Clin Exp Immunol 185(1):22–32
186. Wu BX, Zhao LD, Zhang X (2019) CXCR4 and CXCR5 orchestrate dynamic germinal center reactions and may contribute to the pathogenesis of systemic lupus erythematosus. Cell Mol Immunol 16(8):724–726
187. Lee HT, Shiao YM, Wu TH, Chen WS, Hsu YH, Tsai SF et al (2010) Serum BLC/CXCL13 concentrations and renal expression of CXCL13/CXCR5 in patients with systemic lupus erythematosus and lupus nephritis. J Rheumatol 37(1):45–52
188. Breitbach ME, Ramaker RC, Roberts K, Kimberly RP, Absher D (2020) Population-specific patterns of epigenetic defects in the B cell lineage in patients with systemic lupus erythematosus. Arthritis Rheumatol 72(2):282–291
189. Brookfield JF (2010) Q&A: promise and pitfalls of genome-wide association studies. BMC Biol 8:41

190. Jiang SH, Athanasopoulos V, Ellyard JI, Chuah A, Cappello J, Cook A et al (2019) Functional rare and low frequency variants in BLK and BANK1 contribute to human lupus. Nat Commun 10 (1):2201
191. Mistry P, Nakabo S, O'Neil L, Goel RR, Jiang K, Carmona-Rivera C et al (2019) Transcriptomic, epigenetic, and functional analyses implicate neutrophil diversity in the pathogenesis of systemic lupus erythematosus. Proc Natl Acad Sci USA 116 (50):25222–25228
192. Figgett WA, Monaghan K, Ng M, Alhamdoosh M, Maraskovsky E, Wilson NJ et al (2019) Machine learning applied to whole-blood RNA-sequencing data uncovers distinct subsets of patients with systemic lupus erythematosus. Clin Transl Immunol 8(12):e01093
193. Panousis NI, Bertsias GK, Ongen H, Gergianaki I, Tektonidou MG, Trachana M et al (2019) Combined genetic and transcriptome analysis of patients with SLE: distinct, targetable signatures for susceptibility and severity. Ann Rheum Dis 78(8):1079–1089
194. Rai R, Chauhan SK, Singh VV, Rai M, Rai G (2016) RNA-seq analysis reveals unique transcriptome signatures in systemic lupus erythematosus patients with distinct autoantibody specificities. PLoS One 11(11):e0166312
195. Theodorou E, Nezos A, Antypa E, Ioakeimidis D, Koutsilieris M, Tektonidou M et al (2018) B-cell activating factor and related genetic variants in lupus related atherosclerosis. J Autoimmun 92:87–92
196. Lowenstein EJ, Daly RJ, Batzer AG, Li W, Margolis B, Lammers R et al (1992) The SH2 and SH3 domain-containing protein GRB2 links receptor tyrosine kinases to ras signaling. Cell 70(3):431–442
197. Jang IK, Cronshaw DG, Xie LK, Fang G, Zhang J, Oh H et al (2011) Growth-factor receptor-bound protein-2 (Grb2) signaling in B cells controls lymphoid follicle organization and germinal center reaction. Proc Natl Acad Sci U S A 108(19):7926–7931
198. Ackermann JA, Radtke D, Maurberger A, Winkler TH, Nitschke L (2011) Grb2 regulates B-cell maturation, B-cell memory responses and inhibits B-cell Ca2+ signalling. EMBO J 30(8):1621–1633
199. Xu M, Liu Y, Li X, Cheng C, Liu Y, Dong W et al (2019) Evaluation of genetic susceptibility between systemic lupus erythematosus and GRB2 gene. Sci Rep 9(1):10335
200. Scharer CD, Blalock EL, Barwick BG, Haines RR, Wei C, Sanz I et al (2016) ATAC-seq on biobanked specimens defines a unique chromatin accessibility structure in naive SLE B cells. Sci Rep 6:27030
201. Garrett-Sinha LA, Kearly A, Satterthwaite AB (2016) The role of the transcription factor ets1 in lupus and other autoimmune diseases. Crit Rev Immunol 36(6):485–510
202. Beltran Ramirez O, Mendoza Rincon JF, Barbosa Cobos RE, Aleman Avila I, Ramirez BJ (2016) STAT4 confers risk for rheumatoid arthritis and systemic lupus erythematosus in Mexican patients. Immunol Lett 175:40–43
203. Palomino-Morales RJ, Rojas-Villarraga A, Gonzalez CI, Ramirez G, Anaya JM, Martin J (2008) STAT4 but not TRAF1/C5 variants influence the risk of developing rheumatoid arthritis and systemic lupus erythematosus in Colombians. Genes Immun 9(4):379–382
204. Remmers EF, Plenge RM, Lee AT, Graham RR, Hom G, Behrens TW et al (2007) STAT4 and the risk of rheumatoid arthritis and systemic lupus erythematosus. N Engl J Med 357(10):977–986
205. Saelee P, Kearly A, Nutt SL, Garrett-Sinha LA (2017) Genome-wide identification of target genes for the key B cell transcription factor Ets1. Front Immunol 8:383
206. Kim SJ, Caton M, Wang C, Khalil M, Zhou ZJ, Hardin J et al (2008) Increased IL-12 inhibits B cells' differentiation to germinal center cells and promotes differentiation to short-lived plasmablasts. J Exp Med 205(10):2437–2448
207. Dam EM, Habib T, Chen J, Funk A, Glukhova V, Davis-Pickett M et al (2016) The BANK1 SLE-risk variants are associated with alterations in peripheral B cell signaling and development in humans. Clin Immunol 173:171–180
208. Schickel JN, Pasquali JL, Soley A, Knapp AM, Decossas M, Kern A et al (2012) Carabin deficiency in B cells increases BCR-TLR9 costimulation-induced autoimmunity. EMBO Mol Med 4 (12):1261–1275
209. Ruer-Laventie J, Simoni L, Schickel JN, Soley A, Duval M, Knapp AM et al (2015) Overexpression of Fkbp11, a feature of lupus B cells, leads to B cell tolerance breakdown and initiates plasma cell differentiation. Immun Inflamm Dis 3(3):265–279
210. Brodie EJ, Infantino S, Low MSY, Tarlinton DM (2018) Lyn, lupus, and (B) lymphocytes, a lesson on the critical balance of kinase signaling in immunity. Front Immunol 9:401
211. Simoni L, Delgado V, Ruer-Laventie J, Bouis D, Soley A, Heyer V et al (2018) Trib1 is overexpressed in systemic lupus erythematosus, while it regulates immunoglobulin production in murine B cells. Front Immunol 9:373

B Cell-Targeted Therapies in Systemic Lupus Erythematosus

3

Fabien B. Vincent, William A. Figgett, and Margaret L. Hibbs

Abstract

Most high-profile clinical trials of targeted therapies, including those targeting B cells, have failed in systemic lupus erythematosus (SLE). Currently, only one biological therapy has been approved for this condition, belimumab, a monoclonal antibody neutralizing a key B cell survival factor, B cell activating factor of the tumour necrosis factor (TNF) family (BAFF). However, results with this therapeutic are modest and it is not widely used and thus, the search for better therapeutics continue. The reasons for such across the board failure, which has become the norm in SLE clinical trials, as opposed to the numerous therapeutic successes in other autoimmune diseases, appear multifactorial and include a potential combination of suboptimal target selection, patient stratification, use of concomitant immunosuppressive drugs and/or outcome measures as primary endpoint. Lessons can be learned, as much from successful as from failed clinical trials, and clinicians now have a careful focus on optimisation of the study design of future trials via better target identification, patient selection, and use of high-quality outcome measures. In this Chapter, we will review the current status of B cell-targeted therapies in SLE, including targeting strategy achieved via B cell depletion (CD20), through negative regulation of the B cell antigen receptor (BCR) signalling (CD22, Fc gamma receptor (FcγR)IIb), or via neutralizing one or several components of the BAFF/a proliferation-inducing ligand (APRIL) system.

F. B. Vincent (✉)
Rheumatology Research Group, Centre for Inflammatory Diseases, School of Clinical Sciences at Monash Health, Monash University, Clayton, Victoria 3168, Australia
e-mail: fabien.vincent@monash.edu

W. A. Figgett
Department of Microbiology and Immunology at the Peter Doherty Institute for Infection and Immunity, University of Melbourne, Melbourne, VIC 3000, Australia
e-mail: w.figgett@garvan.org.au

Kinghorn Centre for Clinical Genomics, Garvan Institute of Medical Research, Darlinghurst, NSW 2010, Australia

M. L. Hibbs
Department of Immunology and Pathology, Central Clinical School, Alfred Research Alliance (ARA), Monash University, 89 Commercial Road, Melbourne, Victoria 3004, Australia
e-mail: margaret.hibbs@monash.edu

Introduction

The past two decades have witnessed the emergence of clinical trials focused on B cells in systemic lupus erythematosus (SLE). B cell-

A. Hoi (ed.), *Pathogenesis of Systemic Lupus Erythematosus*,
https://doi.org/10.1007/978-3-030-85161-3_3

targeting can be achieved via B cell depletion (CD20), negative regulation of B cell antigen receptor (BCR) signalling (CD22, Fc gamma receptor (FcγR)IIb), or via neutralizing one or several components of the B cell activating factor of the tumour necrosis factor (TNF) family (BAFF)/a proliferation-inducing ligand (APRIL) system. Among many failed high-profile clinical trials, the phase 3 trials of belimumab, a monoclonal antibody (mAb) targeting BAFF, stands out, as being the first, and currently sole, biological to be approved in SLE (1, 2) (Table 3.1). At least 16 clinical trials are ongoing in adult and paediatric SLE and lupus nephritis (LN), where 15 are targeting CD20 and/or the BAFF/APRIL system (BAFF or BAFF receptor (BAFF-R)) (Table 3.1). In this chapter, we will first provide an overview of the tools used as primary endpoint in these clinical trials, which is essential information in order to understand the outcomes from those trials.

Disease Activity Assessment and Scoring System: Towards a Treat-to-Target (T2T) Strategy in SLE

In the following sub-sections, several scoring systems will be mentioned, such as the SLE Disease Activity Index 2000 (SLEDAI-2K) (3), Safety of Estrogen in Lupus Erythematosus National Assessment-SLEDAI (SELENA-SLEDAI), British Isles Lupus Assessment Group (BILAG), SLE Responder Index (SRI) or the BILAG-based Combined Lupus Assessment (BICLA). These validated scores are used in clinical trials and research settings to assess the activity of the disease in patients with SLE. These are weighted composite scores encompassing several clinical and laboratory parameters, attempting to capture the wide range of possible symptoms in SLE and give an overall activity estimate. Unfortunately, these scores are not perfect, and more research is needed to develop better tools to assess disease activity and severity in SLE. Among tools currently in development, a new scoring system to assess the state of low disease activity in SLE is very promising: the Lupus Low Disease Activity State (LLDAS) (4). This score has been developed by the Asia Pacific Lupus Collaboration (APLC) led by Monash University researchers Prof. Eric Morand and A/Prof. Alberta Hoi, and has now been validated and used by various groups across the world (5–7). If approved by the FDA, it may be used as a primary endpoint in future SLE clinical trials, becoming the first validated T2T approach in SLE.

BAFF-Specific Neutralising Therapy

Belimumab

Belimumab is a fully human recombinant immunoglobulin (Ig)$G_1\lambda$ mAb targeting the cytokine BAFF, originally reported to only inhibit its soluble form (8). However, recent work showed that its trans-membrane form is also antagonised by belimumab in vitro (9, 10). Belimumab-neutralised BAFF cannot transduce signals through its cognate receptors: BAFF-R, transmembrane activator and cyclophilin ligand interactor (TACI) and B cell maturation antigen (BCMA). This, in turn, leads to compromised B cell survival, which is thought to reduce the generation of new autoreactive B cells, although existing memory B cells and plasma cells (PC) may be more difficult to remove (11).

Two successful phase 3 clinical trials, BLISS-52 (N = 865) and BLISS-76 (N = 819), made history achieving FDA approval in March 2011 (1, 2). In those phase 3 clinical trials, the primary endpoint was met (SRI-4), significant improvement on the risk of severe flare and corticosteroid sparing effect were observed, and safety and tolerability profiles were considered acceptable. This agent's therapeutic effects included a reduction in anti-double stranded DNA (dsDNA) autoantibodies (autoAbs) and increase in complement component 3 (C3) and 4 (C4) levels, as well as reduction in numbers of B cells and some B cell subsets (naïve, PC, activated, but not memory). Of note, in BLISS-76, primary endpoint was met only with the dose of 10 mg/kg

Table 3.1 Ongoing clinical trials with B cell-targeted therapy in SLE

	Drug	Target	Condition	Trial ID	Phase	Study design					Status
						Sample size	Study period (weeks)	RCT	Background SOC	Main outcome measures	
Monotherapy	rituximab	CD20	Musculoskeletal and Mucocutaneous SLE	NCT03054259[1]	2	30	26	Yes	CS	BICLA, SRI	Recruiting
			Moderate-to-Severe SLE	NCT04127747	4	110	104	No	N/A	remission (score not specified)	Recruiting
	obinutuzumab	CD20	ISN/RPS 2003 Class III or IV LN	NCT02550652	2	126	52	Yes	MMF/MPA	Protocol-defined CRR, overall response	Active, not recruiting
			ISN/RPS 2003 Class III or IV LN	NCT04221477	3	250	52 to 80	Yes	MMF, CS	CRR, overall renal response, SLEDAI-2K, biological measure (including autoAbs, complements, B cells), PK	Recruiting
	KPG-818	CRBN E3 ubiquitin ligase complex	Mild-to-moderate SLE	NCT04643067	1b/2a	64	16	Yes	AM, CS, IS	PK, SELENA-SLEDAI, CLASI	Not yet recruiting
	belimumab	BAFF	Early SLE	NCT03543839	4	30	104	Yes	N/A	Biological measure (including B cell subsets), SRI, LLDAS, remission	Recruiting
			Paediatric SLE	NCT01649765	2	93	52	Yes	Yes, not specified	SRI, PRINTO/ACR Juvenile SLE Response Evaluation criteria, Parent's Global Assessment, SELENA-SLEDAI	Active, not recruiting
			Paediatric SLE	NCT04179032	2	30	68	No	NSAIDs, AM, CS, IS	PK, PD, biological measure (including autoAbs, complements, B cells and B cell subsets)	Recruiting
			Seropositive SLE	NCT01705977	4	4019	52	Yes	NSAIDs, AM, CS, IS	Death, adverse events, CS tapering	Active, not recruiting
			SLE	NCT04515719	4	334	52	Yes	AM, CS, IS	Flare, SELENA-SLEDAI, BILAG	Not yet recruiting
			SLE	NCT04447053	4	80	48	No	NSAIDs, AM, CS, IS	Biological measure (including autoAbs, complements, T cell subsets, TCR sequencing and RNA expression of the variable region), cognitive function	Not yet recruiting
	ianalumab (VAY736)	BAFF-R	SLE	NCT03656562[2]	2	120	29 to 53	Yes	AM, CS, IS	SRI-4, flare, PK	Recruiting
	telitacicept (RC18)	BAFF + APRIL	SLE	NCT04082416	3	318	52	Yes	NSAIDs, AM, CS, IS	SRI, SELENA-SLEDAI, CS taper, flare	Recruiting
	AMG 570	BAFF & ICOSL	SLE	NCT04058028	2	300	52	Yes	AM, CS, IS	SRI-4, BILAG, LLDAS, CLASI, tender & swollen joint count, PK,	Recruiting

(continued)

Table 3.1 (continued)

	Drug	Target	Condition	Trial ID	Phase	Study design					Status
						Sample size	Study period (weeks)	RCT	Background SOC	Main outcome measures	
Combination therapy	belimumab + rituximab	BAFF + CD20	SLE	NCT03312907	3	292	52 to 104	Yes	AM, CS, NSAIDs	SLEDAI-2K, clinical SLEDAI-2K	Active, not recruiting
			Severe SLE	NCT03747159	2	30	104	No	CS, MMF	Biological measure (including autoAbs, B cell subsets, NET formation), SLEDAI	Recruiting

Data extracted from https://clinicaltrials.gov. (30/11/2020)

[1] biosimilar GP2013

[2] The clinical trial also investigates iscalimab (CFZ533) targeting CD40

ACR: American college of rheumatology; AM: antimalarial; autoAbs: autoantibodies; BICLA: the British Isles Lupus Assessment Group-based Combined Lupus Assessment; BAFF: B cell-activating factor from the tumor necrosis factor family; BAFF-R: BAFF receptor; BILAG: the British Isles Lupus Assessment Group; CD: cluster of differentiation; CLASI: cutaneous lupus erythematosus disease area and severity index; CRR: complete renal response; CS: corticosteroid; ICOSL: inducible T cell costimulator ligand; IS: immunosuppressants; ISN/RPS: international society of nephrology/renal pathology society; LLDAS: lupus low disease activity state; LN: lupus nephritis; MMF: mycophenolate mofetil; MPA: mycophenolic acid; N/A: not available; NSAIDs: non-steroidal anti-inflammatory drugs; PD: pharmacodynamics; PK: pharmacokinetics; PRINTO: Paediatric Rheumatology International Trials Organization; RCT: randomised blinded placebo-controlled clinical trial; SELENA-SLEDAI: Safety of Estrogen in Lupus Erythematosus National Assessment-SLE Disease Activity Index; SLE: systemic lupus erythematosus; SLEDAI: systemic lupus erythematosus disease activity index; SOC: Standard of care; SRI: SLE responder index; TCR: T cell receptor but not 1 mg/kg (2). Interestingly, in pooled analysis from BLISS-52 and BLISS-76 trials, a significant improvement in fatigue and health-related quality of life at week 52 were observed in patients receiving belimumab compared to placebo (12). Another *post-hoc* analysis suggested benefit for belimumab-treated patients regarding the use of corticosteroids (13).

It is worth noting that patients with severe forms of neurological and nephritis manifestations were excluded from these two trials. However, *post-hoc* analysis from the pooled BLISS-52 and BLISS-76 data suggests benefit by preventing renal manifestations (14, 15), as well as ameliorating musculoskeletal and mucocutaneous manifestations (14). Potential predictive markers of clinical response to belimumab were identified as a baseline high disease activity score (SELENA-SLEDAI $\geq$ 10), immunological activity as per low complement and anti-dsDNA positivity as well as the use of corticosteroids (16). Importantly, another successful phase 3 RCT of belimumab (BLISS-LN, N = 448) has recently been published, demonstrating significant clinical benefits in LN (17). Based on this trial, belimumab has been approved as a LN therapeutic biological agent by the FDA in late 2020.

Another phase 3 RCT of belimumab (i.v. 10 mg/kg only) in 677 patients with SLE conducted in North East Asia (18) produced positive results in line with BLISS-52 and BLISS-76; primary endpoint (SRI-4) was met, and significant efficacy was observed on risk of severe flare and corticosteroid use (18).

Pooled data from two open-label long-term follow-up studies (NCT00724867 and NCT00712933) subsequent to both BLISS-52 and BLISS-76 suggests low organ damage accrual in patients over 5 years of treatment with belimumab (19). An 8-year follow-up study of the BLISS-52 and BLISS-76 trials confirmed low damage accrual for long-term belimumab-treated SLE patients (20).

Further development followed with the approval of a subcutaneous self-injectable form of this agent, based on a successful phase 3 RCT of belimumab (BLISS-SC) in SLE patients, also meeting its primary endpoint (SRI-4) (21) in line with the BLISS-52 and BLISS-76 trials.

Belimumab has also been evaluated in paediatric SLE, now being approved in USA, Europe and Japan for children of 5 years of age and older. A phase 2 RCT (PLUTO, N = 93) evaluated this agent in children from 5 to 17 years of age suffering from SLE without severe LN, and reported clinical benefits in line with adult phase 3 RCT, also using SRI-4 as the primary endpoint (22). The prevalence of childhood-onset SLE is much lower than adulthood-onset SLE, explaining why this trial design did not permit statistical analysis to be performed due to the expected small sample size.

Prevalence of anti-drug antibodies (ADAb) was reported at 0.2% in belimumab-treated patient versus 6% in placebo in one phase 3 RCT (18), while no LN patients (17) and no children (22) receiving this agent tested positive for ADAb.

Belimumab is being further investigated in ongoing clinical trials as a biological monotherapy in adult (phase 4, N = 4) and paediatric (phase 2, N = 2) SLE (Table 3.1).

Tabalumab

Tabalumab (LY2127399) is another biological therapy that neutralizes both the soluble and trans-membrane forms of BAFF. It is a fully human IgG_4 mAb. Two Phase 1 clinical trials were published together, with one study (A) pooling patients with SLE (N = 6) and rheumatoid arthritis (RA) (N = 23), and the second (B) only investigating the drug in RA patients (N = 24) [23]. Focusing here on study A, the drug was well tolerated, and a decrease in total $CD20^+$ B cells was observed with no difference in this pattern between SLE and RA patients [23]. However, sample size was too small for any meaningful statistical analysis interpretations.

The phase 3 ILLUMINATE-1 (1164 moderate-to-severe SLE patients; 52 weeks) did not meet its primary (SRI-5 at week 52) and key secondary endpoints (steroid sparing effect, time to severe flare, fatigue, depression, suicide) [24]. However, it is worth noting that patients who had doses of antimalarial or immunosuppressant drugs decreased during this trial were considered non-responders. When performing sensitivity analysis, removing this non-responder definition, the subsequent 'modified' primary endpoint SRI-5 at week 52 was met for tabalumab dose of 120 mg every four weeks compared to placebo. The biological effect of tabalumab includes significant increase in C3 and C4, and decrease in anti-dsDNA autoAbs and Ig (IgG, IgA, IgM) levels and total and naïve B cells [24]. The drug was considered safe and well tolerated.

In the phase 3 ILLUMINATE-2 clinical trial, studying 1124 moderate-to-severe SLE patients for 52 weeks, primary efficacy endpoint (SRI-5 at week 52) was met with tabalumab dose of 120 mg every two weeks compared to placebo [25]. Importantly, as opposed to ILLUMINATE-1 study design, patients who experienced decreases in antimalarial or immunosuppressant drug dose were not deemed non-responders in trial design. As for ILLUMINATE-1 [24], key secondary endpoints were not met (steroid sparing effect, time to severe flare, fatigue). The drug was considered safe and well tolerated, but as opposed to ILLUMINATE-1 [24], rates of depression and suicidality were higher in tabalumab-treated individuals [25]. Biological effects were also noted, including significant decrease in anti-dsDNA autoAbs and Ig (G, A, M) levels and total B cell count, and increase in C3 and C4 levels.

In phase 3 ILLUMINATE-1 and -2 studies, patients with severe active LN were excluded [24, 25], and 219 (10%) patients had renal involvement as assessed by the SELENA-SLEDAI [26]. Data from both trials related to efficacy of tabalumab on renal involvement in SLE, as another predefined secondary endpoint, were published separately [26]. No significant difference in renal function (serum creatine levels, estimated glomerular filtration rate (eGFR)), renal SLE disease activity as assessed by proteinuria (urine protein/creatine ratio (UPCR)), renal flare, and renal adverse event were observed between tabalumab and placebo groups [26].

It is worth mentioning that the observed increased levels of serum BAFF in SLE patients receiving tabalumab in both phase 3 clinical trials could be considered counterintuitive [24, 25], but is explained by the fact that the assay used can detect the complex of BAFF and tabalumab. Of note, baseline whole blood gene expression profile (Affymetrix Human Transcriptome Array 2.0; NanoString) was not predictive of response to tabalumab at 52 weeks (performed in a subset of 1760 from 2288 SLE patients from pooled phase 3 clinical trials) [27]. Interestingly, baseline interferon (IFN) signature in these patients was not affected by 1 year of tabalumab treatment [27].

In the light of the insufficient efficacy of tabalumab in the two phase 3 studies [24, 25], Eli Lilly and Company decided not to progress with the development for SLE. Hence, one open-label long-term safety and efficacy phase 3 study (NCT01488708) and another phase 3 study evaluating different modes of injection (NCT02041091) were terminated. There are currently no recruiting or active clinical trial of tabalumab in SLE (*Source* https://clinicaltrials.gov/).

No increase in ADAb was observed following drug administration in the phase 1 trial [23]. In phase 3 ILLUMINATE-1 and -2 studies [24, 25], low rates of ADAb were observed, with no neutralising Ab noted in tabalumab-treated groups (1 patient with neutralising Ab in placebo group in ILLUMINATE-2), while intriguingly higher in the placebo group in ILLUMINATE-1.

Blisibimod

The peptibody blisibimod (also known as AMG 623, A-623), a fusion protein encompassing a human IgG_1 Fc portion linked to a high-affinity tetrameric BAFF binding domain, is another anti-BAFF targeting therapy, inhibiting both soluble and trans-membrane forms of this cytokine [28].

Same as for belimumab, the mode of action of blisibimod is to prevent BAFF from binding to its three cognate receptors. Along with a satisfactory safety and tolerability profile, the Phase 1a and 1b clinical trials of this drug in 116 SLE patients (1a: 54; 1b: 62) with a stable or inactive mild disease showed a reduction in total and $IgD^{+}CD27^{-}$ naïve B cells, with higher relative level of $IgD^{-}CD27^{+}$ memory B cells [29].

In the large phase 2b PEARL-SC study of 547 moderate-to-severe SLE patients, showing a satisfactory safety and tolerability profile, the primary endpoint comparing efficacy of pooled blisibimod groups versus pooled placebo groups using SRI-5 at week 24 was not met. However, SRI-5 was met for the highest tested dose of blisibimod (200 mg per week) compared to pool placebo groups at week 20 [30]. The authors concluded that SLE patients with high disease activity at baseline, as defined by SELENA-SLEDAI $\geq$ 10 and receiving corticosteroids, are more likely to respond to this drug at the highest dose, when using more stringent SRI threshold (SRI-7 or SRI-8). A significant effect on patient-reported fatigue was also noted in patients receiving higher dose of blisibimod compared to placebo [31]. In post-hoc analysis, improvement in fatigue score was higher in SRI-5 responder compared to non-responders [31]. A biological effect was observed with increase in C3 and C4 levels, and reduction in anti-dsDNA autoAbs, IgG and IgM levels as well as total B cells numbers. Bearing in mind that patients with severe LN were excluded from this trial, subset analysis on patients with proteinuria at baseline showed significant reduction in proteinuria with blisibimod compared to placebo, suggesting potential benefit of this drug in LN patients [30]. An open-label long-term safety phase 2 clinical trial (NCT01305746) for SLE patients who completed the PEARL-SC trial has been completed, but no results are either published or disclosed on the US clinical trials website (source: https://clinicaltrials.gov/).

Using knowledge learned from phase 2 [30], investigators of the phase 3 clinical trial of blisibimod (CHABLIS-SC1) only tested the highest dose of blisibimod (200 mg per week) versus placebo in a large cohort of 442 seropositive SLE patients with high disease activity score, using a higher threshold of SRI (SRI-6) at week 52 as primary endpoint [32]. While some secondary endpoints were met, including significant steroid sparing effect, and effect on proteinuria consistent with the phase 2 outcomes [30], the primary efficacy endpoint was not met. Biological effect included significant increase in C3 and C4 levels, non-significant decreases in anti-dsDNA autoAb levels, and significant decrease in IgM, IgG and IgA. A non-significant decrease in total B cells, significant decrease in $CD20^{+}CD19^{+}IgD^{+}CD27^{-}$ naïve and $CD19^{+}CD38^{+}CD138^{+}$ activated B cells (no statistical analysis data provided), and significant increase in $CD19^{+}CD27^{+}$ memory B cells were also noted. The CHABLIS-SC2 phase 3 clinical trial (NCT02074020) has been withdrawn, while CHABLIS7.5 phase 3 clinical trial (NCT02514967) has been terminated by Anthera Pharmaceuticals. Currently, there are no recruiting or active clinical trials of blisibimod in SLE (*Source* https://clinicaltrials.gov/).

It is noteworthy that in Phase 1a/1b and 2b of blisibimod, meaningful assessment of ADAb was not permitted due to the non-optimised screening assay used [29, 30], and not reported in phase 3 [32].

Neutralisation of Both BAFF and APRIL

Atacicept

Atacicept is a fusion protein consisting of the extra-cellular domain of TACI fused to a human IgG Fc domain, which can bind BAFF, APRIL and BAFF-APRIL heterotrimers [33]. This recombinant form of TACI has also been investigated in SLE [34]. Two phase 1b clinical trials reported acceptable safety and tolerability of atacicept while showing biological effect including reduction in $CD19^{+}$ total and $CD19^{+}CD27^{-}IgD^{+}$ mature B cells, as well as Ig (IgA, IgG, IgM) levels [35–37]. Analysis of the effect of atacicept on C3 and anti-dsDNA autoAbs levels was not meaningful, as most patients were negative for anti-dsDNA auto Abs and had normal C3 levels at baseline visit [35].

A phase 2/3 clinical trial studying atacicept in combination with mycophenolate mofetil (MMF) and corticosteroids in active LN was terminated prematurely after the observation of increased infection rates and decreased serum IgG levels, after enrolling only six patients amongst whom four received atacicept [38]. While a drop in serum IgG levels was actually observed before atacicept exposure, incriminating MMF as a potential cause, these data raise the question of the potential benefit of blocking both BAFF and APRIL in LN. Another phase 1 trial of atacicept in LN patients on a stable dose of MMF was terminated by the EMD Serono for safety issues and time constraints (NCT01369628).

Another phase 2/3 clinical trial (APRIL-SLE randomised trial) was conducted using atacicept in patients with moderate-to-severe SLE, excluding moderate-to-severe glomerulonephritis (GN) and severe central nervous system (CNS) disease, and reported no difference in flare rate or time to first flare between atacicept 75 mg and placebo [39]. Data on atacicept 150 mg suggested clinical benefits on those two flare assessments of higher dosage compared to placebo; however, this arm was terminated due to the occurrence of two deaths (infection) [39]. Of particular interest is the post-hoc analyses of this trial showing that elevated baseline of both BAFF and APRIL serum levels were associated with greater treatment response to atacicept, suggesting serum BAFF and APRIL as potential biomarkers to stratify SLE patients for atacicept therapy [40].

Although the phase 2b clinical trial on atacicept in SLE (ADDRESS II) did not meet its primary efficacy endpoint (SRI-4 at week 24), it showed promising clinical benefit in the arm atacicept 75 mg versus placebo [41]. Moreover, in sensitivity analysis, significant clinical efficacy, as assessed by SRI-4 index, for both atacicept 75 and 150 mg arms were noted when using first treatment injection as baseline instead of screening visit or when performing subset analysis on serologically active SLE patients, or patients with high disease activity at screening visit (only significant for atacicept 150 mg arm). Incidence of severe flares also decreased in both atacicept arms as compared to placebo. No steroid sparing effect of atacicept was noted at week 24. Biological effect included increased C3 and C4 levels, and decreased anti-dsDNA autoAbs, Ig (A, G, M) levels (no statistical test and *P*-value reported). No investigation of treatment effect on B cell population was reported. Of particular interest is the post-hoc analysis of this ADDRESS II trial showing that significantly more SLE patients treated with atacicept 150 mg with high disease activity at screening visit attained the T2T endpoint LLDAS at week 24 compared to placebo [42]. The phase 2 clinical trial ADDRESS has been withdrawn by EMD Serono (NCT01440231), while the open-label long-term safety extension phase 2 trial of the ADDRESS II study was terminated due to a shortage of drug supply (NCT02070978). No atacicept clinical trials are currently registered to study its efficiency in either LN or SLE (source: https://clinicaltrials.gov/).

We and others have shown that urinary BAFF and APRIL were both significantly higher in patients with active renal SLE, suggesting that both BAFF and APRIL may play a role in LN, and that patients with LN may benefit from BAFF ± APRIL-targeting therapy [43, 44]. Whether urinary BAFF and/or APRIL measurement may help to stratify SLE patients who are more likely to respond to atacicept and/or belimumab, particularly in LN, remains to be determined [43]. Of note, it has been reported that serum BAFF is not a predictive biomarker for belimumab therapy effectiveness [45], but may be more promising as a potential predictive biomarker for atacicept therapy [40].

It is noteworthy that immunogenicity was low with only three patients on the atacicept 150 mg arm as positive for ADAb post-treatment (24 weeks after last injection) in the phase 2/3 clinical trial [39]; however, the authors did not report if these were neutralising ADAb [39]. Eight and three patients receiving atacicept developed ADAb at week 24 in ADDRESS II [41] and APRIL-SLE randomised trials [39], respectively. In two phase Ib trials, ADAb were not detectable in any treated SLE patient [35, 36].

Telitacicept

One large 48-week Phase 2b RCT has investigated telitacicept (RC18), a recombinant fusion protein comprising the extra-cellular domain of TACI that neutralises both BAFF and APRIL, in 249 seropositive moderate-to-severe SLE patients (NCT02885610) [46]. This study met its primary endpoint (SRI-4), and one Phase 3 RCT of telitacicept in SLE is now ongoing, with no results available as of today (Table 3.1).

Targeting BAFF-R: Ianalumab

As shown in Table 3.1, a B cell depleting mAb targeting BAFF-R, ianalumab (VAY736), is currently under investigation in a Phase 2 clinical trial in SLE. At present, there is no published data regarding this trial, nor any results posted on the clinicaltrials.gov website and no data was presented at the 2020 European League Against Rheumatism (EULAR) or American College of Rheumatology (ACR) conferences. A recent small phase 2 RCT of this biological agent in primary Sjogren's syndrome (pSS) showed some encouraging results particularly improving fatigue, along with an acceptable safety profile [47].

B Cell-Depleting Anti-CD20 Agents

Rituximab

Rituximab is a chimeric IgG1κ mAb that depletes B cells by antagonising the CD20 surface antigen, acting mainly via its cytotoxicity, namely complement-dependent cytotoxicity (CDC) and antibody-dependent cellular cytotoxicity (ADCC), and direct apoptosis [48]. It is worth mentioning that mature human PC are $CD20^{-}$ [11]. Two pivotal 52-week phase 3 RCT trials, EXPLORER [49] and LUNAR [50], did not meet their primary and secondary efficacy endpoints, while uncontrolled studies including two small phase 1/2 (N = 18 [51]; N = 15 [52]) and one phase 2 trial of rituximab in SLE (N = 34) [53], as well as other small open-label studies have suggested potential clinical benefits and acceptable safety profile in SLE (including refractory SLE) [54–61], including CNS SLE [55] and LN [54, 57, 58, 61].

While in the phase 2/3 EXPLORER trial investigating rituximab in 257 patients with moderate-to-severe extrarenal SLE, primary endpoint was not met [49], pre-specified subgroup analysis revealed potential benefit for African American/Hispanic SLE patients. In the phase 3 LUNAR trial investigating rituximab in 144 class III/IV proliferative LN patients, primary endpoint was not met either [50]. Of particular interest in the exploratory analysis, however, is the observed significantly higher proportion of patients with improved proteinuria with rituximab at week 78 compared to placebo, and a potential steroid-sparing effect [50]. In both EXPLORER and LUNAR trials, rituximab induced a B cell depletion in treated patients, as well as significant decreases in anti-dsDNA autoAbs and increases of both C3 and C4 levels [49, 50]. Safety and tolerability were deemed acceptable in both phase 3 trials [49, 50]. The design of those two RCT of rituximab in SLE has been criticised, raising the issue of substantial concomitant immunosuppression with corticosteroid and immunosuppressants. One phase 2 and one phase 4 clinical trial of rituximab in SLE are currently ongoing (Table 3.1).

Prevalence of ADAb was 64.7% in rituximab-treated patients (including 35.3% with high titre ADAb) [51] and 26.7% [52] in the phase 1/2 trials, and was higher in rituximab-treated patients versus placebo in both EXPLORER (26% vs 3.4%, respectively) [49], and LUNAR (15.1% vs 8.5%, respectively) trials [50].

Ocrelizumab

Ocrelizumab, an anti-CD20 recombinant humanised mAb, is another B cell-depleting agent which has been tested in two Phase 3 clinical trials in SLE and proliferative LN. The phase 3 RCT (BELONG, NCT00626197) of ocrelizumab in association with corticosteroids and either MMF or cyclophosphamide in LN was terminated due to serious adverse events (infections) particularly in patients receiving concomitant MMF, among whom a higher proportion also received baseline IV corticosteroids $\geq$ 1g [62]. This safety profile mirrors the higher rate of infection observed in the phase 2/3 clinical trial of atacicept in combination with MMF and corticosteroids in active LN, also prematurely terminated [38]. The second phase 3 RCT (BEGIN, NCT00539838) in moderate-to-severe SLE patients with no moderate-to-severe GN was also prematurely terminated.

Negative Regulators of BCR Signalling

Epratuzumab

Epratuzumab, a recombinant humanised IgG_1 mAb targeting the B cell-restricted (predominantly expressed on naïve and transitional B cells, but not expressed on PC) CD22, has been tested in clinical trials since 2006. Its target, CD22 (also known as Siglec-2), is a surface-expressed transmembrane sialoglycoprotein, part of the BCR complex acting as a negative regulator of BCR signalling [63]. Its mechanism of action differs from rituximab, being less cytotoxic, particularly characterised by no CDC while ADCC was lower as compared to rituximab, albeit significant [48]. Rather, its mechanism of action is thought to be via BCR complex signalling modulation. Epratuzumab has also been shown to downregulate proinflammatory cytokine expression (TNF, IL-6) by activated total B cells in vitro upon BCR crosslinking $\pm$ co-stimulation by TLR9 agonist CpG in HD and SLE patients, while no influence on either IL-10 expression by activated B cells or IL-10-producing B cells was noted [64]. Another group showed that epratuzumab treatment of $CD10^{-}CD27^{-}$ human tonsillar B cells in vitro upon co-stimulation with TLR7 agonist R848 and BCR engagement led to a significant increase in IL-10 expression and a trend toward downregulation of IL-6 expression (65). Of significant interest is the observation that epratuzumab led to downregulate *PRDM1* mRNA expression (encoding for Blimp-1 [66]) in $CD10^{-}CD27^{-}$ human tonsillar B cells upon co-stimulation with R848 $\pm$ BCR crosslinking in vitro, suggesting that this drug may negatively regulate B cell differentiation into Ab-secreting PC. Finally, along with the demonstrated CD22 downregulation via its internalisation after crosslinking with epratuzumab, as well as its capacity to induce CD22 phosphorylation [67], epratuzumab has been reported to induce downregulation of surface antigen CD19, CD21 and CD79b on B cells of SLE patients via trogocytosis by FcγR-expressing monocytes, NK cells and granulocytes [68].

In a phase 2 open label single centre non-randomised clinical trial of epratuzumab on a small cohort of 14 SLE patients, promising clinical benefit as assessed by the BILAG score, along with expected reduction in total B cells were noted [69]. No obvious change in Ig (IgG, IgA, IgM), C3 and autoAb (anti-nuclear Ab (ANA) & anti-dsDNA) levels were observed [69]. A more detailed B cell subsets analysis has subsequently been performed in PBMC samples obtained from SLE patients enrolled in this trial [69], revealing that $CD27^{+}$ memory B cell and $CD27^{++}$ plasmablast (PB) were not significantly affected by epratuzumab treatment, while a significant reduction in $CD27^{-}$ (most likely naïve and transitional) B cell subsets was noted [70]. In vitro studies report that anti-Ig-stimulated B cell proliferation is antagonised by epratuzumab in SLE but not healthy controls (HC) when co-stimulated by CD40L (T cell dependent) or CpG (T cell independent) [70]. Epratuzumab has indeed been shown to particularly target $CD27^{-}$ (naïve and transitional) B cells, B cell subsets where CD22 surface expression is higher as opposed to $CD27^{+}$ memory B cells [71].

Two phase 3 trials of epratuzumab (ALLEVIATE-1 [NCT00111306] and -2 [NCT00383214]) were terminated due to drug supply issues. Exploratory analysis has been performed with a modified primary endpoint (BILAG with no treatment failure at week 12) on data pooled from 90 SLE patients randomised in both ALLEVIATE trials, and data from an open-label ALLEVIATE extension study (NCT00383513) was also analysed [72, 73]. Along with a satisfactory safety profile, this primary efficacy endpoint was not met. Clinical benefits are difficult to estimate due to the early termination of both trials, the relatively small sample size and the delay between the end of ALLEVIATE trials and the open-label ALLEVIATE extension study. It is, however, worth noting that a significant reduction in cumulative dose of corticosteroids was observed at week 24 in patients receiving epratuzumab (360 mg/m^2) when compared to placebo [73]. A reduction in total CD19$^+$ B cells but no difference in Ig levels were noted in epratuzumab-treated patients versus placebo (no statistical test performed) [72].

Although the phase 2b of epratuzumab (EMBLEM) did not meet its primary efficacy endpoint (BICLA at week 12), it did show an acceptable safety profile, and furthermore, some promising clinical benefit was seen with a significantly higher response rate in patients receiving a cumulative dose of 2.4 g of epratuzumab over 4 weeks versus placebo [74]. However, no significant change in Ig (IgG, IgA, IgM) levels were noted, while total CD20$^+$ B cell count was reduced. An open-label extension (OLE) study of this EMBLEM trial, which terminated early, suggests corticosteroid sparing efficacy, and continuing improvement in disease activity as assessed by the BILAG and SLEDAI scores [75].

Two phase 3 trials of epratuzumab (EMBODY 1 and EMBODY 2) both did not meet their primary efficacy endpoint (BICLA at week 48), while having an acceptable safety profile [76]. No significant effect was seen versus placebo for flare, corticosteroid use, disease activity (as assessed by the BILAG, SLEDAI-2 K or modified SRI (post-hoc analysis)), HR QoL, or fatigue. An expected, moderate reduction in total CD19$^+$ B cell number was observed, as well as a modest reduction in serum IgM, but not A and G. No obvious biological effect of epratuzumab was seen on serum autoAbs and complement C3 and C4 levels [76]. Patients were invited to enter an OLE study (EMBODY 4; NCT01408576; study completed, not yet published in peer-review journal). A post-hoc analysis of the EMBODY 1 and 2 trials suggests potential clinical benefit for SLE patients with associated Sjögren's syndrome [77].

A phase 1/2 RCT clinical trial on Japanese patients with active moderate-to-severe SLE showed consistency with other trials on epratuzumab regarding an observed modest reduction in total CD20$^+$ B cell count [78], highlighting that this biologic is not a potent B cell-depleting agent. A trend toward a reduction in serum IgM, but not IgA and IgG, levels was also noted [78].

Appraisal of immunogenicity of epratuzumab revealed no change from baseline pre-treatment assessment in ADAb levels in the phase 2 open label study [69], while 2.1% (4/189) of epratuzumab-treated patients had ADAb at week 12 in the EMBLEM study [74], 3.9% (2/51) in ALLEVIATE trials [72], and 12.5% (2/16) in another phase 1/2 trial [78]. There are currently no recruiting or active clinical trial of epratuzumab in SLE (*Source* https://clinicaltrials.gov/).

FcγRIIb Inhibition: Obexelimab

Obexelimab (XmAb®5871) is a humanised mAb targeting the CD19 surface antigen with the antibody variable domain, while its Fc-engineered domain targets FcγRIIb on B cells with high affinity [79]. As opposed to rituximab, it is not a B cell-depleting agent but it is engineered to target FcγRIIb inhibition [79]. Rather, engaging with FcγRIIb, a negative regulator of BCR signalling, this agent can antagonise B cell activation and proliferation, including B cell expression of T cell costimulatory molecules CD80/CD86 [79]. A phase 2 clinical trial of obexelimab in SLE has recently been completed. While data are yet to be published in a peer-review journal, information reported in an abstract from the 2018 ACR Annual Meeting showed encouraging results with a trend toward meeting its modified (in intention-to-treat (ITT) population analysis) primary endpoint (maintenance of clinical improvement at week 33 via a score using both SELENA-SLEDAI and BILAG indexes), along with an acceptable safety and tolerability profile [80]. In an abstract from the 2019 EULAR conference, new preliminary data from the same trial were presented, where significant improvement in time to flare and percentage of patients who achieved and maintained LLDAS between month 6 and 8 were observed in obexelimab-treated patients versus placebo [81]. It is worth noting that the study design imposed to stop immunosuppressants other than corticosteroids. No ongoing RCT of this agent in SLE is currently registered on https://clinicaltrials.gov/.

B Cell-Targeted Therapy Combination Strategy

While two Phase 2 (including RCT BEAT Lupus [82], trial registration number ISRCTN47873; not registered in https://clinicaltrials.gov/) and one Phase 3 clinical trials investigating the combination of rituximab and belimumab are currently under way in SLE (Table 3.1), a small Phase 2 trial (CALIBRATE) of the same biological combination strategy in LN has recently been published (NCT02260934) [83]. In this trial, patients were initially treated with rituximab, cyclophosphamide and corticosteroids, followed by belimumab treatment and compared to patients without sequential treatment with belimumab. Of particular interest, the safety profile of this biological combination, the primary endpoint of this trial and one of the major fears for combining biologics, was similar to that of the group treated without belimumab. Bearing in mind that this trial primarily aimed to assess the safety profile of combining both biologics in LN, unfortunately no clinical benefit was demonstrated using this biological combination.

References

1. Navarra SV, Guzman RM, Gallacher AE, Hall S, Levy RA, Jimenez RE et al (2011) Efficacy and safety of belimumab in patients with active systemic lupus erythematosus: a randomised, placebo-controlled, phase 3 trial. Lancet 377(9767):721–731
2. Furie R, Petri M, Zamani O, Cervera R, Wallace DJ, Tegzova D et al (2011) A phase III, randomized, placebo-controlled study of belimumab, a monoclonal antibody that inhibits B lymphocyte stimulator, in patients with systemic lupus erythematosus. Arthritis Rheum 63(12):3918–3930
3. Gladman DD, Ibanez D, Urowitz MB (2002) Systemic lupus erythematosus disease activity index 2000. J Rheumatol 29(2):288–291
4. Franklyn K, Lau CS, Navarra SV, Louthrenoo W, Lateef A, Hamijoyo L et al (2016) Definition and initial validation of a Lupus Low Disease Activity State (LLDAS). Ann Rheum Dis 75(9):1615–1621
5. Houssiau FA, Thanou A, Mazur M, Ramiterre E, Gomez Mora DA, Misterska-Skora M et al (2020) IFN-alpha kinoid in systemic lupus erythematosus: results from a phase IIb, randomised, placebo-controlled study. Ann Rheum Dis 79(3):347–355
6. Babaoglu H, Li J, Goldman D, Magder LS, Petri M (2019) Predictors of predominant Lupus Low Disease Activity State (LLDAS-50). Lupus 28 (14):1648–1655
7. Piga M, Floris A, Cappellazzo G, Chessa E, Congia M, Mathieu A et al (2017) Failure to achieve lupus low disease activity state (LLDAS) six months after diagnosis is associated with early damage accrual in Caucasian patients with systemic lupus erythematosus. Arthritis Res Ther 19(1):247
8. Baker KP, Edwards BM, Main SH, Choi GH, Wager RE, Halpern WG et al (2003) Generation and characterization of LymphoStat-B, a human monoclonal antibody that antagonizes the bioactivities of B lymphocyte stimulator. Arthritis Rheum 48 (11):3253–3265
9. Kowalczyk-Quintas C, Chevalley D, Willen L, Jandus C, Vigolo M, Schneider P (2018) Inhibition of Membrane-Bound BAFF by the Anti-BAFF Antibody Belimumab. Front Immunol 9:2698
10. Nicoletti AM, Kenny CH, Khalil AM, Pan Q, Ralph KL, Ritchie J et al (2016) Unexpected Potency Differences between B-Cell-Activating Factor (BAFF) Antagonist Antibodies against Various Forms of BAFF: Trimer, 60-Mer, and Membrane-Bound. J Pharmacol Exp Ther 359(1):37–44
11. Vincent FB, Morand EF, Mackay F (2012) BAFF and innate immunity: new therapeutic targets for systemic lupus erythematosus. Immunol Cell Biol 90 (3):293–303

12. Strand V, Levy RA, Cervera R, Petri MA, Birch H, Freimuth WW et al (2014) Improvements in health-related quality of life with belimumab, a B-lymphocyte stimulator-specific inhibitor, in patients with autoantibody-positive systemic lupus erythematosus from the randomised controlled BLISS trials. Ann Rheum Dis 73(5):838–844
13. van Vollenhoven RF, Petri M, Wallace DJ, Roth DA, Molta CT, Hammer AE et al (2016) Cumulative Corticosteroid Dose Over Fifty-Two Weeks in Patients With Systemic Lupus Erythematosus: Pooled Analyses From the Phase III Belimumab Trials. Arthritis & rheumatology. 68(9):2184–2192
14. Manzi S, Sanchez-Guerrero J, Merrill JT, Furie R, Gladman D, Navarra SV et al (2012) Effects of belimumab, a B lymphocyte stimulator-specific inhibitor, on disease activity across multiple organ domains in patients with systemic lupus erythematosus: combined results from two phase III trials. Ann Rheum Dis 71(11):1833–1838
15. Dooley MA, Houssiau F, Aranow C, D'Cruz DP, Askanase A, Roth DA et al (2013) Effect of belimumab treatment on renal outcomes: results from the phase 3 belimumab clinical trials in patients with SLE. Lupus 22(1):63–72
16. van Vollenhoven RF, Petri MA, Cervera R, Roth DA, Ji BN, Kleoudis CS et al (2012) Belimumab in the treatment of systemic lupus erythematosus: high disease activity predictors of response. Ann Rheum Dis 71(8):1343–1349
17. Furie R, Rovin BH, Houssiau F, Malvar A, Teng YKO, Contreras G et al (2020) Two-Year, Randomized, Controlled Trial of Belimumab in Lupus Nephritis. N Engl J Med 383(12):1117–1128
18. Zhang F, Bae SC, Bass D, Chu M, Egginton S, Gordon D et al (2018) A pivotal phase III, randomised, placebo-controlled study of belimumab in patients with systemic lupus erythematosus located in China, Japan and South Korea. Ann Rheum Dis 77 (3):355–363
19. Bruce IN, Urowitz M, van Vollenhoven R, Aranow C, Fettiplace J, Oldham M et al (2016) Long term organ damage accrual and safety in patients with SLE treated with belimumab plus standard of care. Lupus 25(7):699–709
20. van Vollenhoven RF, Navarra SV, Levy RA, Thomas M, Heath A, Lustine T et al (2020) Long-term safety and limited organ damage in patients with systemic lupus erythematosus treated with belimumab: a Phase III study extension. Rheumatology (Oxford) 59(2):281–291
21. Stohl W, Schwarting A, Okada M, Scheinberg M, Doria A, Hammer AE et al (2017) Efficacy and Safety of Subcutaneous Belimumab in Systemic Lupus Erythematosus: A Fifty-Two-Week Randomized, Double-Blind. Placebo-Controlled Study. Arthritis & rheumatology. 69(5):1016–1027
22. Brunner HI, Abud-Mendoza C, Viola DO, Calvo Penades I, Levy D, Anton J et al (2020) Safety and efficacy of intravenous belimumab in children with systemic lupus erythematosus: results from a randomised, placebo-controlled trial. Ann Rheum Dis 79(10):1340–1348
23. Witcher J, Fleischmann R, Chindalore VL, Hansen RJ, Hu L, Radtke D et al (2016) Pharmacokinetics and safety of single doses of tabalumab in subjects with rheumatoid arthritis or systemic lupus erythematosus. Br J Clin Pharmacol 81(5):908–917
24. Isenberg DA, Petri M, Kalunian K, Tanaka Y, Urowitz MB, Hoffman RW et al (2016) Efficacy and safety of subcutaneous tabalumab in patients with systemic lupus erythematosus: results from ILLUMINATE-1, a 52-week, phase III, multicentre, randomised, double-blind, placebo-controlled study. Ann Rheum Dis 75(2):323–331
25. Merrill JT, van Vollenhoven RF, Buyon JP, Furie RA, Stohl W, Morgan-Cox M et al (2016) Efficacy and safety of subcutaneous tabalumab, a monoclonal antibody to B-cell activating factor, in patients with systemic lupus erythematosus: results from ILLUMINATE-2, a 52-week, phase III, multicentre, randomised, double-blind, placebo-controlled study. Ann Rheum Dis 75(2):332–340
26. Rovin BH, Dooley MA, Radhakrishnan J, Ginzler EM, Forrester TD, Anderson PW (2016) The impact of tabalumab on the kidney in systemic lupus erythematosus: results from two phase 3 randomized, clinical trials. Lupus 25(14):1597–1601
27. Hoffman RW, Merrill JT, Alarcon-Riquelme MM, Petri M, Dow ER, Nantz E et al (2017) Gene Expression and Pharmacodynamic Changes in 1,760 Systemic Lupus Erythematosus Patients From Two Phase III Trials of BAFF Blockade With Tabalumab. Arthritis & rheumatology. 69(3):643–654
28. Hsu H, Khare SD, Lee F, Miner K, Hu YL, Stolina M et al (2012) A novel modality of BAFF-specific inhibitor AMG623 peptibody reduces B-cell number and improves outcomes in murine models of autoimmune disease. Clin Exp Rheumatol 30 (2):197–201
29. Stohl W, Merrill JT, Looney RJ, Buyon J, Wallace DJ, Weisman MH et al (2015) Treatment of systemic lupus erythematosus patients with the BAFF antagonist "peptibody" blisibimod (AMG 623/A-623): results from randomized, double-blind phase 1a and phase 1b trials. Arthritis Res Ther 17:215
30. Furie RA, Leon G, Thomas M, Petri MA, Chu AD, Hislop C et al (2015) A phase 2, randomised, placebo-controlled clinical trial of blisibimod, an inhibitor of B cell activating factor, in patients with moderate-to-severe systemic lupus erythematosus, the PEARL-SC study. Ann Rheum Dis 74(9):1667–1675
31. Petri MA, Martin RS, Scheinberg MA, Furie RA (2017) Assessments of fatigue and disease activity in patients with systemic lupus erythematosus enrolled in the Phase 2 clinical trial with blisibimod. Lupus 26 (1):27–37
32. Merrill JT, Shanahan WR, Scheinberg M, Kalunian KC, Wofsy D, Martin RS (2018) Phase III trial

results with blisibimod, a selective inhibitor of B-cell activating factor, in subjects with systemic lupus erythematosus (SLE): results from a randomised, double-blind, placebo-controlled trial. Ann Rheum Dis 77(6):883–889
33. Vincent FB, Morand EF, Schneider P, Mackay F (2014) The BAFF/APRIL system in SLE pathogenesis. Nat Rev Rheumatol 10(6):365–373
34. Vincent FB, Saulep-Easton D, Figgett WA, Fairfax KA, Mackay F (2013) The BAFF/APRIL system: Emerging functions beyond B cell biology and autoimmunity. Cytokine Growth Factor Rev 24 (3):203–215
35. Pena-Rossi C, Nasonov E, Stanislav M, Yakusevich V, Ershova O, Lomareva N et al (2009) An exploratory dose-escalating study investigating the safety, tolerability, pharmacokinetics and pharmacodynamics of intravenous atacicept in patients with systemic lupus erythematosus. Lupus 18(6):547–555
36. Dall'Era M, Chakravarty E, Wallace D, Genovese M, Weisman M, Kavanaugh A et al (2007) Reduced B lymphocyte and immunoglobulin levels after atacicept treatment in patients with systemic lupus erythematosus: results of a multicenter, phase Ib, double-blind, placebo-controlled, dose-escalating trial. Arthritis Rheum 56(12):4142–4150
37. Nestorov I, Papasouliotis O, Pena Rossi C, Munafo A (2010) Pharmacokinetics and immunoglobulin response of subcutaneous and intravenous atacicept in patients with systemic lupus erythematosus. J Pharm Sci 99(1):524–538
38. Ginzler EM, Wax S, Rajeswaran A, Copt S, Hillson J, Ramos E et al (2012) Atacicept in combination with MMF and corticosteroids in lupus nephritis: results of a prematurely terminated trial. Arthritis Res Ther 14(1):R33
39. Isenberg D, Gordon C, Licu D, Copt S, Rossi CP, Wofsy D (2015) Efficacy and safety of atacicept for prevention of flares in patients with moderate-to-severe systemic lupus erythematosus (SLE): 52-week data (APRIL-SLE randomised trial). Ann Rheum Dis 74(11):2006–2015
40. Gordon C, Wofsy D, Wax S, Li Y, Pena Rossi C, Isenberg D (2017) Post Hoc Analysis of the Phase II/III APRIL-SLE Study: Association Between Response to Atacicept and Serum Biomarkers Including BLyS and APRIL. Arthritis & rheumatology. 69(1):122–130
41. Merrill JT, Wallace DJ, Wax S, Kao A, Fraser PA, Chang P et al (2018) Efficacy and Safety of Atacicept in Patients With Systemic Lupus Erythematosus: Results of a Twenty-Four-Week, Multicenter, Randomized, Double-Blind, Placebo-Controlled, Parallel-Arm. Phase IIb Study. Arthritis & rheumatology. 70(2):266–276
42. Morand EF, Isenberg DA, Wallace DJ, Kao AH, Vazquez-Mateo C, Chang P, et al. Attainment of treat-to-target endpoints in SLE patients with high disease activity in the atacicept phase 2b ADDRESS II study. Rheumatology (Oxford). 2020.
43. Vincent FB, Kandane-Rathnayake R, Hoi AY, Slavin L, Godsell JD, Kitching AR et al (2018) Urinary B-cell-activating factor of the tumour necrosis factor family (BAFF) in systemic lupus erythematosus. Lupus 27(13):2029–2040
44. Phatak S, Chaurasia S, Mishra SK, Gupta R, Agrawal V, Aggarwal A, Misra R, Urinary B cell activating factor (BAFF) and a proliferation-inducing ligand (APRIL): potential biomarkers of active lupus nephritis. Clinical & Experimental Immunology. 2017;187(3):376–382. https://doi.org/10.1111/cei.12894
45. Stohl W, Hiepe F, Latinis KM, Thomas M, Scheinberg MA, Clarke A et al (2012) Belimumab reduces autoantibodies, normalizes low complement levels, and reduces select B cell populations in patients with systemic lupus erythematosus. Arthritis Rheum 64 (7):2328–2337
46. Wu D, Li J, Xu D, Wang W, Li L, Fang J, et al. A Human Recombinant Fusion Protein Targeting B Lymphocyte Stimulator (BlyS) and a Proliferation-Inducing Ligand (APRIL), Telitacicept (RC18), in Systemic Lupus Erythematosus (SLE): Results of a Phase 2b Study [abstract]. Arthritis Rheumatol. 2019;71 (suppl 10).
47. Dorner T, Posch MG, Li Y, Petricoul O, Cabanski M, Milojevic JM et al (2019) Treatment of primary Sjogren's syndrome with ianalumab (VAY736) targeting B cells by BAFF receptor blockade coupled with enhanced, antibody-dependent cellular cytotoxicity. Ann Rheum Dis 78(5):641–647
48. Carnahan J, Stein R, Qu Z, Hess K, Cesano A, Hansen HJ et al (2007) Epratuzumab, a CD22-targeting recombinant humanized antibody with a different mode of action from rituximab. Mol Immunol 44(6):1331–1341
49. Merrill JT, Neuwelt CM, Wallace DJ, Shanahan JC, Latinis KM, Oates JC et al (2010) Efficacy and safety of rituximab in moderately-to-severely active systemic lupus erythematosus: the randomized, double-blind, phase II/III systemic lupus erythematosus evaluation of rituximab trial. Arthritis Rheum 62 (1):222–233
50. Rovin BH, Furie R, Latinis K, Looney RJ, Fervenza FC, Sanchez-Guerrero J et al (2012) Efficacy and safety of rituximab in patients with active proliferative lupus nephritis: the Lupus Nephritis Assessment with Rituximab study. Arthritis Rheum 64 (4):1215–1226
51. Looney RJ, Anolik JH, Campbell D, Felgar RE, Young F, Arend LJ et al (2004) B cell depletion as a novel treatment for systemic lupus erythematosus: a phase I/II dose-escalation trial of rituximab. Arthritis Rheum 50(8):2580–2589
52. Tanaka Y, Yamamoto K, Takeuchi T, Nishimoto N, Miyasaka N, Sumida T et al (2007) A multicenter phase I/II trial of rituximab for refractory systemic lupus erythematosus. Mod Rheumatol 17(3):191–197
53. Tanaka Y, Takeuchi T, Miyasaka N, Sumida T, Mimori T, Koike T et al (2016) Efficacy and safety of

rituximab in Japanese patients with systemic lupus erythematosus and lupus nephritis who are refractory to conventional therapy. Mod Rheumatol 26(1):80–86
54. Iaccarino L, Bartoloni E, Carli L, Ceccarelli F, Conti F, De Vita S et al (2015) Efficacy and safety of off-label use of rituximab in refractory lupus: data from the Italian Multicentre Registry. Clin Exp Rheumatol 33(4):449–456
55. Tokunaga M, Saito K, Kawabata D, Imura Y, Fujii T, Nakayamada S et al (2007) Efficacy of rituximab (anti-CD20) for refractory systemic lupus erythematosus involving the central nervous system. Ann Rheum Dis 66(4):470–475
56. Leandro MJ, Cambridge G, Edwards JC, Ehrenstein MR, Isenberg DA (2005) B-cell depletion in the treatment of patients with systemic lupus erythematosus: a longitudinal analysis of 24 patients. Rheumatology (Oxford) 44(12):1542–1545
57. Moroni G, Gallelli B, Sinico RA, Romano G, Sinigaglia L, Messa P (2012) Rituximab versus oral cyclophosphamide for treatment of relapses of proliferative lupus nephritis: a clinical observational study. Ann Rheum Dis 71(10):1751–1752
58. Vigna-Perez M, Hernandez-Castro B, Paredes-Saharopulos O, Portales-Perez D, Baranda L, Abud-Mendoza C et al (2006) Clinical and immunological effects of Rituximab in patients with lupus nephritis refractory to conventional therapy: a pilot study. Arthritis Res Ther 8(3):R83
59. Jonsdottir T, Gunnarsson I, Risselada A, Henriksson EW, Klareskog L, van Vollenhoven RF (2008) Treatment of refractory SLE with rituximab plus cyclophosphamide: clinical effects, serological changes, and predictors of response. Ann Rheum Dis 67(3):330–334
60. Fernandez-Nebro A, de la Fuente JL, Carreno L, Izquierdo MG, Tomero E, Rua-Figueroa I et al (2012) Multicenter longitudinal study of B-lymphocyte depletion in refractory systemic lupus erythematosus: the LESIMAB study. Lupus 21 (10).1063–1076
61. Condon MB, Ashby D, Pepper RJ, Cook HT, Levy JB, Griffith M et al (2013) Prospective observational single-centre cohort study to evaluate the effectiveness of treating lupus nephritis with rituximab and mycophenolate mofetil but no oral steroids. Ann Rheum Dis 72(8):1280–1286
62. Mysler EF, Spindler AJ, Guzman R, Bijl M, Jayne D, Furie RA et al (2013) Efficacy and safety of ocrelizumab in active proliferative lupus nephritis: results from a randomized, double-blind, phase III study. Arthritis Rheum 65(9):2368–2379
63. Monroe JG (2006) ITAM-mediated tonic signalling through pre-BCR and BCR complexes. Nat Rev Immunol 6(4):283–294
64. Fleischer V, Sieber J, Fleischer SJ, Shock A, Heine G, Daridon C et al (2015) Epratuzumab inhibits the production of the proinflammatory cytokines IL-6 and TNF-alpha, but not the regulatory cytokine IL-10, by B cells from healthy donors and SLE patients. Arthritis Res Ther 17:185
65. Giltiay NV, Shu GL, Shock A, Clark EA (2017) Targeting CD22 with the monoclonal antibody epratuzumab modulates human B-cell maturation and cytokine production in response to Toll-like receptor 7 (TLR7) and B-cell receptor (BCR) signaling. Arthritis Res Ther 19(1):91
66. Kallies A, Hasbold J, Fairfax K, Pridans C, Emslie D, McKenzie BS et al (2007) Initiation of plasma-cell differentiation is independent of the transcription factor Blimp-1. Immunity 26(5):555–566
67. Carnahan J, Wang P, Kendall R, Chen C, Hu S, Boone T et al (2003) Epratuzumab, a humanized monoclonal antibody targeting CD22: characterization of in vitro properties. Clin Cancer Res 9(10 Pt 2):3982S-S3990
68. Rossi EA, Goldenberg DM, Michel R, Rossi DL, Wallace DJ, Chang CH (2013) Trogocytosis of multiple B-cell surface markers by CD22 targeting with epratuzumab. Blood 122(17):3020–3029
69. Dorner T, Kaufmann J, Wegener WA, Teoh N, Goldenberg DM, Burmester GR (2006) Initial clinical trial of epratuzumab (humanized anti-CD22 antibody) for immunotherapy of systemic lupus erythematosus. Arthritis Res Ther 8(3):R74
70. Jacobi AM, Goldenberg DM, Hiepe F, Radbruch A, Burmester GR, Dorner T (2008) Differential effects of epratuzumab on peripheral blood B cells of patients with systemic lupus erythematosus versus normal controls. Ann Rheum Dis 67(4):450–457
71. Daridon C, Blassfeld D, Reiter K, Mei HE, Giesecke C, Goldenberg DM et al (2010) Epratuzumab targeting of CD22 affects adhesion molecule expression and migration of B-cells in systemic lupus erythematosus. Arthritis Res Ther 12(6):R204
72. Wallace DJ, Gordon C, Strand V, Hobbs K, Petri M, Kalunian K et al (2013) Efficacy and safety of epratuzumab in patients with moderate/severe flaring systemic lupus erythematosus: results from two randomized, double-blind, placebo-controlled, multicentre studies (ALLEVIATE) and follow-up. Rheumatology (Oxford) 52(7):1313–1322
73. Strand V, Petri M, Kalunian K, Gordon C, Wallace DJ, Hobbs K et al (2014) Epratuzumab for patients with moderate to severe flaring SLE: health-related quality of life outcomes and corticosteroid use in the randomized controlled ALLEVIATE trials and extension study SL0006. Rheumatology (Oxford) 53 (3):502–511
74. Wallace DJ, Kalunian K, Petri MA, Strand V, Houssiau FA, Pike M et al (2014) Efficacy and safety of epratuzumab in patients with moderate/severe active systemic lupus erythematosus: results from EMBLEM, a phase IIb, randomised, double-blind, placebo-controlled, multicentre study. Ann Rheum Dis 73(1):183–190
75. Wallace DJ, Hobbs K, Clowse ME, Petri M, Strand V, Pike M et al (2016) Long-Term Safety and Efficacy of Epratuzumab in the Treatment of

Moderate-to- Severe Systemic Lupus Erythematosus: Results From an Open-Label Extension Study. Arthritis Care Res (Hoboken). 68(4):534–543

76. Clowse ME, Wallace DJ, Furie RA, Petri MA, Pike MC, Leszczynski P et al (2017) Efficacy and Safety of Epratuzumab in Moderately to Severely Active Systemic Lupus Erythematosus: Results From Two Phase III Randomized, Double-Blind. Placebo-Controlled Trials. Arthritis & rheumatology. 69 (2):362–375
77. Gottenberg JE, Dorner T, Bootsma H, Devauchelle-Pensec V, Bowman SJ, Mariette X et al (2018) Efficacy of Epratuzumab, an Anti-CD22 Monoclonal IgG Antibody, in Systemic Lupus Erythematosus Patients With Associated Sjogren's Syndrome: Post Hoc Analyses From the EMBODY Trials. Arthritis & rheumatology. 70(5):763–773
78. Tsuru T, Tanaka Y, Kishimoto M, Saito K, Yoshizawa S, Takasaki Y et al (2016) Safety, pharmacokinetics, and pharmacodynamics of epratuzumab in Japanese patients with moderate-to-severe systemic lupus erythematosus: Results from a phase 1/2 randomized study. Mod Rheumatol 26(1):87–93
79. Horton HM, Chu SY, Ortiz EC, Pong E, Cemerski S, Leung IW et al (2011) Antibody-mediated coengagement of FcgammaRIIb and B cell receptor complex suppresses humoral immunity in systemic lupus erythematosus. J Immunol 186(7):4223–4233
80. Merrill JT, June J, Koumpouras F, Machua W, Khan MF, Askanase A, et al. Top-Line Results of a Phase 2, Double-Blind, Randomized, Placebo-Controlled Study of a Reversible B Cell Inhibitor, XmAb®5871, in Systemic Lupus Erythematosus (SLE) [abstract]. Arthritis & rheumatology. 2018;70 (suppl 10).
81. Merrill J, June J, Koumpouras F, Machua W, Khan MF, Askanase A et al (2019) PHASE 2, Double-blind, randomized, placebo-controlled study of a reversible b cell inhibitor, XMAB®5871, in systemic lupus erythematosus (SLE)). Ann Rheum Dis 78:761–762
82. Jones A, Muller P, Dore CJ, Ikeji F, Caverly E, Chowdhury K, et al. Belimumab after B cell depletion therapy in patients with systemic lupus erythematosus (BEAT Lupus) protocol: a prospective multicentre, double-blind, randomised, placebo-controlled, 52-week phase II clinical trial. BMJ Open. 2019;9(12):e032569.
83. Atisha-Fregoso Y, Malkiel S, Harris KM, Byron M, Ding L, Kanaparthi S, et al. Phase II Randomized Trial of Rituximab Plus Cyclophosphamide Followed by Belimumab for the Treatment of Lupus Nephritis. Arthritis & rheumatology. 2020.

Type I Interferons and the Perpetuation of a Loss of Tolerance

4

Kathryn Connelly and Alberta Hoi

Abstract

Dysregulation of the type I interferon (IFN) system has emerged as a key pathway involved in the perpetuation of autoimmunity and inflammation in systemic lupus erythematosus (SLE). As our understanding of this pathway and its role in disease continues to grow, clinical application through type I IFN based disease biomarkers and targeted therapeutics are emerging. This is the first of two chapters covering the role of type I interferons in lupus and will focus on the basics of the type I IFN system, in particular the role of IFN alpha produced by plasmacytoid dendritic cells and relevant downstream effects that can contribute to the initiation and propagation of immune pathology in SLE.

K. Connelly · A. Hoi (✉)
Rheumatology Research Group, Centre for Inflammatory Diseases, School of Clinical Sciences At Monash Health, Monash University, Clayton, VIC 3168, Australia
e-mail: alberta.hoi@monash.edu

K. Connelly
e-mail: Kathryn.connelly@monash.edu

K. Connelly · A. Hoi
Department of Rheumatology, Monash Health, Clayton, VIC 3168, Australia

A. Hoi
Department of Rheumatology, Austin Health, Heidelberg, VIC 3084, Australia

Biology of the Type I IFN System

Interferons are a family of soluble cytokines with pleiotropic immune effects. Their most well-known function, first documented over 50 years ago, is in the innate host defence against viral pathogens [1]. However, a more extensive role as an immune effector in a range of clinical settings has subsequently been described, including antiproliferative and immunomodulatory effects. Inappropriate or persistent activation of the type I IFN system has also been linked to deleterious effects in a number of disease states, including autoimmune and inflammatory disorders such as SLE.

In humans there are three distinct families of IFN—type I, type II and type III—which are distinguished by the receptors via which they signal (Table 4.1). Types and subtypes of IFN also differ in precise chromosomal location, signal transduction pathways and downstream effects, although significant overlap exists [2]. The type I IFNs form the largest family and are the best characterised in relation to SLE pathogenesis, in particular IFN alpha (IFNα). Other IFN subsets may also play a role in SLE, however their associations with disease are less well established and translation to the clinical setting remains preliminary [3].

A. Hoi (ed.), *Pathogenesis of Systemic Lupus Erythematosus*,
https://doi.org/10.1007/978-3-030-85161-3_4

Table 4.1 The interferon family

Interferon	Subtypes	Receptor	Major sources
Type I	IFNα (13 human subtypes), IFNβ, IFNε, IFNκ, IFNω, IFNδ, IFNτ, IFNζ	IFNAR1 IFNAR2	Plasmacytoid dendritic cells Macrophages Fibroblasts Epithelial cells
Type II	IFNγ	IFNGR1 IFNGR2	T cells Natural killer cells
Type III	IFNλ1-4	IFNλR1 IL10R2	Epithelial cells Dendritic cells

Induction of IFN Alpha Production by Plasmacytoid Dendritic Cells

Many cells can produce type I IFNs, however the generation of IFNα by plasmacytoid dendritic cells (pDCs) represents a major source and the most relevant pathway in SLE. pDCs have been described as "professional" producers of IFNα due to their capacity to generate large quantities of this cytokine when activated. At rest pDCs are predominantly found within blood and lymphoid organs, however when stimulated migrate to sites of inflammation [4]. In addition to IFNα production, activated pDCs also secrete other pro-inflammatory proteins and play a role in antigen presentation [4]. In contrast, IFN beta (IFNβ) is produced by multiple cell types including pDCs, phagocytes, epithelial cells and fibroblasts [5].

The stimulus for type I IFN production usually occurs via the detection of pathogen-associated molecular patterns (PAMPs) by pattern recognition receptors (PRRs) which may be cell surface, endosomal or cytoplasmic [6]. As well as viral and bacterial products, endogenous nucleic acids in various forms can also trigger PRRs, with implications for loss of tolerance and the development of autoimmunity. The activation of PRRs and downstream signalling pathways culminate in the phosphorylation of IFN regulatory factors (IRFs) which positively regulate genes encoding type I IFNs [7]. Major pathways resulting in type I IFN production and those most relevant to SLE pathogenesis are detailed below. The particular route of production that is activated depends on the cell type and particular PRRs expressed and the location of stimulating nucleic acids within the cell.

In the pDC, stimulation of both endosomal and cytosolic PRRs can lead to the activation of downstream IRFs and type I IFN production (Fig. 4.1). Endosomal TLR7/8 senses RNA which triggers activation of IRF7 and IRF5, while TLR9 senses DNA-containing unmethylated CpG motifs and results in activation of IRF7 [8]. When activated these TLRs recruit myeloid differentiation primary response protein 88 (MyD88) which in turn interacts with interleukin-1 receptor associated kinase 1 and 4 (IRAK1 and IRAK4) resulting in a signalling complex that phosphorylates IRF7 and IRF5 [5]. Retinoic acid-inducible gene I (RIG1) and melanoma differentiation-associated gene 5 (MDA5) are cytosolic RNA-detecting PRRs which also activate IRF7 via interaction with mitochondrial antiviral signalling protein (MAVS) [9]. Once activated, IRF5 and 7 translocate to the nucleus and stimulate transcription of IFNα (mainly) and IFNβ. IRF5 also stimulates transcription of other pro-inflammatory cytokines [7, 10].

Additional PRRs, signalling pathways and IRFs are also activated in other cell types such as monocytes/macrophages, endothelial cells and dendritic cells to produce type I IFNs, in particular IFNβ. For example endosomal TLR3 (stimulated by double stranded RNA) signals through TIR-domain-containing adaptor-inducing IFNβ (TRIF), TBK1 (TANK-binding kinase 1) and IKKε (IκB kinase ε) to activate IRF3 and IFNβ production [5]. Other cytosolic

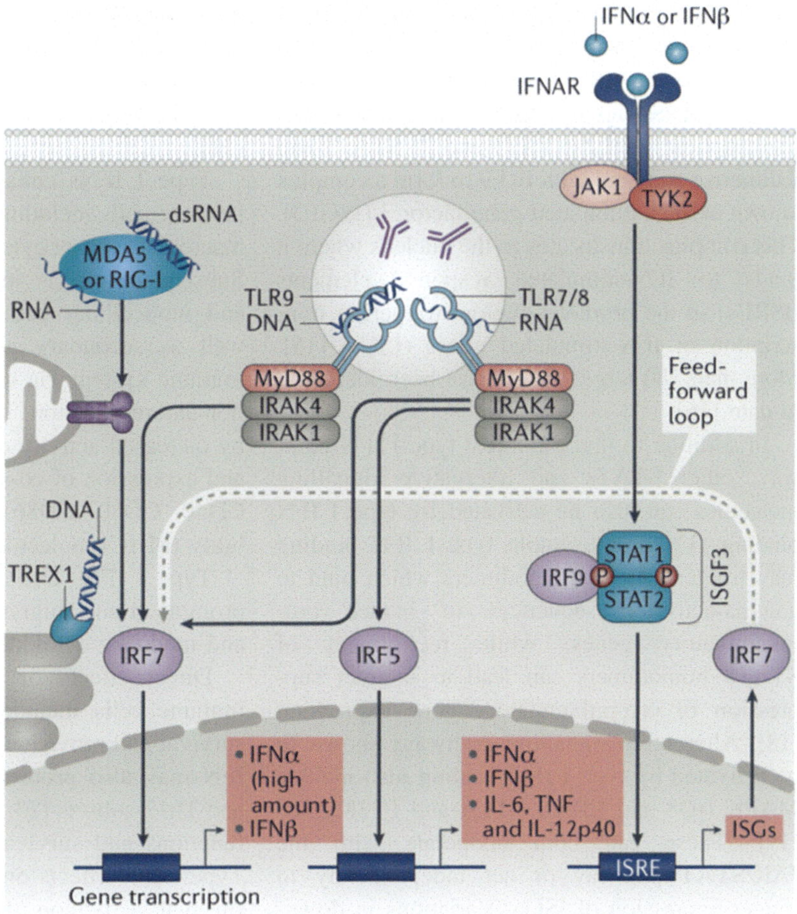

Fig. 4.1 **Type I IFN production by the plasmacytoid dendritic cell.** Major intracellular pathways by which type I IFN is produced by pDC. Activation of endosomal toll-like receptors leads differential activation of the IRF pathways depending on the specific TLR engaged. IRF5 and IRF7 translocates into the nucleus and induce transcription of IFN. Binding of type I interferon to the type I interferon receptor (IFNAR) results in activation of the canonical Janus kinase (JAK)–signal transducer and activator of transcription (STAT) pathway that results in transcription of type I interferon stimulated genes (ISGs). From Muskardin, Nat Rev Rheum 2018 (5)

nucleic acid sensors such as RNA helicases, DNA-dependent activator of IFN-regulatory factor (DAI) and cytosolic GAMP synthase (cGAS) have also been shown to play a role in type I IFN production [11]. In the case of cGAS, stimulation by DNA (both microbial and self-DNA) generates cGAMP which is recognised by stimulator of IFN genes (STING) leading to IRF3 phosphorylation and thus IFNβ production [12]. In endothelial cells and macrophages, signalling through tumour necrosis factor receptors (TNFR) can also induce IFNβ via IRF1 [13].

Type I IFN Signalling Pathways

Type I IFNs signal through binding a common type I IFN receptor (interferon α/β receptor; IFNAR) which is expressed on all nucleated cell types [14]. The receptor consists of two transmembrane proteins, IFNAR1 and IFNAR2, which dimerise upon cytokine binding triggering intracellular signalling.

The most well-defined type I IFN signalling pathway is depicted in Fig. 4.1 and characterised by phosphorylation of cytosolic janus kinase 1 (JAK1) and tyrosine kinase 2 (TYK2) in response to IFNAR engagement, which in turn phosphorylates signal transducer and activator of transcription 1 and 2 (STAT1 and 2). STAT1 and 2 dimerise and bind with IRF9 to form a complex known as IFN-stimulated gene factor 3 (ISGF3). This complex translocates to the nucleus where it binds to IFN-stimulated response elements (ISREs) in the promoter region, triggering transcription of IFN-stimulated genes (ISGs) [15]. More than 380 target genes have been identified to date [16].

In addition to this canonical type I IFN pathway, other STATs and alternative signalling molecules can also be activated by type I IFN binding [17]. For example type I IFN binding may trigger STAT1 homodimers which bind to gamma-activated sequences to induce pro-inflammatory genes, while recruitment of STAT3 homodimers can lead to indirect suppression of pro-inflammatory gene expression [18]. Alternative signalling pathways shown to be activated by type I IFN binding also include MAPK (p38 and ERK), NFkB and P13K/AKT [19]. These can both cooperate with the JAK/STAT pathway or act independently to trigger expression of ISGs. Variations in downstream signalling pathways may occur due to differential engagement of IFNAR by type I IFNs inducing different conformational changes and thus triggering different signalling pathways [20]. Alternative signalling pathways may also be differentially activated depending on the cell type and environmental context [15], and this may play a role in the regulation of overall type I IFN activity and effects.

Downstream Effects of Type I IFN Activation

Type I IFN pathway activation and the subsequent upregulation of ISG transcription has numerous downstream effects. In addition to direct antimicrobial effects that limit spread of infection, IFNα and IFNβ also have effects on a range of immune cells, modulating and bridging both the innate and adaptive immune response [21].

Type I IFNs enhance the activity of innate immune cells including natural killer (NK) cells, macrophage/monocytes and dendritic cells. Subsequent effects include increased cytolysis and intracellular pathogen destruction [21], as well as secondary activation of the adaptive immune system via enhanced antigen presentation and recognition. This latter effect is mediated by increased activity of antigen presenting cells, and expression of co-stimulatory molecules (e.g. CD40, CD80, CD86) and major histocompatibility (MHC) molecules [19].

Type I IFNs also induce chemokines which promote tissue migration of inflammatory cells and modulate cytokine secretion [19].

Direct effects of type I IFN on adaptive immune cells include enhanced CD4 + T cell survival and provision of B cell help [22]. Type I IFN may also promote differentiation into Th1 and Th17 subsets [23, 24]. CD8 + T cell effector functions and survival are also promoted [22]. Type I IFN effects on B cells include increased survival, activation, differentiation and antibody production [22] and IFNα has been shown to increase B-lymphocyte stimulator (BLyS) [25]. Collectively these effects promote antigen-specific T and B cell responses and memory cell generation.

Type I IFNs also have effects on non-immune cell types. For example type I IFN signalling has been shown to negatively affect haemopoietic stem cell differentiation in the bone marrow [26], impair function of endothelial cells [27] and increase reactivity of microglia in the nervous system [28].

While the major downstream consequences of type I IFN are vital in host immunity, the same effects also hold the potential to promote loss of tolerance and exacerbate immune pathology such as SLE in susceptible individuals.

Regulation of Type I IFN Activity

Type I IFN activity is tightly regulated, balancing its important role in host defence with the potential harms of prolonged or inappropriate activation. In healthy individuals there is only very small constitutive IFNα production and complex cellular and molecular mechanisms regulate activity of the type I IFN system. IFN production and the outcome of IFNAR engagement (i.e. activation of signalling pathways and ISG transcription) are influenced by numerous factors including cell type, genetic context, phase of the immune response, and the influence of surrounding cells and local cytokine milieu [15, 29–31].

Cell-specific factors play an important role in type I IFN production. For example pDCs have the capacity to produce large amounts of IFNα rapidly due to several factors, including high constitutive production of IRF7 [4] and endosomal retention of the MyD88-IRF7 complex [10]. Furthermore a self-amplification mechanism occurs, whereby factors involved in type I IFN signalling, such as IRF7, are in fact ISGs themselves, which facilitates a feed-forward means of producing further type I IFNs [32]. Termination of IFNα production occurs once pathogen eradication is achieved, and pDCs become temporarily refractory to new stimuli, mediated by inhibition and degradation of transcription factors and signal transducers [33].

Cellular interactions between type I IFN producing cells and other cell types, and the influence of other cytokine signalling pathways also modulate the production of type I IFNs. For example, pDCs are influenced by NK cells, B cells and activated T cells which have stimulatory effects on IFNα production via a range of mechanisms, while monocytes/macrophages have an inhibitory effect by releasing reactive oxygen species and prostaglandin E2 [11, 29] (Fig. 4.2). Activity of other inflammatory pathways also influence type I IFN activity. For example tumour necrosis factor alpha (TNFα) can have a suppressive effect on type I IFN activity (and vice versa) [34]. This bears particular clinical relevance given the availability of TNFα inhibitors which can upregulate ISGs and in some patients trigger a reversible lupus-like syndrome [34].

Diversity in the biological consequences of type I IFN production is also controlled by variations in IFNAR engagement, downstream signalling pathways and consequent gene transcription. At the receptor level, down-regulation of type I IFN activity may be achieved by internalisation and degradation of IFNAR [18], a process regulated by phosphatases such as protein-tyrosine phosphatase 1B (PTPB1) [35]. Differential binding of the IFNAR receptor by type I IFNs also contribute to variations in downstream outcomes [36]. Furthermore a range of post-translational modifications, including phosphorylation, ubiquitylation, acetylation and methylation targeting various components of the signalling pathway can modify type I IFN activity [37]. This includes an important role for negative regulators in the control of type I IFN activity, such as ubiquitin-specific protease (USP)-18 [38] and suppressor of cytokine signalling (SOCS) [39] which target particular JAK and STAT proteins. Epigenetic regulation including histone modification, DNA methylation and non-coding RNAs adds a further level of control impacting ISG expression [37].

Association of Type I IFN With SLE Pathogenesis

Multiple lines of evidence support a link between type I IFN activity and SLE. In patients with established SLE, early studies identified serum levels of IFNα to be elevated compared to healthy controls [40]. This has subsequently been replicated in various SLE populations using different laboratory techniques, with between 50–90% of patients consistently demonstrating elevated IFNα. It has also been demonstrated that levels of IFNα markedly rise in the year prior to the development of overt disease [41]. Furthermore, family members of lupus patients have also been shown to have higher serum IFNα

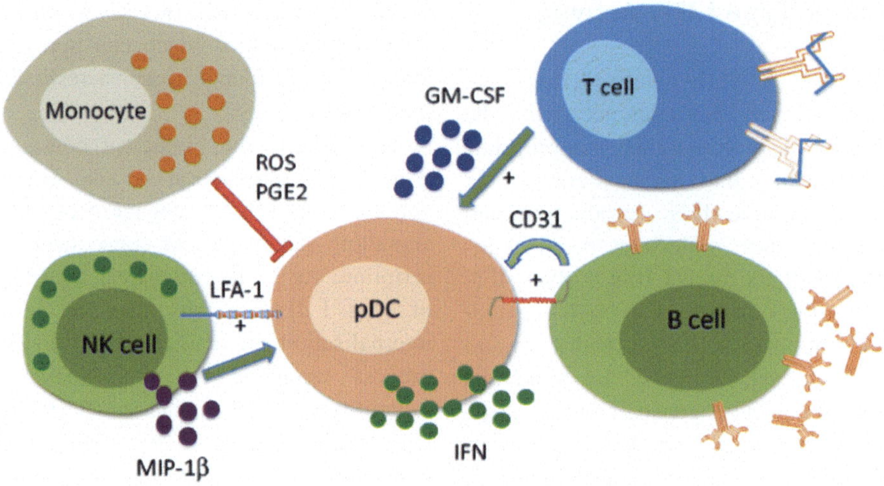

Fig. 4.2 Inducers and regulators of IFNα production by pDCs. There are a number of mechanisms by which IFN is induced by pDC. The most well studied is the presence of immune complexes which become endocytosed by pDC. A range of cells can stimulate further production by pDC via process of costimulation. From Eloranta et al. J Molecular Med 2016

suggesting it may be a heritable trait predisposing to SLE [42]. This notion is supported by genetic studies which have identified numerous risk genes linked to the type I IFN system [43].

It has also been observed that therapeutic administration of exogenous type I interferon, as has been used in hepatitis C, multiple sclerosis and certain malignancies, can cause a reversible lupus syndrome as a complication of treatment, supporting a role in disease initiation [44]. Conversely, traditional treatments used in SLE such as hydroxychloroquine which improves disease activity and outcomes, has been shown to reduce the production of IFNα by pDCs [45]. Studies in lupus animal models also support the importance of the association; for example in NZB/W lupus prone mice, administration of IFNα has been shown to promote SLE manifestations, while IFNAR deficiency leads to disease amelioration [46, 47].

The "Interferon Signature"

Further evidence of association between the type I IFN system and SLE is the consistent finding of a distinct pattern of upregulated ISGs measurable in the peripheral blood of a significant proportion of individuals with SLE. Known as the "interferon signature" this marker of type I IFN activation was initially detected with microarray analysis and subsequently confirmed with real-time polymerase chain reaction (rtPCR) [48–52]. Essentially mRNA transcripts of selected ISGs measured whole blood, peripheral blood mononuclear cells or tissue are presumed to reflect gene expression and hence type I IFN activity. A select group of genes has been consistently replicated in studies from multiple centres.

A peripheral blood IFN signature has been detected in approximately 50–75% of adult patients with SLE and an even greater proportion of paediatric SLE patients [48–52]. It may predict development of SLE in at-risk individuals [53]. Phenotypic associations have been described, including associations with both past and active renal disease and links with a range of autoantibodies [54]. The IFN signature has also been consistently associated with high disease activity in cross-sectional studies, although studies of longitudinal correlations of the IFN signature with disease activity over time have been inconsistent [55, 56].

Measurement and interpretation of the IFN signature is complicated by the probable contribution of multiple IFN types to the signature due to the significant overlap in genes induced by different members of the IFN family [57, 58] and differential expression of genes in different immune cells. It also remains unclear how the predictive value and associations observed in cohort studies can be applied at an individual patient level. Despite these challenges, use of the IFN signature as a disease and treatment response biomarker is of significant interest. Other gene signatures, such as neutrophil and plasmablast signatures [59], as well as alternate means of measuring type I IFN activity are also under investigation [5].

Dysregulation of Type I IFN in SLE

Clearly there are many lines of evidence which point to the association between type I IFN and SLE. Increasingly the underlying mechanisms by which dysregulated type I IFN activity occurs, and subsequently promotes disease initiation and perpetuation are being identified.

Stimulation of Type I IFN Production in SLE

pDCS represent the major source of type I IFNs in SLE. In addition to activation of PRRs by exogenous material as previously described, nucleic acids from endogenous sources can also trigger pathways of type I IFN activation and contribute to inappropriate or persistent type I IFN production in patients with SLE [60]. Figure 4.2 illustrates the range of possible inducers of IFN production that may be active in SLE [61]. As well as excess activation, impaired clearance mechanisms of endogenous nucleic acids have also been implicated, leading to prolonged exposure of self-antigens to the immune system promoting formation of immune complexes. Increased apoptosis, deficiencies in clearance of apoptotic debris, complement system dysfunction and reduced phagocyte function have all been described in SLE [62, 63].

Immune complexes containing nucleic acid binding proteins and auto-antibodies are one of the important endogenous stimuli for type I IFN production in SLE [64]. These complexes bind to Fc-γ-RIIa receptors on pDCs and are endocytosed, with the nucleic acid component interacting with TLR7/9 resulting in stimulation of type I IFN pathways and IFNα production [65]. Other PRRs which lead to type I IFN production, such as cGAS, RIG-1 and MDA5, may also be activated in a subset of SLE patients, but mechanisms and significance are yet to be as well elucidated [66, 67].

Neutrophil extracellular traps (NETs) are another source of nuclear material which activate pDCs and promote high levels of TLR9-dependent type I IFN production and can also promote autoantibody formation and activate autoreactive B cells in SLE [68, 69]. NETs are produced when neutrophils undergo a cell death pathway known as NETosis, involving the extrusion of nuclear material in a net-like structure designed to entrap pathogens. A reduced capacity to degrade NETs has been described in some patients with SLE [69]. Mitochondrial DNA which can be released via NETosis is particularly interferonogenic and may also induce type I IFN production via the cytosolic cGAS/STING pathway [70]. A novel mechanism of mitochondrial DNA release triggering type I IFN production may also occur as a result of voltage-dependent anion channel oligomers forming pores in the outer mitochondrial membrane [71].

Transposable elements (movable DNA sequences) such as long interspersed nuclear element type I (LINE-1) may also induce type I IFN production. For example, it has been shown LINE-1 mRNA transcripts are increased in lupus nephritis kidneys and can trigger pDC and monocyte production of type I IFN [72].

Genetic and Environmental Factors

Type I IFN activity is influenced by genetic factors and is a polygenic trait. Genes relevant to type I IFN activity make up a significant proportion of the identified risk genes for SLE in genome wide association studies. Many variants associated with SLE correspond with increased activity of the type I IFN pathway [73]. For example, IRF5 is a consistently identified genetic risk factor in SLE, and is involved in type I IFN production, as well as promoting pro-inflammatory cytokines and B cell responses. IRF5 risk variants have been associated with higher type I IFN activity in the presence of anti-RNA-binding protein or anti-dsDNA autoantibodies in SLE [74]. However, the exact function of most risk alleles is not well understood, and the pathogenic associations of an individual genetic risk variant likely also depends on other contextual factors. A STAT4 risk variant, for example, has been associated in SLE patients with increased responses to IFNα in activated T cells, while in healthy individuals the same responses were found to be decreased [74].

Environmental associations with SLE have also been linked to type I IFN activity. The best characterised of these is the association with ultraviolet (UV) light, which is a well-known trigger of flares in some patients with SLE. It has been shown that UV light can trigger keratinocyte apoptosis and the exposure of auto-antigens. This promotes auto-antibody and immune complex formation which in turn stimulates type I IFN production by pDCs in the skin. UV light can also induce reactive oxygen species which cause DNA damage and exposure of further endogenous nucleic acids [75]. A recent murine study has also demonstrated that acute skin exposure to UV-B light can trigger ISG expression in both skin and blood, further supporting the link between UV light, type I IFN activity and SLE [76].

Disturbed Feedback Mechanisms

As described in the first part of this chapter, a broad range of regulatory measures influence the production of type I IFNs and its downstream effects. It is estimated that up to 10% of our genes have the potential to be regulated by type I IFNs [77], and hence consequences of type I IFN activation may vary significantly and is context-dependent.

Regulatory mechanisms limiting type I IFN activity may be impaired in SLE. For example, the capacity of monocytes to suppress IFNα production by pDCs appears to be reduced in patients with SLE compared to healthy individuals [78]. Conversely stimulatory influences of other cell types such as activated T cells, B cells, NK cells and platelets may enhance type I IFN production in SLE (Fig. 4.2) [61]. Priming of immune cells by exposure to basal levels of IFN may also contribute to enhanced type I IFN responses upon subsequent stimulation in SLE. This may be mediated by increased STAT-1 activation and increased expression of TLRs [79]. Epigenetic regulation is likely also important, with findings that differential methylation of ISGs is a feature of SLE and may correspond with active disease [80]. There still remains much to be elucidated however, regarding the exact means by which type I IFN escapes regulatory mechanisms to perpetuate autoimmunity and inflammation in SLE.

Type I IFN Dysregulation and the Promotion of SLE

In the context of SLE type I IFNs likely exert pathogenic influence via several mechanisms. General effects of type I IFN activation on cells of the immune system relevant to SLE are summarised in Table 4.2.

Innate immune activity is modulated by type I IFN and includes enhanced cytotoxicity and

Table 4.2 Immune effects of type I IFN activation relevant to SLE pathogenesis

Immune cell	Effects of type I IFN activation
Dendritic cells	Induces monocyte differentiation into dendritic cells [82] Enhances MHC expression and antigen presentation [94]
NK cells	Enhances cytotoxicity and cytokine production [95]
Monocytes/ Macrophages	Impairs clearance mechanisms [81] Increases co-stimulatory molecule expression, antigen presentation [82]
Neutrophils	Promotes NET formation [68]
T cells	Promotes CD4 + differentiation and survival [83] Enhances CD8 + T cell cytotoxicity and memory cell formation [96] Suppresses regulatory T cell activity [97]
B cells	Increased levels of BLyS [25] Promotes activation and plasma cell differentiation [22] Increases antibody synthesis and isotype class switching [87]

modulation of cytokine and chemokine release, relevant to various pathways of tissue inflammation and damage in SLE. Type I IFNs may also contribute to auto-antigen persistence, via reducing splenic marginal zone macrophage function which reduces clearance of apoptotic cells [81] and by priming neutrophils for apoptosis in response to autoantibodies, promoting the release of NETs [68].

Bridging the innate and adaptive immune systems, type I IFN also augments antigen presentation by promoting differentiation and maturation of antigen presenting cells and upregulating T cell co-stimulatory capacity [82], contributing to increased presentation of self-antigens. Type I IFNs also have direct effects on T cell activity stimulating differentiation of naïve T cells into helper T cells [83] and the development of CD8 + memory T cells [84], while suppressing regulatory T cell activity [85]. Together these actions promote expansion of autoreactive T cells [22].

Type I IFNs also have numerous direct and indirect effects on B cells relevant to SLE pathogenesis. It reduces the activation threshold of B cells [86], promotes B cell differentiation, isotype switching [87] and memory B cell formation [88]. Many of these effects B cell promoting effects are mediated by increased production of B lymphocyte stimulating factor (BLyS), an important B cell survival factor and therapeutic target in SLE. Increased BLyS production and release occurs in response to IFNα [25] and is associated with increased B cell activity and differentiation of autoreactive B cells into antibody secreting plasma cells [89].

Type I IFN activation may also have organ-specific implications in SLE. For example type I IFNs have been shown to promote loss and inhibit maturation of renal podocytes, which bears relevance to lupus nephritis pathophysiology [81]. In the skin IFNα may promote apoptosis of keratinocytes leading to the release of nuclear antigens and perpetuation of cutaneous SLE manifestations [90]. Modulatory effects of type I IFN on the vascular system may also be relevant to SLE, where patients are observed to have accelerated atherosclerosis and high rates of cardiovascular morbidity. Type I IFNs have been shown to impair endothelial function via multiple mechanisms [27, 91], and may also alter platelet function [92] and increase lipid uptake [93] promoting atherogenesis.

Conclusion

The type I IFN system is well recognised as a key pathway in SLE pathophysiology. Our current model of understanding involves the production of excess auto-antigens from a number of sources, of which a consequence is stimulation of type I IFN production by pDCs. This, in combination with impaired clearance mechanisms promotes chronic type I IFN system activation, which has pleiotropic effects on the immune system that may perpetuate autoimmunity and promote the development of a lupus clinical syndrome in a genetically predisposed individual. Although highly complex and not yet fully understood, these mechanisms lend support to the ongoing investigation of type I IFN based biomarkers, as well as promising therapies which modify type I IFN activity in SLE.

References

1. Isaacs A, Lindenmann J (1957) Virus interference. I. The interferon. Proc R Soc Lond B Biol Sci 147 (927):258–67
2. Pestka S, Krause CD, Walter MR (2004) Interferons, interferon-like cytokines, and their receptors. Immunol Rev 202:8–32
3. Chasset F, Arnaud L (2018) Targeting interferons and their pathways in systemic lupus erythematosus. Autoimmun Rev 17(1):44–52
4. Swiecki M, Colonna M (2015) The multifaceted biology of plasmacytoid dendritic cells. Nat Rev Immunol 15(8):471–485
5. Muskardin TLW, Niewold TB (2018) Type I interferon in rheumatic diseases. Nat Rev Rheumatol 14 (4):214–228
6. Gurtler C, Bowie AG (2013) Innate immune detection of microbial nucleic acids. Trends Microbiol 21 (8):413–420
7. Jensen MA, Niewold TB (2015) Interferon regulatory factors: critical mediators of human lupus. Transl Res J Lab Clin Med 165(2):283–295
8. Kawai T, Akira S (2010) The role of pattern-recognition receptors in innate immunity: update on Toll-like receptors. Nat Immunol 11(5):373–384
9. Kato H, Takeuchi O, Sato S, Yoneyama M, Yamamoto M, Matsui K et al (2006) Differential roles of MDA5 and RIG-I helicases in the recognition of RNA viruses. Nature 441(7089):101–105
10. Tamura T, Yanai H, Savitsky D, Taniguchi T (2008) The IRF family transcription factors in immunity and oncogenesis. Annu Rev Immunol 26:535–584
11. Eloranta ML, Rönnblom L (2016) Cause and consequences of the activated type I interferon system in SLE. J Mol Med (Berl) 94(10):1103–1110
12. Sun L, Wu J, Du F, Chen X, Chen ZJ (2013) Cyclic GMP-AMP synthase is a cytosolic DNA sensor that activates the type I interferon pathway. Sci (N.Y., NY) 339(6121):786–791
13. Yarilina A, Park-Min KH, Antoniv T, Hu X, Ivashkiv LB (2008) TNF activates an IRF1-dependent autocrine loop leading to sustained expression of chemokines and STAT1-dependent type I interferon-response genes. Nat Immunol 9 (4):378–387
14. Mesev EV, LeDesma RA, Ploss A (2019) Decoding type I and III interferon signalling during viral infection. Nat Microbiol 4(6):914–924
15. Platanias LC (2005) Mechanisms of type-I- and type-II-interferon-mediated signalling. Nat Rev Immunol 5(5):375–386
16. Schoggins JW, Wilson SJ, Panis M, Murphy MY, Jones CT, Bieniasz P et al (2011) A diverse range of gene products are effectors of the type I interferon antiviral response. Nature 472(7344):481–485
17. van Boxel-Dezaire AH, Rani MR, Stark GR (2006) Complex modulation of cell type-specific signaling in response to type I interferons. Immunity 25 (3):361–372
18. Ivashkiv LB, Donlin LT (2014) Regulation of type I interferon responses. Nat Rev Immunol 14(1):36–49
19. Hervas-Stubbs S, Perez-Gracia JL, Rouzaut A, Sanmamed MF, Le Bon A, Melero I (2011) Direct effects of type I interferons on cells of the immune system. Clin Cancer Res Official J Am Assoc Cancer Res 17 (9):2619–2627
20. Piehler J, Thomas C, Garcia KC, Schreiber G (2012) Structural and dynamic determinants of type I interferon receptor assembly and their functional interpretation. Immunol Rev 250(1):317–334
21. McNab F, Mayer-Barber K, Sher A, Wack A, O'Garra A (2015) Type I interferons in infectious disease. Nat Rev Immunol 15(2):87–103
22. Le Bon A, Thompson C, Kamphuis E, Durand V, Rossmann C, Kalinke U et al (2006) Cutting edge: enhancement of antibody responses through direct stimulation of B and T cells by type I IFN. J Immunol 176(4):2074–8
23. Alunno A, Bartoloni E, Bistoni O, Nocentini G, Ronchetti S, Caterbi S et al (2012) Balance between regulatory T and Th17 cells in systemic lupus erythematosus: the old and the new. Clin Dev Immunol 2012:823085
24. Brinkmann V, Geiger T, Alkan S, Heusser CH (1993) Interferon alpha increases the frequency of interferon gamma-producing human CD4+ T cells. J Exp Med 178(5):1655–1663
25. Lopez P, Scheel-Toellner D, Rodriguez-Carrio J, Caminal-Montero L, Gordon C, Suarez A (2014) Interferon-alpha-induced B-lymphocyte stimulator expression and mobilization in healthy and systemic lupus erthymatosus monocytes. Rheumatol (Oxford) 53(12):2249–2258
26. Honda K, Yanai H, Takaoka A, Taniguchi T (2005) Regulation of the type I IFN induction: a current view. Int Immunol 17(11):1367–1378
27. Thacker SG, Zhao W, Smith CK, Luo W, Wang H, Vivekanandan-Giri A et al (2012) Type I interferons

modulate vascular function, repair, thrombosis, and plaque progression in murine models of lupus and atherosclerosis. Arthritis Rheum 64(9):2975–2985
28. Bialas AR, Presumey J, Das A, van der Poel CE, Lapchak PH, Mesin L et al (2017) Microglia-dependent synapse loss in type I interferon-mediated lupus. Nature 546(7659):539–543
29. Bengtsson AA, Ronnblom L (2017) Role of interferons in SLE. Best Pract Res Clin Rheumatol 31 (3):415–428
30. Ng CT, Mendoza JL, Garcia KC, Oldstone MB (2016) Alpha and Beta Type 1 interferon signaling: passage for diverse biologic outcomes. Cell 164 (3):349–352
31. Sharma S, Jin Z, Rosenzweig E, Rao S, Ko K, Niewold TB (2015) Widely divergent transcriptional patterns between SLE patients of different ancestral backgrounds in sorted immune cell populations. J Autoimmun 60:51–58
32. Levy DE, Marie I, Smith E, Prakash A (2002) Enhancement and diversification of IFN induction by IRF-7-mediated positive feedback. J Interferon Cytokine Res Official J Int Soc Interferon Cytokine Res 22(1):87–93
33. Swiecki M, Colonna M (2010) Unraveling the functions of plasmacytoid dendritic cells during viral infections, autoimmunity, and tolerance. Immunol Rev 234(1):142–162
34. Cantaert T, Baeten D, Tak PP, van Baarsen LG (2010) Type I IFN and TNFalpha cross-regulation in immune-mediated inflammatory disease: basic concepts and clinical relevance. Arthritis Res Ther 12 (5):219
35. Carbone CJ, Zheng H, Bhattacharya S, Lewis JR, Reiter AM, Henthorn P et al (2012) Protein tyrosine phosphatase 1B is a key regulator of IFNAR1 endocytosis and a target for antiviral therapies. Proc Natl Acad Sci USA 109(47):19226–19231
36. Stanifer ML, Pervolaraki K, Boulant S (2019) Differential regulation of Type I and Type III interferon signaling. Int J Mol Sci 20(6)
37. Chen K, Liu J, Cao X (2017) Regulation of type I interferon signaling in immunity and inflammation: a comprehensive review. J Autoimmun 83:1–11
38. Malakhova OA, Kim KI, Luo JK, Zou W, Kumar KG, Fuchs SY et al (2006) UBP43 is a novel regulator of interferon signaling independent of its ISG15 isopeptidase activity. EMBO J 25 (11):2358–2367
39. Piganis RA, De Weerd NA, Gould JA, Schindler CW, Mansell A, Nicholson SE et al (2011) Suppressor of cytokine signaling (SOCS) 1 inhibits type I interferon (IFN) signaling via the interferon alpha receptor (IFNAR1)-associated tyrosine kinase Tyk2. J Biol Chem 286(39):33811–33818
40. Ytterberg SR, Schnitzer TJ (1982) Serum interferon levels in patients with systemic lupus erythematosus. Arthritis Rheum 25(4):401–406
41. Munroe ME, Lu R, Zhao YD, Fife DA, Robertson JM, Guthridge JM et al (2016) Altered type II interferon precedes autoantibody accrual and elevated type I interferon activity prior to systemic lupus erythematosus classification. Ann Rheum Dis 75 (11):2014–2021
42. Niewold TB, Hua J, Lehman TJ, Harley JB, Crow MK (2007) High serum IFN-alpha activity is a heritable risk factor for systemic lupus erythematosus. Genes Immun 8(6):492–502
43. Bronson PG, Chaivorapol C, Ortmann W, Behrens TW, Graham RR (2012) The genetics of type I interferon in systemic lupus erythematosus. Curr Opin Immunol 24(5):530–537
44. Borg FA, Isenberg DA (2007) Syndromes and complications of interferon therapy. Curr Opin Rheumatol 19(1):61–66
45. Sacre K, Criswell LA, McCune JM (2012) Hydroxychloroquine is associated with impaired interferon-alpha and tumor necrosis factor-alpha production by plasmacytoid dendritic cells in systemic lupus erythematosus. Arthritis Res Ther 14(3):R155
46. Liu Z, Bethunaickan R, Huang W, Lodhi U, Solano I, Madaio MP et al (2011) Interferon-alpha accelerates murine systemic lupus erythematosus in a T cell-dependent manner. Arthritis Rheum 63 (1):219–229
47. Santiago-Raber ML, Baccala R, Haraldsson KM, Choubey D, Stewart TA, Kono DH et al (2003) Type-I interferon receptor deficiency reduces lupus-like disease in NZB mice. J Exp Med 197(6):777–788
48. Baechler EC, Batliwalla FM, Karypis G, Gaffney PM, Ortmann WA, Espe KJ et al (2003) Interferon-inducible gene expression signature in peripheral blood cells of patients with severe lupus. Proc Natl Acad Sci USA 100(5):2610–2615
49. Bennett L, Palucka AK, Arce E, Cantrell V, Borvak J, Banchereau J et al (2003) Interferon and granulopoiesis signatures in systemic lupus erythematosus blood. J Exp Med 197(6):711–723
50. Crow MK, Kirou KA, Wohlgemuth J (2003) Microarray analysis of interferon-regulated genes in SLE. Autoimmunity 36(8):481–490
51. Han GM, Chen SL, Shen N, Ye S, Bao CD, Gu YY (2003) Analysis of gene expression profiles in human systemic lupus erythematosus using oligonucleotide microarray. Genes Immun 4(3):177–186
52. Kirou KA, Lee C, George S, Louca K, Papagiannis IG, Peterson MG et al (2004) Coordinate overexpression of interferon-alpha-induced genes in systemic lupus erythematosus. Arthritis Rheum 50 (12):3958–3967
53. Md Yusof MY, Psarras A, El-Sherbiny YM, Hensor EMA, Dutton K, Ul-Hassan S et al (2018) Prediction of autoimmune connective tissue disease in an at-risk cohort: prognostic value of a novel two-score system for interferon status. Ann Rheum Dis 77 (10):1432–1439

54. Kirou KA, Lee C, George S, Louca K, Peterson MG, Crow MK (2005) Activation of the interferon-alpha pathway identifies a subgroup of systemic lupus erythematosus patients with distinct serologic features and active disease. Arthritis Rheum 52 (5):1491–1503
55. Landolt-Marticorena C, Bonventi G, Lubovich A, Ferguson C, Unnithan T, Su J et al (2009) Lack of association between the interferon-alpha signature and longitudinal changes in disease activity in systemic lupus erythematosus. Ann Rheum Dis 68 (9):1440–1446
56. Petri M, Singh S, Tesfasyone H, Dedrick R, Fry K, Lal P et al (2009) Longitudinal expression of type I interferon responsive genes in systemic lupus erythematosus. Lupus 18(11):980–989
57. Chiche L, Jourde-Chiche N, Whalen E, Presnell S, Gersuk V, Dang K et al (2014) Modular transcriptional repertoire analyses of adults with systemic lupus erythematosus reveal distinct type I and type II interferon signatures. Arthritis Rheumatol (Hoboken, NJ) 66(6):1583–1595
58. Karonitsch T, Feierl E, Steiner CW, Dalwigk K, Korb A, Binder N et al (2009) Activation of the interferon-gamma signaling pathway in systemic lupus erythematosus peripheral blood mononuclear cells. Arthritis Rheum 60(5):1463–1471
59. Banchereau R, Hong S, Cantarel B, Baldwin N, Baisch J, Edens M et al (2016) Personalized immunomonitoring uncovers molecular networks that stratify lupus patients. Cell 165(3):551–565
60. Mustelin T, Lood C, Giltiay NV (2019) Sources of pathogenic nucleic acids in systemic lupus erythematosus. Front Immunol 10:1028
61. Ronnblom L, Leonard D (2019) Interferon pathway in SLE: one key to unlocking the mystery of the disease. Lupus Sci Med 6(1):e000270
62. Mahajan A, Herrmann M, Munoz LE (2016) Clearance deficiency and cell death pathways: a model for the pathogenesis of SLE. Front Immunol 7:35
63. Leffler J, Ciacma K, Gullstrand B, Bengtsson AA, Martin M, Blom AM (2015) A subset of patients with systemic lupus erythematosus fails to degrade DNA from multiple clinically relevant sources. Arthritis Res Ther 17:205
64. Lovgren T, Eloranta ML, Bave U, Alm GV, Ronnblom L (2004) Induction of interferon-alpha production in plasmacytoid dendritic cells by immune complexes containing nucleic acid released by necrotic or late apoptotic cells and lupus IgG. Arthritis Rheum 50(6):1861–1872
65. Bave U, Magnusson M, Eloranta ML, Perers A, Alm GV, Ronnblom L (2003) Fc gamma RIIa is expressed on natural IFN-alpha-producing cells (plasmacytoid dendritic cells) and is required for the IFN-alpha production induced by apoptotic cells combined with lupus IgG. J Immunol 171(6):3296–302
66. An J, Durcan L, Karr RM, Briggs TA, Rice GI, Teal TH et al (2017) Expression of cyclic GMP-AMP synthase in patients with systemic lupus erythematosus. Arthritis Rheumatol (Hoboken, NJ) 69(4):800–807
67. Shao WH, Shu DH, Zhen Y, Hilliard B, Priest SO, Cesaroni M et al (2016) Prion-like aggregation of mitochondrial antiviral signaling protein in lupus patients is associated with increased levels of Type I interferon. Arthritis Rheumatol (Hoboken, NJ) 68 (11):2697–2707
68. Garcia-Romo GS, Caielli S, Vega B, Connolly J, Allantaz F, Xu Z et al (2011) Netting neutrophils are major inducers of type I IFN production in pediatric systemic lupus erythematosus. Sci Transl Med 3 (73):73ra20
69. Gestermann N, Di Domizio J, Lande R, Demaria O, Frasca L, Feldmeyer L et al (2018) Netting neutrophils activate autoreactive B cells in lupus. J Immunol 200(10):3364–71
70. Lood C, Blanco LP, Purmalek MM, Carmona-Rivera C, De Ravin SS, Smith CK et al (2016) Neutrophil extracellular traps enriched in oxidized mitochondrial DNA are interferogenic and contribute to lupus-like disease. Nat Med 22(2):146–153
71. Kim J, Gupta R, Blanco LP, Yang S, Shteinfer-Kuzmine A, Wang K et al (2019) VDAC oligomers form mitochondrial pores to release mtDNA fragments and promote lupus-like disease. Science (New York, NY) 366(6472):1531–1536
72. Mavragani CP, Sagalovskiy I, Guo Q, Nezos A, Kapsogeorgou EK, Lu P et al (2016) Expression of long interspersed nuclear element 1 retroelements and induction of Type I interferon in patients with systemic autoimmune disease. Arthritis Rheumatol (Hoboken, NJ) 68(11):2686–2696
73. Ghodke-Puranik Y, Niewold TB (2013) Genetics of the type I interferon pathway in systemic lupus erythematosus. Int J Clin Rheumatol 8(6)
74. Niewold TB, Kelly JA, Kariuki SN, Franek BS, Kumar AA, Kaufman KM et al (2012) IRF5 haplotypes demonstrate diverse serological associations which predict serum interferon alpha activity and explain the majority of the genetic association with systemic lupus erythematosus. Ann Rheum Dis 71(3):463–468
75. Barbhaiya M, Costenbader KH (2014) Ultraviolet radiation and systemic lupus erythematosus. Lupus 23(6):588–595
76. Skopelja-Gardner S AJ, Sun X, Tanaka L, Tai J, Hermanson P, Kawasumi M, Elkon K (2019) UV light induces acute Type I interferon production in the skin and blood which is cGAS dependent [abstract]. Arthritis Rheumatol (Hoboken, NJ) 71
77. Schoggins JW (2019) Interferon-stimulated genes: what do they all do? Ann Rev Virol 6 (1):567–584
78. Eloranta ML, Lovgren T, Finke D, Mathsson L, Ronnelid J, Kastner B et al (2009) Regulation of the interferon-alpha production induced by RNA-containing immune complexes in plasmacytoid dendritic cells. Arthritis Rheum 60(8):2418–2427

79. Kalliolias GD, Ivashkiv LB (2010) Overview of the biology of type I interferons. Arthritis Res Ther 12 (Suppl 1):S1
80. Imgenberg-Kreuz J, Carlsson Almlof J, Leonard D, Alexsson A, Nordmark G, Eloranta ML et al (2018) DNA methylation mapping identifies gene regulatory effects in patients with systemic lupus erythematosus. Ann Rheum Dis 77(5):736–743
81. Li H, Fu YX, Wu Q, Zhou Y, Crossman DK, Yang P et al (2015) Interferon-induced mechanosensing defects impede apoptotic cell clearance in lupus. J Clin Investig 125(7):2877–2890
82. Blanco P, Palucka AK, Gill M, Pascual V, Banchereau J (2001) Induction of dendritic cell differentiation by IFN-alpha in systemic lupus erythematosus. Science (New York, NY) 294 (5546):1540–1543
83. Chan VS, Nie YJ, Shen N, Yan S, Mok MY, Lau CS (2012) Distinct roles of myeloid and plasmacytoid dendritic cells in systemic lupus erythematosus. Autoimmun Rev 11:890–897
84. Curtsinger JM, Valenzuela JO, Agarwal P, Lins D, Mescher MF (2005) Type I IFNs provide a third signal to CD8 T cells to stimulate clonal expansion and differentiation. J Immunol (Baltimore, Md: 1950) 174(8):4465–9
85. Le Buanec H, Gougeon ML, Mathian A, Lebon P, Dupont JM, Peltre G et al (2011) IFN-alpha and CD46 stimulation are associated with active lupus and skew natural T regulatory cell differentiation to type 1 regulatory T (Tr1) cells. Proc Natl Acad Sci USA 108(47):18995–19000
86. Vazquez MI, Catalan-Dibene J, Zlotnik A (2015) B cells responses and cytokine production are regulated by their immune microenvironment. Cytokine 74(2):318–326
87. Le Bon A, Schiavoni G, D'Agostino G, Gresser I, Belardelli F, Tough DF (2001) Type i interferons potently enhance humoral immunity and can promote isotype switching by stimulating dendritic cells in vivo. Immunity 14(4):461–470
88. Fleischer SJ, Giesecke C, Mei HE, Lipsky PE, Daridon C, Dorner T (2014) Increased frequency of a unique spleen tyrosine kinase bright memory B cell population in systemic lupus erythematosus. Arthritis Rheumatol (Hoboken, NJ) 66(12):3424–3435
89. Sjostrand M, Johansson A, Aqrawi L, Olsson T, Wahren-Herlenius M, Espinosa A (2016) The expression of BAFF is controlled by IRF transcription factors. J Immunol (Baltimore, Md : 1950) 196 (1):91–6
90. Zahn S, Rehkamper C, Ferring-Schmitt S, Bieber T, Tuting T, Wenzel J (2011) Interferon-alpha stimulates TRAIL expression in human keratinocytes and peripheral blood mononuclear cells: implications for the pathogenesis of cutaneous lupus erythematosus. Br J Dermatol 165(5):1118–1123
91. Denny MF, Thacker S, Mehta H, Somers EC, Dodick T, Barrat FJ et al (2007) Interferon-alpha promotes abnormal vasculogenesis in lupus: a potential pathway for premature atherosclerosis. Blood 110 (8):2907–2915
92. Scherlinger M, Sisirak V, Richez C, Lazaro E, Duffau P, Blanco P (2017) New insights on platelets and platelet-derived microparticles in systemic lupus erythematosus. Curr Rheumatol Rep 19(8):48
93. Li J, Fu Q, Cui H, Qu B, Pan W, Shen N et al (2011) Interferon-alpha priming promotes lipid uptake and macrophage-derived foam cell formation: a novel link between interferon-alpha and atherosclerosis in lupus. Arthritis Rheum 63(2):492–502
94. Montoya M, Schiavoni G, Mattei F, Gresser I, Belardelli F, Borrow P et al (2002) Type I interferons produced by dendritic cells promote their phenotypic and functional activation. Blood 99(9):3263–3271
95. Lee CK, Rao DT, Gertner R, Gimeno R, Frey AB, Levy DE (2000) Distinct requirements for IFNs and STAT1 in NK cell function. J Immunol (Baltimore, Md : 1950) 165(7):3571–7
96. Kolumam GA, Thomas S, Thompson LJ, Sprent J, Murali-Krishna K (2005) Type I interferons act directly on CD8 T cells to allow clonal expansion and memory formation in response to viral infection. J Exp Med 202(5):637–650
97. Bacher N, Raker V, Hofmann C, Graulich E, Schwenk M, Baumgrass R et al (2013) Interferon-α suppresses cAMP to disarm human regulatory T cells. Cancer Res 73:5647–5656

Therapeutic Modulation of the Interferon Pathway in Systemic Lupus Erythematosus

5

Shereen Oon

Abstract

Interferons (IFNs), in particular type I IFN, have emerged as key pathogenic cytokines in systemic lupus erythematosus (SLE). The previous Chapter outlined the biology of type I IFN, and evidence for its contribution to disease pathogenesis in SLE. This Chapter will explore the evidence for therapeutic targeting of the IFN pathway in preclinical and clinical trial studies. There are multiple potential avenues to target the IFN pathway therapeutically (Fig. 5.1)—these include neutralizing the IFNs themselves, interfering with stimulators of IFN production such as nucleic acids, and altering the cellular machinery involved in IFN production (such as Toll-like receptors (TLRs), and the Janus Kinase/Signal Transducer and Activator of Transcription (JAK/STAT) signalling pathway). Later phase clinical trials of these agents have mostly used composite disease activity measures as their primary outcomes, and are conducted in patients with at least moderate disease activity (excluding those with severe lupus nephritis and central nervous system lupus), with standard of care (including immunosuppressive agents and anti-malarials) as placebo. Secondary outcome measures often include clinical parameters, steroid sparing effects, and the ability of the drug to suppress an IFN gene signature.

Therapeutic Targeting of Type I IFN

Therapeutics that neutralize type I IFN (Table 5.1) are the furthest progressed in the therapeutic pipeline. These drugs have targeted either type I IFN itself (anti-IFNα monoclonal antibodies (mAbs), or an IFN-kinoid vaccine) or the type I IFN receptor (IFNAR). The results of clinical trials of these agents have been mixed, with the anti-IFNAR mAbs potentially appearing the most promising.

The potential benefit of therapeutically targeting type I IFN or its receptor was first seen in murine models. Injection of poly I:C, a potent inducer of type I IFN, exacerbated disease in the B6 *lpr* murine model [1]. IFNAR deficiency has been found to ameliorate disease in a variety of murine lupus models [1–4], as have anti-IFNAR mAbs [5].

S. Oon (✉)
Department of Rheumatology, The Royal Melbourne Hospital, Parkville, VIC, Australia
e-mail: shereen.oon@unimelb.edu.au

S. Oon
Department of Rheumatology, St Vincent's Hospital, Fitzroy, VIC, Australia

S. Oon
Department of Medicine at St Vincent's Hospital, The University of Melbourne, Parkville, VIC, Australia

A. Hoi (ed.), *Pathogenesis of Systemic Lupus Erythematosus*,
https://doi.org/10.1007/978-3-030-85161-3_5

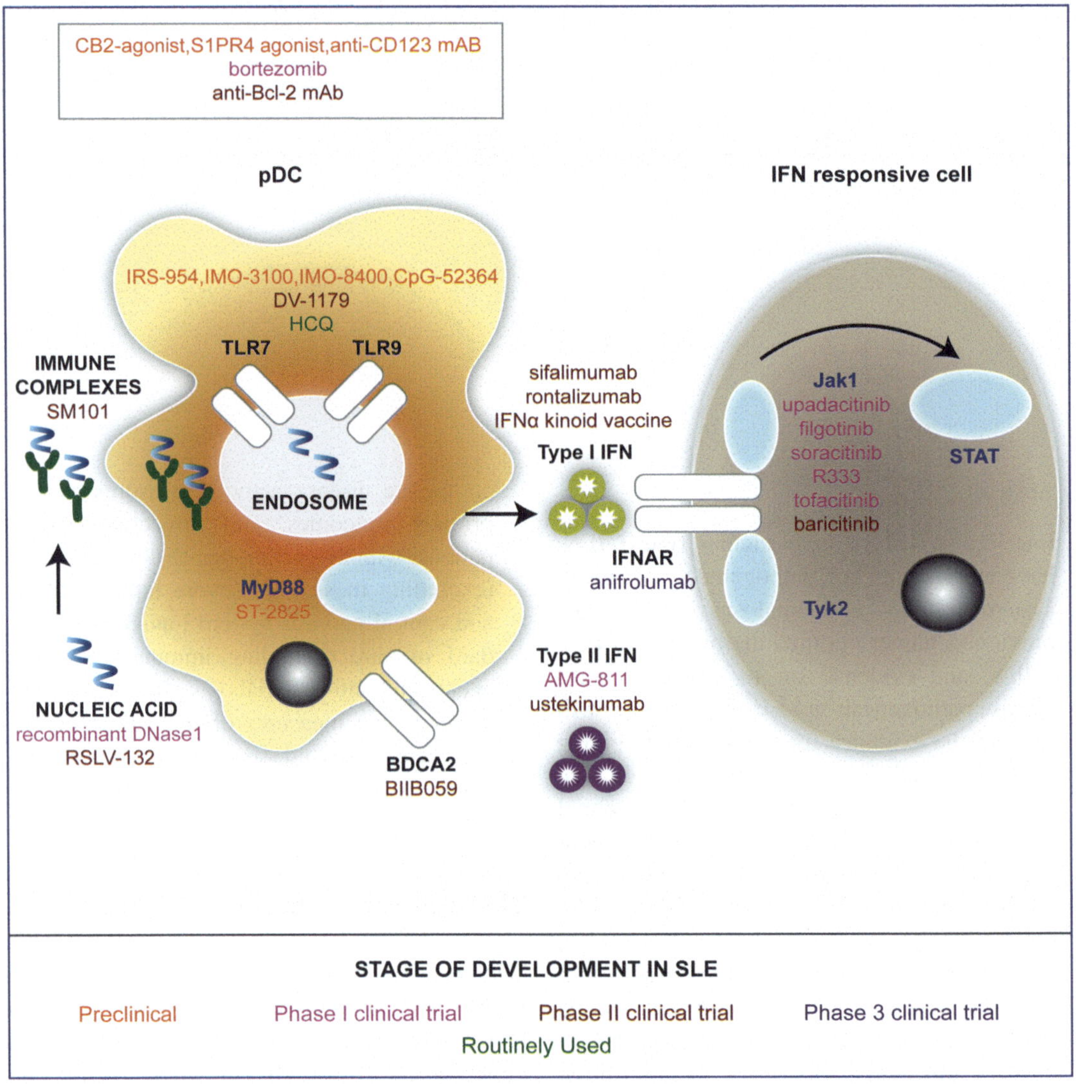

Fig. 5.1 Therapeutics targetting the interferon system in systemic lupus erythematosus. Therapeutics targeting various aspects of the interferon pathway are in different stages of development, ranging from those that target the IFN producing cell, the pDC, to various parts of the IFN signaling machinery. (BDCA—blood dendritic cell antigen, IFN—interferon, JAK—Janus Kinase, pDC—plasmacytoid dendritic cell, STAT—Signal Transducer and Activator of Transcription, TLR—toll-like receptor)

Anifrolumab, a fully human, $IgG_1\kappa$ mAb which targets the type I interferon receptor subunit 1, and thus has an advantage of neutralising *all* type I IFN subtypes, is the only type I IFN targeting therapeutic which has progressed to phase III trials [6, 7]. Of these two phase III trials, one (TULIP-2) met its primary endpoint [6], with 47.8% of patients in the anifrolumab 300 mg treatment group achieving BICLA response at week 52 of the trial compared to 31.5% in the placebo group (16.3% difference, 95% CI (confidence interval) 6.3–26.3%, p = 0.001). The BICLA (BILAG (British Isles Lupus Assessment Group)-based Combined Lupus Assessment) is a BILAG-based [8] composite response measure defined as having all of

Table 5.1 Clinical trials of type I IFN targeting therapies in SLE

Therapeutic	Trial phase/pt pop[n]	Primary and key secondary outcomes	Effect on IFN signature
Anifrolumab **Anti-IFNAR mAb** ***(TULIP-2) (6)*** 300mg IV q4w (52w)	**Phase III RCT** n=362 Mod-severe SLE (SLEDAI-2K≥6)	**O1:** w52 BICLA[A] response favoured anifrolumab - 47.8% vs placebo 31.5% (difference 16.3%; 95% CI 6.3-26.3%, p=0.001). In high IFN-signature subgroup (83.3%), response higher in anifrolumab (48%) vs placebo (37.7%) group (difference 17.3%, 95% CI 6.5-28.2%, p=0.002) **O2:** greater response in anifrolumab group for skin disease[1] (49% vs 25%, p=0.04 at week 12) vs joint disease[2] (42.2% vs 37.5%, p=0.55), steroid sparing effect[3] (51.5% vs 30.2%, p=0.01, between w40-52), lower BILAG-2004-based annualized flare rate (0.43 vs 0.64, p=0.08) **AE:** higher rate of herpes zoster in anifrolumab group (7.2% vs 1.1%)	Suppressed – early (w12) and sustained (to w52)
Anifrolumab **Anti-IFNAR mAb** ***(TULIP-1) (7)*** 300mg IV q4w or 150mg q4w (52w)	**Phase III RCT** n=457 Mod-severe SLE (SLEDAI-2K≥6)	**O1:** w52 SRI-4[B] response similar between anifrolumab 300mg group and placebo (36% vs 40%, difference -4.2%, 95% CI -14.2-5.8%, p=0.41). IFN high subpop[n] (82.2%): similar SRI-4 response (anifrolumab 36% vs placebo 39%) **O2:** numerically higher (37% vs 27%) higher w52 BICLA[A] response in anifrolumab 300mg group, skin disease[1] (42% vs 25%, at w12), steroid sparing effect[3] (41% vs 32%, between w40-52), BILAG-based annualised flare rate (0.60 vs 0.72). **AE**: higher rates of herpes zoster in anifrolumab groups (6% in 300mg, 5% in 150mg) vs placebo (1.1%)	Suppressed – early (w12) and sustained (to w52)
Anifrolumab **Anti-IFNAR mAb** ***(MUSE) (13, 15)*** 1000mg IV q4w or 300mg IV q4w (52w)	**Phase II RCT** n=305 Mod-severe SLE (SLEDAI-2K≥6, BILAG-2004≥1A or ≥2B), PGA≥1)	**O1:** w24 SRI-4[B] with sustained reduction in oral corticosteroids from w12-w24 (<10mg/d or ≤ week 1 dosage) achieved in more anifrolumab treated pts vs placebo (1000mg 28.8%, 300mg 34.3%, placebo 17.6%, p=0.063/0.014 respectively) IFN high subpop[n] (75%): greater response in anifrolumab groups (1000mg 28.2%, 300mg 36%, placebo 13.2%, p=0.029/0.004) IFN low subpop[n] – no difference in response. **O2:** greater percentage of anifrolumab-treated (vs placebo) patients achieved SRI-4[B], BICLA[A], steroid reduction[3] (1000mg 31.7%, 300mg 56.4%, placebo 26.6%, p=0.595/0.001, at w52), skin disease improvement[1] (1000mg 58.3%,	Suppressed – sustained from w24-w52.

(continued)

Table 5.1 (continue)

		300mg 63.0%, placebo 30.8%, p=0.077/0.013, at w52), and joint disease improvement[4] (1000mg 64.6%, 300mg 69.6%, placebo 48.6%, p=0.156/0.038, at w52), trend towards improvement in well-being scales (FACIT-F[C], SF-36[D]) Subgroup analysis suggested better response for skin than joints. **AE:** dose dependent increase in herpes zoster reactivation, URTIs and influenza infection.	
Anifrolumab **Anti-IFNAR mAb** **(14)** Dose escalation – 100mg, 300mg, 1000mg IV	**Phase II open label** n=17 Japanese Mod-severe SLE	Non-linear pharmacokinetic profile after first and last dose. Similar AEs between dose cohorts. Generally well tolerated – most common AEs were URTIs and nasopharyngitis.	Suppressed – dose dependent
Anifrolumab **Anti-IFNAR mAb** **(18)** 150mg, 300mg SC q2w (52w)	**Phase II RCT** n=36 IFN-signature high, active skin disease, SLE	Non-linear pharmacokinetics. **O1:** w12 neutralization of IFN-signature in 88% (150mg), 90.7% (300mg) and 18.5% (placebo). **O2:** Greater reduction in w52 CLASI score in anifrolumab groups (-10.2 (150mg), -13.2 (300mg), -6.3 (placebo). Low immunogenicity (ADA in 2 patients, low titre), no injection site reactions. **AE:** 85.2% (anifrolumab) vs 77.8% (placebo) ≥1AE, all 10 SAEs were in 6 anifrolumab patients.	Suppressed – early (w12), sustained (to w52).
Anifrolumab **Anti-IFNAR mAb** **(19)** 300mg (initially 1000mg) IV q4w, 3Y	**Phase II** **Open label extension (post MUSE)** n=218 SLE	**O1**: n=139 completed 3Y treatment. SAE rate 8.56/100 PY. Most common AE of special interest was herpes zoster (1.92/100 PY). 1 death due to pneumonia. 2/218 ADA positive at any time during treatment (both persistently positive). **O2**: sustained efficacy for SLEDAI (mean change -2.1 [SD=3.5] from baseline to Week 168)	Suppressed – rapid, sustained.
Anifrolumab **Anti-IFNAR mAb** **(17)**	**Phase I RCT** n=30 Healthy volunteers	SC dose approximately 87% of intravenous dose equivalent SC formulation well tolerated, minimal injection site reactions. Most common AEs URTIs and dry throat. No SAEs.	Not assessed – low IFN signature in healthy volunteers

(continued)

Table 5.1 (continue)

300mg SC, 300mg IV, 600mg SC, single dose			
Anifrolumab (anti-IFNAR) and sifalimumab (anti-IFNα) (28) Dose escalation Sifalimumab – 1, 3, 10mg/kg IV q4w, 100mg SC q2w, 600mg, 1200mg IV q4w (365d) Anifrolumab – 100mg, 300mg, 1000mg IV q4w (169d)	**Phase II x 2** **Open label** Sifalimumab: n=30 Anifrolumab: n=17 Japanese patients with SLE	IFN-high pop[n] (97% of sifalimumab treated patients, 88% of anifrolumab treated patients). Superior suppression of 21-gene IFN-signature with anifrolumab compared to sifalimumab (maximum median suppression of IFN-signature 97% with 1000mg IV anifrolumab cohort, compared to 61% with 1200mg IV sifalimumab cohort). No major safety issues, however small study numbers.	Suppressed – dose dependent.
Sifalimumab **Anti-IFNα mAb (20)** 200mg, 600mg, 1200mg IV, D1, 15, 29, then every 28d (52w)	**Phase IIb RCT** n=431 Mod-severe SLE (SLEDAI-2K≥6, BILAG-2004≥1A or ≥2B, PGA≥1)	**O1:** w52 SRI-4[B] met more frequently in sifalimumab-treated groups (placebo 45.4%, 200mg 58.3%, 600mg 56.5%,1200mg 59.8%; p=0.057/0.094/0.031) IFN-high pop[n]: w52 SRI-4[B] met more frequently in sifalimumab groups (placebo 42%, 200mg 57.5%, 600mg 50.0%, 1200mg 57.5%, p=0.042/0.264/0.038) IFN-low pop[n]: high placebo response rate (60%), numerically greater w52 SRI-4 response in 600mg (85%) and 1200mg (70%) treatment groups; small numbers limited analysis. **O2:** Improvements in skin disease[1], joint disease[4], and BICLA[A] at w52. **AE:** Similar rates of AEs and SAEs between drug and placebo, except herpes zoster reactivation more frequent with sifalimumab (1 reactivation, 1 ophthalmic) – 0.9% placebo vs 5.9% sifalimumab (highest in 1200mg group)	Not stated
Sifalimumab	**Phase II** **Open label**	**AEs:** Relatively high discontinuation rate due to AEs (10% in Stage I, 28.6% in Stage II).	Remained at baseline or

(continued)

Table 5.1 (continue)

Anti-IFNα mAb **(21)** Dose escalation Stage 1: 1mg/kg, 3mg/kg, 10mg/kg IV, 100mg SC q6w, 600mg, 1200mg IV q6w (52w) Stage 2: same as Stage 1 or 600mg IV q4w (157w)	**Japanese** Stage 1: n=30 Stage 2: n=21 Mod-severe SLE (SELENA-SLEDAI[E]≥6, BILAG≥1A or ≥2B)	Most common AEs were urinary tract infections, nausea and headache. ADA: 2/30 (Stage 1), 4/21 (Stage 2) **Efficacy**: Majority of patients (3/5 in 3mg/kg IV, 3/5 in 10mg/kg IV, and all other patients) did not have a BILAG flare.	below during Stage 1. No clear trend between gene expression levels and serum sifalimumab levels
Sifalimumab **Anti-IFNα mAb** **(22)** Dose escalation 0.3, 1, 3, 10mg/kg q2w (26w)	**Phase Ib RCT** n=161 Mod-severe SLE (mean SLEDAI 11.0)	Linear pharmacokinetics IFN-high subpop[n] (75.2%) **AEs:** most frequent in sifalimumab group – SLE flare, urinary tract infection, nausea, hypokalaemia, nasopharyngitis, diarrhoea, headache. Headache and viral infection more frequent at higher sifalimumab doses. 5 deaths – 2 from sepsis in 10mg/kg group. Trends towards improvements in disease activity	Suppressed – dose dependent, incomplete
Sifalimumab **Anti-IFNα mAb** **(23)** Dose escalation 0.3, 1, 3, 10, 30mg/kg single dose	**Phase Ia RCT + open label extension** n=33 blinded phase n=17 open label extension Mod SLE	Linear pharmacokinetics IFN-high subpop[n] (58%) **AEs**: similar AEs between groups, no increase in viral infections or reactivation. Similar and low conversion rates in sifalimumab and placebo groups for CMV, EBV, HPV, HSV-1 and HSV-2. No ADAs **Efficacy** Trends towards improvement in various disease activity measures. Decrease in IFNa/b inducible proteins at D14 post dose in skin biopsy.	Suppressed – dose dependent
Sifalimumab **Anti-IFNα mAb** **(24)** Dose escalation	**Phase Ia RCT + open label extension** n=62	IFN-mod to high subpop[n] (60%) 7/8 patients with positive IFNα/β gene signatures in both skin and blood showed similar trend of IFN-gene neutralization D14 after receiving treatment.	Suppressed – dose dependent, as early as D1 post treatment,

(continued)

Table 5.1 (continue)

0.3, 1, 3, 10, 30mg/kg single dose	Mild-mod SLE with cutaneous involvement	Decrease in CD4+ T cell infiltrates and IFN-producing cells D14 post dose, in addition to staining for IFN-inducible proteins (ISG-15, HERC-5, IP-10) Gene signatures induced by other cytokines such as TNFα, IL-1 and GM-CSF also suppressed in some patients.	sustained (signature not returned to baseline levels D84 post dosing)
Rontalizumab **Anti-IFNα mAb** ***(ROSE) (25)*** Part 1: 750mg q4w IV (20w) Part 2: 300mg q2w SC (22w)	**Phase II RCT** n=238 Mod-severe SLE (BILAG≥1A OR ≥2B)	**O1:** No significant difference between treatment groups (pooled IV and SC or separately) compared to placebo (pooled treatment difference 3.5%, p=0.60) in w24 BILAG Index Response[F]. IFN-high subpop[n] (76%) – no significant difference in primary endpoint between treatment and placebo. **O2:** w24 SRI-4[B] – no significant difference in overall population, however IFN-low patients appeared more responsive (pooled treatment difference 31.1%, 95% CI 8.9-51%, p=0.03) **AE:** Comparable between treatment and placebo groups.	
Rontalizumab **Anti-IFNα mAb** **(26)** Dose escalation 0.3, 1, 3, 10mg/kg IV; 1, 3mg/kg SC single dose	**Phase I RCT** n=60 Mild SLE (mean≥SD SELENA-SLEDAI[E] 3.4≥2.7)	Similar rates of AEs between treatment and placebo groups. More higher grade SAEs in rontalizumab-treated groups (25% vs 1% placebo). IFN-high subpop[n] (50%) No changes in serum levels of IFN-inducible proteins even after repeated doses of rontalizumab.	Suppressed – dose dependent, incomplete
AGS-009 **Anti-IFNα mAb** **(27)** 0.01, 0.1, 0.6, 3, 10, 30mg/kg IV single dose	**Phase I RCT** n=25 Mild-mod SLE	No AGS-009 related SAEs IFN-high subpop[n] (81%)	Suppressed for doses > 0.6mg/kg, trend towards dose dependent suppression.

(continued)

Table 5.1 (continue)

IFN-kinoid vaccine (30) Day 0, 7, 28, w12, w24 (36w)	**Phase IIb RCT** Mod-severe SLE with positive IFN gene signature	**O1:** Coprimary endpoints w36 – neutralisation of IFN signature and BICLA[A] (modified by mandatory corticosteroid tapering). Anti-IFN-a2b antibodies induced in 91% of patients. Modified BICLA not significantly different between treatment (41%) and placebo groups. **O2:** Significant corticosteroid sparing effect from w28 onwards, significant difference in w36 LLDAS[G] **AE:** generally well tolerated	Suppressed
IFN-kinoid vaccine (29) Dose escalation 30µg, 60µg, 120µg, 240µg (3-4 doses)	**Phase I/II RCT** n=28 Mild-mod SLE (SLEDAI-2K 4-10)	Anti-IFNα antibodies induced in all patients, in a mostly dose-dependent fashion. IFNα-neutralizing activity seen in some patients with doses ≥60µG. IFN-high subpop[n] (81%) – produced higher levels of anti-IFNα antibodies than IFN-signature negative patients. Only 2 SAEs – both SLE flares (one placebo, one 240µg drug group).	Suppressed – decrease correlated with anti-IFNα Ab titre

Shaded = trials positive for clinical primary outcome (clinical outcome)

Ab—antibody; ADA—anti drug antibody; AE—adverse event; BICLA—BILAG (British Isles Lupus Assessment Group)-based Composite Lupus Assessment; CI—confidence interval; CLASI—Cutaneous Lupus Disease Area and Severity Index; D/d—day; DB—double blind; IFN—interferon; IFNAR—type I interferon receptor; IV—intravenous; mAb—monoclonal antibody; mod—moderate; O1—primary outcome; O2—key secondary outcomes; PGA—physicians' global assessment; pop[n]—population; pt—patient; q4w—4 weekly; PY—patient years; RCT—randomised controlled trial; SAE—serious adverse events; SC—subcutaneous; SD—standard deviation; SLEDAI—systemic lupus erythematosus disease activity index; SRI—systemic lupus erythematosus responder index; URTI—upper respiratory tract infection; w—week, Y—year

[1]patients with at least moderately active skin disease (CLASI $\geq$ 10) at baseline, with a reduction of $\geq$50% in CLASI at specified timepoint

[2]patients with $\geq$6 swollen and tender joints at baseline, with a reduction of $\geq$50% at w52

[3]patients receiving prednisone or equivalent $\geq$10 mg/day at baseline, with sustained reduction to $\leq$7.5 mg/day at specified timepoint(s)

[4]patients with $\geq$8 swollen and tender joints at baseline, with a reduction of $\geq$50% at w52

[A]BICLA—(a) a reduction of all severe (BILAG-2004 A) or moderately severe (BILAG-2004 B) disease activity at baseline to lower levels (BILAG-2004 B, C, or D and C or D, respectively) and no worsening in other organ systems (defined as $\geq$1 new BILAG-2004 A item or $\geq$2 new BILAG-2004 B items), (b) no worsening in SLEDAI-2K score from baseline, (c) no worsening in the PGA score by $\geq$0.3 points from baseline

[B]SRI-4—(a) $\geq$4-point reduction in SLEDAI-2K, (b) no new BILAG-2004 A or >1 new BILAG-2004 B organ domain scores, and (c) <0.3 point increase in PGA from baseline

[C]FACIT-Fatigue (Functional Assessment of Chronic Illness Therapy-Fatigue)—13-question questionnaire assessing fatigue in previous 7 days

[D]SF-36—validated questionnaire assessing HRQOL (Health Related Quality of Life and Well Being) during previous 4 weeks in 8 domains (physical functioning, role-physical, bodily pain, general health, vitality, social functioning, role-emotional and mental health)

[E]SELENA-SLEDAI—Safety of Estrogens in Lupus Erythematosus National Assessment version of the SLEDAI [36]

[F]BILAG Index Response—reduction of all BILAG A domains at baseline to BILAG B or better, and all BILAG B domains to BILAG C or better, with no new BILAG A or more than one new BILAG B, without treatment failure due to additional therapies

[G]LLDAS—Lupus Low Disease Activity State [35]—(1) SLEDAI-2K $\leq$ 4, with no activity in major organ systems (renal, central nervous system, cardiopulmonary, vasculitis, fever) and no haemolytic anaemia or gastrointestinal activity, (2) no new lupus disease activity compared with the previous assessment, (3) PGA < 1, (4) current prednisolone (or equivalent) dose $\leq$7.5 mg daily, and (5) well tolerated standard maintenance doses of immunosuppressive drugs and approved biologic agents

(a) a reduction of all severe (BILAG-2004 A) or moderately severe (BILAG-2004 B) disease activity at baseline to lower levels (BILAG-2004 B, C, or D and C or D, respectively) and no worsening in other organ systems (defined as ≥ 1 new BILAG-2004 A item or ≥ 2 new BILAG-2004 B items), (b) no worsening in SLEDAI-2K [9] score from baseline, and c) no worsening in the PGA (Physician Global Assessment) score by ≥ 0.3 points from baseline. There were additional positive results in key secondary endpoint analyses. A better response for skin, as compared to joint, disease was suggested by a significant difference in patients with at least moderately active skin disease (CLASI 10; CLASI—Cutaneous Lupus Disease Area and Severity Index [10]) achieving a reduction of $\geq 50\%$ in week 12 CLASI with anifrolumab as treatment compared to placebo (49% vs 25%; adjusted difference 24%, 95% CI 4.3–43.6%, $p = 0.04$). A steroid sparing effect was also seen, with more patients receiving prednisolone or equivalent at a dose of ≥ 10 mg/day at baseline achieving a sustained reduction to ≤ 7.5 mg/day between weeks 40 and 52 of the trial in those who were taking anifrolumab (51.5%) as compared to placebo (30.2%; adjusted difference 21.2%, 95% CI 6.8–35.7%, $p = 0.01$). Anifrolumab was also seen to suppress the IFN gene signature in a sustained fashion in the 83.3% of patients who had a high baseline interferon gene signature. Similar overall rates of adverse events were seen between treatment and placebo groups; of note, herpes zoster infections occurred more frequently in anifrolumab-treated patients (7.2%) compared to placebo (1.1%), potentially reflecting the consequences of antagonising the anti-viral effects of type I IFNs, although all cases were cutaneous and resolved without drug discontinuation.

The second phase III trial (TULIP-1) [7] did not meet its primary endpoint, the SRI-4 (SLE Responder Index-4 [11]), a different endpoint to the TULIP-2 trial. The SRI-4 is also a composite response measure utilised in a number of SLE clinical trials, most notably the phase III belimumab trials (which yielded moderately positive results), and is defined as (a) $\geq$ 4-point reduction in SLEDAI-2K, (b) <1 new BILAG-2004 A or <2 new BILAG-2004 B organ domain scores, and (c) <0.3 point increase in PGA from baseline (with no use of restricted medications or discontinuation of investigational product). Similar proportions of patients achieved an SRI-4 at week 52 between those who received anifrolumab 300 mg (36%) and placebo (40%) (difference −4.2%, 95% CI −14.2–5.8%, $p = 0.41$). The TULIP-1 trial also tested a 150 mg dose of anifrolumab, which also showed no additional efficacy compared to placebo (difference −2.6%, 95% CI −14.7–9.6%). A potential reason for the difference in primary outcome achievement between the two TULIP trials is the difference in primary endpoints. The BICLA is BILAG-based outcome measure, as opposed to the SRI-4, which is a SLEDAI-based measure. The BILAG is able to register partial improvements within an organ system, whereas the SLEDAI can only track complete improvements [12], rendering the BICLA a potentially more sensitive measure of change than the SRI-4. Indeed, week 52 BICLA responses in the TULIP-1 trial were achieved by numerically more patients treated with anifrolumab 300 mg (37%) than placebo-treated patients (27%; difference 10.1%, 95% CI 0.6–19.7%). In the TULIP-1 trial, similar to the TULIP-2 trial, multiple secondary endpoints showed positive results (Table 5.1), including an effect on skin disease, a steroid sparing effect, and a decrease in BILAG-based annualized flare rates. Again, rates of herpes zoster were higher in anifrolumab-treated groups (6% in 300 mg group, 5% in 150 mg group) compared to placebo (2%), however were mostly mild to moderate (one severe), and all resolved with anti-viral treatment.

Earlier phase trials of anifrolumab had also shown promising results. In the phase IIb (MUSE) trial of anifrolumab [13] both tested doses of anifrolumab (1000 mg and 300 mg) had shown benefit in achieving the primary endpoint of a week 24 SRI-4 with sustained corticosteroid reduction between weeks 12 and 24 (to <10 mg/d or less than week 1 dosage). A greater effect size was seen in those with a high baseline IFN-signature, however a difference in response was not seen in those with a low IFN-signature; the latter was due to a greater response to placebo, potentially identifying the IFN high group as more likely to benefit from the addition of anifrolumab to standard therapy. The lower dose of anifrolumab (300 mg) showed greater numerical response rates at week 52 for many endpoints, potentially indicating a plateau effect in dosing as the degree of IFN gene signature inhibition was similar with both tested doses. This finding is in contrast to an open label phase II trial of anifrolumab in Japanese patients [14], in which a dose-dependent suppression of a 21-gene IFN signature was seen between a dose of 300 mg and 1000 mg (a 100 mg dose did not adequately suppress the gene signature). In the MUSE trial, multiple secondary endpoints were also met, including week 52 SRI-4, BICLA, and corticosteroid reduction at week 52; an improvement both in skin *and* joint disease was also seen, although the joint disease outcome measure used 8 swollen and tender joints at baseline compared to 6 in the TULIP-2 trial. A subsequent post-hoc analysis of the MUSE trial [15], which analysed responses by SLEDAI-2K, BILAG and organ-specific disease measures (CLASI, swollen and tender joint counts) suggested that the benefit of anifrolumab treatment was more robust for skin than joint disease and that there may be improvement even for IFN-signature low patients, as benefit for anifrolumab was seen when measures of disease activity that captured partial improvements (such as the BILAG) were used. Anifrolumab was also seen to reduce IFN-induced chemokines associated with B, T and other immune cell migration, and proteins associated with endothelial activation that were previously dysregulated [16]. Anifrolumab appeared to increase the rate of herpes zoster reactivations, and upper respiratory tract and influenza infections (although the majority of the reported influenza cases were not confirmed by laboratory testing or application of strict clinical criteria) in the MUSE trial, in dose-dependent fashion; the open label phase II Japanese trial similarly found upper respiratory tract infections to be common, but did not find that adverse effects were influenced by anifrolumab dose [14].

Although the intravenous (IV) formulation of anifrolumab showed efficacy relative to placebo in the phase II and III trials, subcutaneous administration of biologic agents is often preferred by patients. A phase I study comparing 300 and 600 mg subcutaneous, to 300 mg of intravenous anifrolumab [17], found that subcutaneous dosing at 300 mg was equivalent to 87% of intravenous dosing, and similarly well tolerated. A subsequent phase II pharmacokinetic/pharmacodynamic (PK/PD) study of subcutaneous dosing for 52 weeks of a dose of 150 or 300 mg in IFN-signature high patients with active skin disease found that the PK/PD profile of subcutaneous dosing was consistent with previous studies using IV administration, and that the subcutaneous dosing showed low immunogenicity and a comparable safety profile to previous IV studies [18]. A phase II trial of anifrolumab is currently underway in lupus nephritis (TULIP-LN1—NCT02547922), as is a long term, placebo-controlled extension study post phase III TULIP trials (TULIP SLE LTE—NCT02794285). A prior open-label extension study post phase II MUSE trial [19] had shown that anifrolumab was generally well tolerated with sustained efficacy seen over a 3 year period.

Mixed results were seen in the trials of anti-IFNα monoclonal antibodies, sifalimumab and rontalizumab. Sifalimumab, a fully human, $IgG_1\kappa$ mAb, that binds and neutralizes the majority of IFNa subtypes, met its primary endpoint in a multi-centre phase II trial [20]. In this trial, the week 52 SRI-4 was met more frequently in sifalimumab-treated groups (58.3% for 200 mg dose, 56.5% for 600 mg, and 59.8% for 1200 mg dose) compared to placebo (45.5%, $p = 0.057$, 0.094 and 0.031 respectively). Numerically greater week 52 SRI-4 responses were seen for the 600 mg and 1200 mg doses even for IFN-signature-low patients, however smaller numbers in this group limited analysis. Improvements in skin and joint disease, and BICLA at week 52, were also seen. Again, a higher rate of herpes zoster reactivation (including 1 ophthalmic case) was evident. A relatively smaller open-label phase II dose escalation study of sifalimumab in Japanese patients [21] did not report any herpes zoster reactivations, however did report a relatively high discontinuation rate due to adverse events (28.6% in the long term extension phase). Interestingly, this study found that a 21-gene IFN-signature remained at baseline level or below during the treatment stage of the study, and there appeared to be no clear trend between the level of gene expression and serum concentration of sifalimumab. Earlier phase I studies [22–24] had shown an early, dose dependent and sustained suppression of the IFN-gene signature. Two of the phase I studies [23, 24] also showed a decrease in IFN-inducible proteins in biopsies of affected skin with sifalimumab treatment. In contrast to sifalimumab, the phase II trial of rontalizumab, a humanized IgG1 mAb which neutralizes all IFNα subtypes, did not show a difference in primary endpoint compared to placebo [25]. The primary endpoint in this trial was different to the sifalimumab trial, being a week 24 BILAG Index Response (defined as a reduction of all BILAG A domains at baseline to BILAG B or better, and all BILAG B domains to BILAG C or better, with no new BILAG A or more than one new BILAG B, without treatment failure due to additional therapies). Interestingly, a difference was seen in the IFN-signature low patients for the key secondary endpoint of SRI-4 at week 24 (31.1%, 95% CI 8.9–51%, $p = 0.03$), but not for the IFN-signature high patients. This may have been due to relatively higher mean trough concentrations of rontalizumab in the IFN-low group compared to the IFN-high group, however comparable attenuation of IFN-regulated genes was seen between the two groups. As with the phase I trials of sifalimumab, and the phase I [26] and II trials of rontalizumab, the phase I trial of a different anti-IFNα mAb, also AGS-009 [27], saw a dose dependent suppression of the IFN-gene signature in those with a high expression of IFN-inducible genes at baseline. However, the suppression was incomplete, potentially reflecting the contribution of other type I IFNs (such as IFNβ) to the IFN gene signature. This finding suggests therefore that blockade of the receptor may be more efficacious—indeed, in two small phase II trials in Japanese patients, neutralization of the IFNAR with anifrolumab, rather than IFNα alone with sifalimumab, led to a superior suppression of the IFN gene signature [28].

An IFN-α-kinoid vaccine has also undergone phase I/II [29] and IIb [30] clinical trials. This drug, composed of inactivated IFNα coupled to a carrier protein, keyhole limpet hemocynanin, has been shown in human IFNα-transgenic mice to induce polyclonal antibodies that neutralize all subtypes of human IFNα (without neutralizing IFNβ or IFNγ) [31]. Low titres of autoantibodies to IFNα have been found in SLE patient serum [32]. The advantages of this therapeutic approach compared to a mAb is the ability of polyclonal antibodies to neutralize *all* IFNα subtypes, and the relative cost and ease with which they are produced. The antibody was shown in an SLE-like murine flare model driven by recombinant adenovirus-expressing IFNα5 to delay disease progression and death [33]. A phase I/II trial [29] showed that the vaccine was able to induce anti-IFNα antibodies in SLE patients, with a greater production of these antibodies in IFN-signature high patients. A decrease in expression of B-cell associated gene transcripts was also seen [34]. A recently published phase IIb trial [30], with co-primary endpoints of neutralisation of the IFN signature and modified BICLA (modified by mandatory corticosteroid tapering) saw anti-IFN antibodies induced in the majority (91%) of patients. However, though the IFN gene signature was suppressed, the modified BICLA was not significantly different between treatment and placebo groups. Benefit in secondary endpoints were however seen, with a significant corticosteroid sparing effect from week 28 onwards, and a significant difference in week 36 Lupus Low Disease Activity State (LLDAS) [35].

Therapeutic Targeting of Type II and III IFN

Although the strongest evidence exists for the contribution of type I IFN to SLE pathogenesis, therapeutic targeting of other IFN types (type II and III) has also been explored (Table 5.2).

Table 5.2 Clinical trials of therapeutics targeting type II IFN, and plasmacytoid dendritic cells in SLE

Therapeutic	Trial phase/pt pop^n	Primary and key secondary outcomes	Effect on IFN signature
Type II IFN			
AMG-811 **Anti-IFNγ mAb** **(37)** Single dose, 2, 6, 20, 60 or 180mg SC or 60mg IV	**Phase I RCT** n=26 Mild-moderate, stable SLE	**AEs:** Generally well tolerated, no deaths, no withdrawals due to AEs. 1 SAE in 60mg SC group – pyelonephritis. **Efficacy:** No significant treatment-related changes in BILAG global scores and SELENA-SLEDAI scores. No significant treatment-related changes over time in C3, C4, CH50, anti-dsDNA antibody levels. Decreased CXCL10 (IFNγ-upregulated protein) levels by day 15 post treatment.	Suppressed – dose-dependent
AMG-811 **Anti-IFNγ mAb** **(49)** Multiple dose escalation SLE: 6, 20, 60mg SC SLE+LN: 20, 60mg SC q4w (12w)	**Phase I RCT** n=56 SLE +/- lupus nephritis (LN) class III or IV	**AEs:** 2 discontinuations in LN group (1 placebo, 1 treatment group) – due to AE. Majority had AEs, mostly grade 2 or 3. No deaths. **Efficacy:** Dose-related reduction in mean serum CXCL10 (IFNγ-upregulated protein) with treatment – reduction less pronounced in SLE with LN.	Suppressed – less pronounced and not sustained in SLE with LN.
AMG-811 **Anti-IFNγ mAb** **(48)** Single dose 180mg SC, 2 phase crossover	**Phase I RCT** n=16 Discoid lupus erythematosus (DLE) +/- SLE	**AEs:** No deaths or withdrawals due to AEs SAEs only in treatment group, none in placebo. **Efficacy:** Reduced serum CXCL10 (IFNγ-upregulated protein) with treatment. No significant difference in CLASI scores, physician's or patient's self-assessed skin disease.	Suppressed in blood and lesional skin.
Ustekinumab	**Phase II RCT** n=102	**O1:** w24 SRI-4 achieved by greater percentage in ustekinumab group, compared to placebo (percentage difference 28%, 95% CI 10-47, p=0.006).	Ustekinumab response

(continued)

Table 5.2 (continued)

Anti-IL-12/23 (p40) mAb (50, 51) Dose adjusted IV ustekinumab, then 90mg SC q8w	Moderate-severe SLE (SLEDAI-2K≥6, BILAG-2004≥1A or ≥2B)	**AEs** Similar rates of infections between ustekinumab (45%) and placebo (50%) groups. No deaths, treatment-emergent opportunistic infections, herpes zoster, tuberculosis, malignancies, between w0-24.	associated with decrease in IFNγ and IFNγ gene signatures (not seen in non-responders, or placebo)
pDCs			
BIIB059 **Anti-BDCA2 mAb** ***(LILAC) (86, 87)*** SLE: 450mg SC q4w (20w) CLE: 50mg, 150mg, 450mg SC q4w	**Phase II RCT** n=120 SLE, n=132 CLE	**O1:** SLE: Statistically significant difference in total active joint count from baseline to week 24 (least squares mean change -15.0 (standard error 1.2), p=0.037, compared to placebo -11.6 (1.3)). CLE: Statistically significant difference in % change from baseline in CLASI-A score in BIIB059-treated patients vs placebo.	Not stated
BIIB059 **Anti-BDCA2 mAb (85)** Part 1: dose ascending, 0.05-20mg/kg IV or 50mg SC in healthy volunteers Part 2: 20mg/kg IV in SLE patients	**Phase I RCT** n=54 healthy, n=12 SLE with active skin disease	**O1:** Rapid internalization of BDCA2 on pDCs (90% within 1hr to 2d after administration), with transient reduction of pDC number (50% recovery in average numbers by week 2) Marked reduction (average 70%) in MxA and IFIM3 (IFN-induced proteins) in 6/7 SLE skin lesions at w4. 6/8 pts were CLASI-A responders (at least 4 point reduction from baseline at 24 and/or w12). **AEs:** Part 1: no SAEs. Most common AE – URTI. Part 2: most AEs mild to moderate, most common AE – URTI. One case of herpes zoster of moderate severity. No study discontinuations due to AEs	Suppressed within 24 hours
Bortezomib **Proteasome inhibitor (94)**	**Phase I RCT** n=14	**O1:** Trend to decreased anti-dsDNA titre.	Not stated

(continued)

Table 5.2 (continued)

1.3mg/m² d1, 4, 8, 11 (2 cycles of 28 days)	Moderate-severe refractory SLE (SLEDAI≥6, ant-dsDNA+, PNL≥10mg/d)	No significance difference in SRI-4[A], trend to reduced mild-moderate flares in bortezomib group (no severe flares in either group) **AEs:** Infection rate doubled in bortezomib group. All SAEs in bortezomib group (fever x 3, thrombocytopenia x 1, thrombotic microangiopathy x 1, liver dysfunction x 1)	
Venetoclax **Anti-Bcl-2 mAb** **(97)** Single ascending doses: 10, 30, 90, 180, 300, 500mg Multiple ascending doses (2 cycles): 30, 60, 120, 240, 400, 600mg for 1w, then 3w off per cycle.	**Phase I RCT**	**AEs:** Non dose dependent, slightly higher AE rate in venetoclax groups compared with placebo. Most common AEs – headache, nausea, fatigue. **Efficacy:** Total lymphocyte and B cell depletion seen; no consistent or marked change in neutrophils, natural killer cells, haemoglobin or platelets.	Not stated

Shaded = trials positive for clinical primary outcome (clinical outcome)

AE—adverse event; Bcl-2—B-cell lymphoma 2; BDCA2—blood dendritic cell antigen 2; BILAG—British Isles Lupus Assessment Group Index; CI—confidence interval; CLASI—Cutaneous Lupus Erythematosus Disease Area and Severity Index; CLE—cutaneous lupus erythematosus; d—day; dsDNA—double stranded DNA; hr—hour; IFN—interferon; IFNγ—interferon gamma; IV—intravenous; mAb—monoclonal antibody; O1—primary outcome; pDC—plasmacytoid dendritic cell; PNL—prednisolone; pop^{n}—population; pt—patient; q4w—4 weekly; RCT—randomized controlled trial; SAE—serious adverse event; SC—subcutaneous; SLE—systemic lupus erythematosus; SLEDAI—systemic lupus erythematosus disease activity index; URTI—upper respiratory tract infection; w—week

[A]SRI-4—(a) $\geq$ 4-point reduction in SLEDAI-2K, (b) no new BILAG-2004 A or >1 new BILAG-2004 B organ domain scores, and (c) <0.3 point increase in PGA from baseline

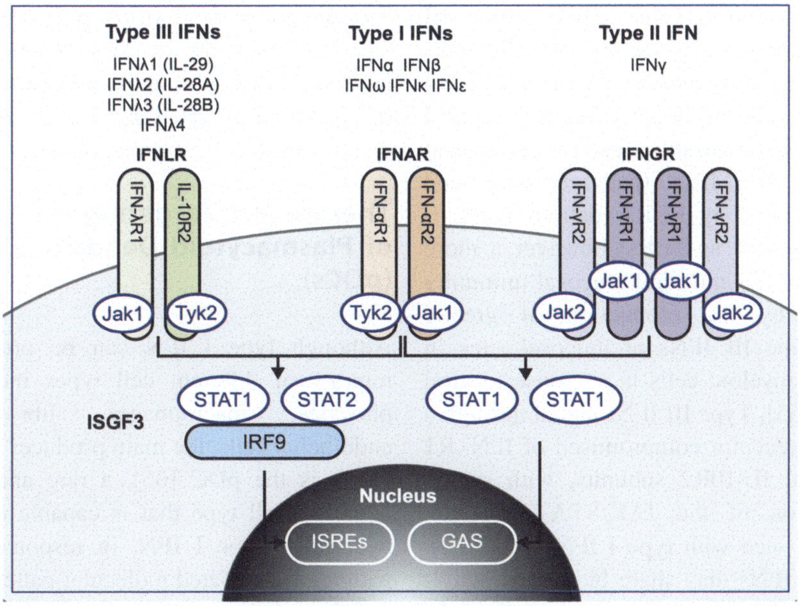

Fig. 5.2 Type I, II, and III interferon signalling pathways. Type I, II and III interferons signal via distinct receptors (IFNAR, IFNGR and IFNLR respectively) with signal transduction mediated through JAK/STAT activation. The downstream signalling pathways of the different interferons overlap, resulting in the production of interferon stimulated genes following activation of transcriptional response elements in the nucleus (GAS—IFNγ activated sites, IFN—interferon, IRF—interferon regulatory factor, ISGF—interferon-stimulated gene factor, ISRE—interferon stimulated response elements, JAK—Janus Kinase, STAT—Signal Transducer and Activator of Transcription)

Type II IFN (IFNγ), signals through a different receptor complex to type I IFNs, the IFNγ receptor (IFNGR), which is expressed by most cell types. Although the IFNγ is distinct to the IFNAR, their downstream signalling pathways overlap (Fig. 5.2), and IFNγ-induced gene signatures have been found to have considerable overlap with those induced by IFNα [37]. Similar to type I IFN, IFNγ administration in humans can induce an SLE-like disease [38]. Serum IFNγ levels are elevated in SLE patient serum [39, 40] and SLE patient T and natural killer (NK) cells produce higher amounts of IFNγ compared to controls [41, 42]; increased IFNγ levels in lymphoid organs of prediseased MRL$^{lpr/lpr}$ mice [43] and IFNγ-producing T cells correlating with autoantibody titres and proteinuria in aged diseased mice has also been seen [44]. Administration of IFNγ accelerated disease, and neutralization decreased autoantibody and BAFF production, myeloid proliferation and T cell-hyperactivation, reducing renal disease in other murine models [45–47]. Phase I trials of an anti-IFNγ mAb, AMG811, have shown treatment to be relatively well tolerated in SLE, SLE with cutaneous lupus, and SLE with lupus nephritis patients, with reductions in biomarkers such as the serum IFNγ signature and IFNγ-upregulated proteins, and the IFN-signature in cutaneous lesions [37, 48, 49]. A positive phase II trial of ustekinumab (an anti-IL-12/23 mAb) in SLE [50] found that ustekinumab-responders had suppressed IFNγ serum levels and an IFNγ gene signature compared to baseline, which did not occur in non-responders or in the placebo group [51]. A phase III trial of ustekinumab in SLE (NCT03517722) has been discontinued due to lack of efficacy in an interim analysis.

Type III IFNs are the most recently discovered IFN type and include IFNλ1 (IL-29), IFNλ2 (IL-28A), IFNλ3 (IL-28B) and IFNλ4. Unlike type I IFN, which is produced mainly by plasmacytoid dendritic cells (pDCs), type III IFNs are produced by a variety of cell types, including pDCs, regulatory T cells (Tregs), Th17 cells, macrophages and hepatocytes. Like type I IFNs, type III IFNs are produced in response to detection of PAMPs (pathogen-associated molecular patterns) by PRRs (pattern recognition receptors) such as TLRs; however a more predominant role in anti-viral mucosal immunity is suggested by the production of greater amounts of type III IFNs at mucosal sites in epithelial and myeloid cells in response to viral infection [52, 53]. Type III IFNs signal through a heterodimeric receptor compromised of IFNλR1 (IL-28RA) and IL-10R2 subunits, with subsequent activation of the JAK/STAT cascade, similar to that seen with type I IFNs (Fig. 5.2). Thus, type III IFNs may share biological activities with type I IFNs through the induction of the ISGF3 (interferon stimulated gene factor 3) transcription complex, and have been postulated as a potential explanation for partial responses to type I IFN blockade [54]. However, type III IFNs have a narrower range of effects than type I IFNs, due to the limited expression of its receptor on epithelial, and some immune cells (such as NK cells, pDCs, and dendritic cells) [55–57] and may therefore contribute less to the peripheral blood IFN gene signature in SLE [58].

Evidence for the pathological role of type III IFN in SLE is accumulating. Elevated serum levels have been found in SLE patients compared to healthy controls, in some studies correlating with SLE disease activity measures (such as the SLEDAI, raised anti-dsDNA levels and low complement) [59]; other studies have found IFNλs in active cutaneous lupus lesions [60], and an association with lupus nephritis activity [61]. There are, as yet, no therapeutics that specifically target type III IFNs in clinical trials for SLE. In pre-clinical studies, an anti-CD123 mAb that depletes pDCs was shown to decrease both type I and type III IFN production in vitro [62]. Blockade of the IFNAR or type I IFN itself may also reduce type III IFN, as reduced type III IFN production has been observed in IFNAR-deficient mice [63], and type III IFN has been shown to be induced by type I IFN in vitro [64]. Therapeutic agents which alter the JAK/STAT signalling pathway or TLR function, discussed later in this Chapter, may also reduce the effects of type III IFN signalling.

Therapeutic Targeting of Plasmacytoid Dendritic Cells (pDCs)

Although type I IFN can be produced by a number of different cell types including lymphocytes, macrophages, fibroblasts and endothelial cells, the main producer (on a per cell basis) is the pDC [65], a rare and specialized dendritic cell type that is capable of rapid production of type I IFN, in response to various pathogen-associated molecular patterns (PAMPs) which trigger pattern-recognition receptors (PRRs) such as toll-like receptors 7 and 9 (TLR7 and 9). In SLE, TLR7 and TLR9 are triggered by self nucleic acids contained in immune complexes or neutrophil-extracellular traps (NETS) [66–68].

The pDC has been indirectly implicated in SLE pathogenesis in many in vitro human studies. pDCs have been mostly found to be decreased in SLE peripheral blood [69–73], correlating with disease activity, although there are fewer studies of unaltered [74] or increased [75] pDC numbers compared to healthy donors. Some of these studies have shown a concomitant infiltration of pDCs into nephritic kidneys [69, 70], and other studies have shown increased pDCs in cutaneous lupus lesions [76–78], suggesting the pDCs may be decreased in peripheral blood due to migration to sites of disease activity. Peripheral blood pDCs from SLE patients have been found to be functionally abnormal, with impaired ability to induce regulatory T cell differentiation in vitro [79]. Several murine lupus models have elucidated a direct role for pDCs in SLE pathogenesis. Impairment of pDC function [80, 81] and transient pDC depletion [82, 83] were shown to abolish splenomegaly and myeloid cell expansion, decrease autoantibody levels and ameliorate the development of glomerulonephritis, in these models.

A number of ways to target the pDC therapeutically in SLE have been explored (major clinical trials are listed in Table 5.2), including inhibition of pDC activation and pDC depletion. A mAb that binds BDCA2 (blood dendritic cell antigen 2), which is a pDC-specific cell surface lectin, increased internalization of BDCA2 and CD32 (FcγRIIa) in vitro, which resulted in inhibition of both TLR and SLE immune complex stimulated type I IFN, and other inflammatory cytokine, production [84]. A phase I study of this mAb, BIIB059, in healthy volunteers and SLE patients with active cutaneous disease showed efficacy (decreased IFN gene expression in blood, reduced IFN-induced proteins in cutaneous lesions, and decreased CLASI-A score) and safety [86]. A phase II trial in SLE [86] showed efficacy for the primary endpoint of a week 24 change in total active joint count (sum of tender and swollen joints) from baseline (least squares mean change -15.0 (standard error 1.2) compared to placebo −11.6 (1.3), p = 0.037); a phase II trial in cutaneous lupus also showed efficacy for the primary endpoint of a change in CLASI-A from baseline to week 16 (least squares mean change −43.5 (5.5) for the 450 mg dose group, compared to placebo −14.5 (6.4), p-0.001) [87]. In vitro human studies have shown that cannabinoid receptor 2 (CB2)-specific agonists [88] and S1PR4 (sphingosine-1-phosphate-receptor-4) agonists [89] decrease IFNα production by decreasing pDC activation. In the case of CB2-agonists (CB2 is largely expressed by immune system cells, rather than in the central nervous system), these caused phosphorylation of IRF7 (interferon regulatory factor 7), TBK1 (TANK-binding kinase 1), NFκB and IKKγ (inhibitor of nuclear factor kappa-B kinase subunit gamma), key events in pDC activation, to be suppressed [88]. S1PR4 activation preserved ILT7 (Ig-like transcript 7), a human pDC-specific surface marker, which decreases IFNα through ITAM (immunoreceptor tyrosine-based activation motif)-mediated signalling [89].

Several drugs have been shown to deplete pDCs, resulting in impaired type I IFN production in vitro and in animal models, including Bcl-2 (an anti-apoptotic protein) family antagonists (ABT-737 and ABT-199) [90, 91], the proteasome inhibitor bortezomib [92], an anti-CD123 (pDC-specific cell surface receptor) mAb [62] and a benzenediamine derivative, FC-99 [93]. A recent multi-centre double-blind RCT (randomized controlled trial) of bortezomib did not find a significant difference in SRI-4 compared to placebo, however there was a trend to decreased mild-moderate flares with bortezomib treatment; small numbers may have limited analysis [94]. Non-controlled trials in lupus nephritis [95] and SLE [96] have also suggested benefit, although as occurred in the RCT, a higher rate of adverse effects (including peripheral neuropathy) was seen with bortezomib treatment. Venetoclax, a Bcl-2 inhibitor approved for treatment in acute myeloid leukemia, was generally well tolerated in a phase I trial in women with SLE [97].

Therapeutic Targeting of Nucleic Acids

As outlined in the previous Chapter, nucleic acids, accumulated as a result of impaired clearance of apoptotic debris in SLE, are important stimulators of IFN production, through both TLR and non-TLR dependent pathways; indeed, anti-dsDNA antibodies are routinely used in the clinic to track disease activity. However, attempts to neutralise extracellular nucleic acids therapeutically have met with mixed results (Table 5.3). Although recombinant DNase1 slowed disease progression in a murine lupus model [98], a phase Ib trial in lupus nephritis patients, though well tolerated, showed no change in relevant serum biomarkers [99]. In contrast, targeting RNA has yielded more promising results. *Tlr7* x *RNase* double transgenic mice (which have higher concentrations of serum RNase) had increased survival compared to *Tlr7* Tg mice, associated with reduced T and B cell activation and less IgG and C3 deposition in the kidneys [100]. A phase I trial of a RNase-IgG1-Fc fusion protein, RSLV-132, showed a decrease in RNA autoantibodies and BAFF levels, correlating with reduced disease activity, with a favourable safety profile [101]. A phase IIa study of RSLV-132

Table 5.3 Clinical trials of therapeutics targeting nucleic acids, toll-like receptors and the JAK/STAT pathway in SLE

Therapeutic	Trial phase/pt pop^n	Primary and key secondary outcomes	Effect on IFN signature
Nucleic acids			
Recombinant DNase 1 (99) 25µg/kg, 125µg/kg IV, then 10 SC doses over 40d	**Phase I RCT** n=17 SLE with class III, IV, V LN	**AEs:** No significant drug related adverse events **Efficacy:** No relevant change in serum biomarkers (serum dsDNA antibodies, C3 and C4 levels, circulating immune complexes, serum sIL-2 receptor, IL-6, IL-10, TNFα. Skin biopsy immune complexes unchanged pre- and post treatment.	Not tested
RSLV-132 RNase-Fc fusion protein (102)	**Phase IIa RCT** n=66, moderate-severe SLE with severe CLE (CLASI score≥10)	**Efficacy**: Higher percentage (33%) of RSLV-132 treated patients had CLASI 50 (50% reduction in CLASI score) compared to 23% of placebo patients.	
RSLV-132 RNase-Fc fusion protein (101) Multiple dose ascending – 5 weekly, or 3 bi-weekly doses of 0.3-10.0mg/kg IV	**Phase Ib RCT** n=32 Inactive-mild SLE Positive antibody-bound U1 or Y1 RNA higher than healthy controls	**AEs:** Well tolerated, similar AE frequency between treatment and placebo, mostly grade 1 AEs. **Efficacy:** Modest decrease in RNA autoantibodies in 5 drug-treated subjects (vs 1 in placebo group), correlating with improvements in SELENA-SLEDAI^A Decrease in BAFF levels by average 9% in treatment group (vs 11% increase in placebo group), correlating with improvements in disease activity.	Decreased in 2 patients on treatment.
SM101 FcγRIIb (103) 2 doses, either 6mg/kg or 12mg/kg IV	**Phase IIa RCT** n=61 Mod-severe (SELENA-SLEDAI^A≥6; LN included, anti-dsDNA+ or low C3)	**O1:** w24 SRI-4^B higher in treatment groups (36% for 6mg/kg group, 39% for 12mg/kg group vs 18% placebo) especially in lupus nephritis patients (50% for 6mg/kg and 60% for 12mg/kg, vs 0% placebo) **AEs:** No drug related serious adverse events.	Not stated

(continued)

Table 5.3 (continued)

Toll like receptors			
DV-1179 **TLR7/9 inhibitor** **(119)** Up to 60mg/week SC for 8w	**Phase Ib/IIa** n=52 SLE, positive IFN gene signature	**Safety:** Well tolerated, most common AE was injection site reactions **Efficacy:** Failed to meet primary or secondary pharmacodynamic endpoints (reduction in IFN genes)	Not decreased
JAK/STAT pathway			
Baricitinib **JAK1/2 inhibitor** **(134)** 2mg, 4mg oral daily, 24w	**Phase II RCT** n=314 SLEDAI-2K≥4, or active arthritis or rash	**O1:** w24 SLEDAI-2K resolution of rash or arthritis, significantly greater in baricitinib 4mg (OR 1.8, 95% CI 1.0-3.3, p=0.0414) but not 2mg (OR 1.3, 95% CI 0.7-2.3, p=0.39) Significant difference for 4mg dose compared to placebo for 28-tender joint count and worse joint pain score, but no meaningful change in skin disease measures (CLASI, SLEDAI cutaneous domains) **AEs:** Higher rate of discontinuation due to adverse effects with baricitinib, and higher rate of SAEs including serious infections; one occurrence of DVT in baricitinib 4mg group (30% overall rate of APL Abs in study patients)	Not stated
Tofacitinib **JAK1/3>2 inhibitor** **(139)** 5mg bd oral, 56w	**Phase Ib/IIa** Mild-mod SLE n=30 Stratified by STAT 4 risk allele	**AEs:** Well tolerated, no severe AEs, opportunistic infections or liver function abnormalities. **Efficacy:** Significant increase in HDL-C and HDL particle size in tofacitinib-treated patients at d56 (p=0.006), and trends for improvements in vascular stiffness	Decreased

Shaded = trials positive for clinical primary outcome (clinical outcome)

Abs—antibodies; AE—adverse event; APL—antiphospholipid; CI—confidence interval; CLASI—Cutaneous Lupus Erythematosus Disease Area and Severity Index; CLE—cutaneous lupus erythematosus; d- day; HDL—high density lipoprotein; IFN—interferon; IV—intravenous; LN—lupus nephritis; mod—moderate; OR—odds ratio; pop^{n}—population; pt—patient; RCT—randomised controlled trial; SAEs—serious adverse events, SC—subcutaneous; SLE—systemic lupus erythematosus; SLEDAI—systemic lupus erythematosus disease activity index; TLR—toll-like receptor; w—week

[A]SELENA-SLEDAI—Safety of Estrogens in Lupus Erythematosus National Assessment version of the SLEDAI [36]

[B]SRI-4—(a) $\geq$ 4-point reduction in SLEDAI-2K, (b) no new BILAG-2004 A or >1 new BILAG-2004 B organ domain scores, and (c) <0.3 point increase in PGA from baseline.

(NCT02660944) showed a 50% reduction in CLASI score in 33% of RSLV-132 treated patients compared to 23% of the placebo group [102]. Immune-complex neutralisation has also been trialled, with a phase IIa trial of SM101 (FcγRIIb) showing greater week 24 SRI-4 responses in treatment groups, with an acceptable adverse event profile [103].

Therapeutic Targeting of Toll-Like Receptors (TLRs)

Multiple TLRs (such as TLRs 3, and 7–9) can be activated by PAMPs in various cell types, leading to IFN production, as detailed in the previous Chapter. Similar to other components of the interferon signalling pathway, data from both human and murine studies support the role of TLRs in SLE pathogenesis. In humans, TLR7 and TLR9 m*RNA* expression, correlating with IFNα expression, has been seen in SLE peripheral blood mononuclear cells (PBMCs). TLR7 deficiency has reduced autoantibody production, immunoglobulin levels, lymphocyte activation and glomerulonephritis in various murine lupus models including the BXSB-*Yaa* [104], MRL/*lpr* [105], and the highly interferon dependent pristane-induced model [106]. Reduced production of IFNα and IL-6 induced by immune complexes containing ribonucleoprotein were also seen in murine pDCs deficient in TLR7 [107]. In contrast, the evidence for TLR9 in the pathogenesis of SLE is less clear; in murine lupus models TLR9 deletion *worsened* disease [105, 108], and in human disease, B cells were less able to be activated by the TLR9-agonist ODN-CpG, though there was higher expression of TLR9 in B and dendritic cells correlating with more significant disease activity [109]. TLR8 resembles TLR7 phylogenetically and shares similar ligands (single-stranded RNA). Due to its position on the X chromosome, it has been surmised to be a reason for the female preponderance observed in SLE. However, similar to TLR9, data regarding the role of TLR8 in SLE is conflicting. Support for a pathogenic contribution of TLR8 in SLE was seen in a murine lupus model where the absence of TLR7 and TLR8 together resulted in decreased autoantibody production, although reduced autoantibodies were not seen with combined deletion of TLR7 and TLR9 [110]. Conversely, disease stimulated by TLR7 activation was *augmented* in lupus-prone mice in the absence of TLR8 [111].

The antimalarial drug hydroxychloroquine (HCQ) is a mainstay of SLE treatment, with multiple benefits seen in SLE including decreasing overall disease severity, preventing disease flares [112], increasing survival [113], altering lipid profiles favourably [114], and having anti-thrombotic effects [115]. It is a TLR7/8/9 antagonist, with its activity attributed to reduced endosomal acidification, which is required for TLR activation. More recent evidence suggests that HCQ also binds directly to nucleic acids, causing structural modifications that prevent ligand binding to TLRs [116]. HCQ has been shown to decrease TLR9-stimulated pDC responses to a greater extent than those evoked by TLR7/8 stimulation ex vivo [117] and reduce pDC production of IFNα and TNFα [118]. These benefits of HCQ occur without clinically significant immunosuppression, vindicating the concept of therapeutically targeting TLRs.

Several other drugs that antagonise TLRs or components of their signalling pathways have been tested in SLE, with varying results. IRS-954, a dual TLR7/9 antagonist, ameliorated disease in murine lupus models [119] and reduced DNA/RNA/SLE immune complex-stimulated IFNα production from human pDCs ex vivo [120]. However, although there were no significant side effects, another dual TLR7/9 antagonist, DV-1179, did not reach its IFN-induced gene reduction endpoint in a phase Ib/IIa trial [121]. IMO-3100 (a dual TLR7/9 antagonist), and IMO-8400 (which antagonises TLR7, 8 and 9) have both been found to be effective and safe in phase II trials in psoriasis [122, 123], another autoimmune disease in which IFN may be pathogenic. Both of these drugs have shown efficacy in murine lupus models [124, 125]. ST-2825 is a small molecule inhibitor of MyD88, which is involved in the downstream signalling pathway of TLRs. ST-2825 has been found to decrease TLR-9 stimulated plasma cell and autoantibody generation in human lupus peripheral blood mononuclear cells [126] and reduce pro-inflammatory IL-6 production in mice [127]. Another small molecule inhibitor CpG-52364, which antagonises TLR 7/8/9, was found to have greater efficacy than hydroxychloroquine in pre-clinical animal models [116, 128], however a subsequent phase I trial was discontinued [129].

Therapeutic Targeting of the Janus Kinase (JAK)/STAT Pathway

Each of the four JAK family members, JAK1, JAK2, JAK3 and TYK2, is involved in the signalling cascade of multiple cytokine receptors [130], therefore, a theoretical advantage of modulating the JAK/STAT pathway in SLE is the potential inhibit both IFNs *and* other pathogenic cytokines, such as IL-6 [131].

The role of the JAK/STAT pathway in SLE will be explored in more detail in a subsequent Chapter (*Intracellular targets of JAK/STAT pathways and immunoproteasomes*), however the potential efficacy of modulating the pathway to ameliorate IFN-driven diseases has been seen in studies in the rare monogenic interferonopathies. A compassionate use study of baricitinib (a JAK1/2 inhibitor) in CANDLE (chronic atypical neutrophilic dermatosis with lipodystrophy and elevated temperatures) showed decreased disease manifestations and steroid requirements [132]. In vitro, tofacitinib (a JAK1/3 > 2 inhibitor), baricitinib and ruxolitinib (a JAK1/2 inhibitor) reduced constitutive upregulation of phosphorylated STAT molecules in lymphocytes from patients with SAVI (STING-associated vasculopathy with onset in infancy), with a dose dependent blockade of some interferon-responsive genes [133].

JAK/STAT inhibition in SLE has shown promise, with several drugs having undergone, or are currently undergoing phase II/III clinical trials, including phase III baricitinib in SLE studies (BRAVE I—NCT03616912, BRAVE II—NCT03616964, BRAVE-X—NCT03843125), filgotinib (a JAK-1 selective inhibitor) phase II studies in cutaneous lupus erythematosus (CLE) (NCT03134222) and lupus membranous nephropathy (NCT03285711), and a phase II study of upadacitinib (a selective JAK1 inhibitor) compared to ABBV-105 (a Bruton's tyrosine kinase inhibitor) (NCT03978520). A phase II study of baricitinib in non-renal SLE had shown efficacy for the primary endpoint of a SLEDAI resolution of arthritis *or* rash, with greater efficacy suggested for joint rather than skin disease [134] (Table 5.3). Data from the two phase II filgotinib trials has not been formally presented, although both have appear to have failed to meet their primary endpoints [135]; prior murine studies had suggested efficacy in reducing disease activity, various pro-inflammatory cytokines and a majority of interferon-upregulated genes tested [136]. An earlier phase II trial in non-renal SLE of another JAK1-specific inhibitor, soracitinib, was terminated early due to lack of efficacy. In this trial, two patients who had received the drug also developed the DRESS syndrome (a rare but serious idiosyncratic drug reaction causing rash, eosinophilia and systemic symptoms) and severe, but reversible, abnormalities in liver function [137]. Although mild liver function abnormalities are a recognized side effect of other JAK inhibitors, DRESS has not previously been reported with this class of drugs. The development of a JAK/SYK inhibitor, R333, has also been terminated, after failing to meet its primary endpoint of a week four 50% decrease in active skin lesions in a phase II trial in discoid lupus [138]. Tofacitinib, one of the first JAK inhibitors developed for autoimmune disease treatment in humans, was well tolerated in a phase Ib/IIa trial in SLE, with improvements in plasma lipid profiles and decreased IFN response genes in STAT4 positive patients at baseline [139]. Phase I/II trials focussing on skin disease are currently recruiting (NCT03159936, NCT03288324).

Other Approaches to Therapeutic Targeting of the IFN Pathway in SLE

A number of other approaches to modulating the IFN system have been efficacious in pre-clinical studies. These include the use of microRNAs [140, 141], prostaglandins [142], and IgE [143], and inhibitors of Bruton's tyrosine kinase [144–146], cyclin-dependent kinase 1 [147], IKKβ (inhibitor of nuclear factor kappa B subunit beta) [148, 149], and NETS [150].

Conclusion

A wealth of data has now emerged pointing to interferon as a very promising therapeutic target in SLE, however not all clinical trials of drugs targeting this pathway have been successful. Further understanding of this pathway may allow refinement of future therapeutic exploration. For example, there is evidence from murine models suggesting that targeting the IFN pathway *early* on in the disease may be more effective [5, 82] and the association of different interferon types with particular disease manifestations [151] may allow streamlining of clinical trials of different IFN targeting agents to specific organs to which they are most likely to be effective. With continued examination of this pathway and its therapeutics, interferon-targeting therapies are well placed to become an important part of the future therapeutic armamentarium for SLE.

Acknowledgements Elaine Oon—Graphics.

References

1. Braun D, Geraldes P, Demengeot J (2003) Type 1 interferon controls the onset and severity of autoimmune manifestations in lpr mice. J Autoimmun 20:15–25
2. Santiago-Raber ML, Baccala R, Haraldsson KM, Choubey D, Stewart TA, Kono DH et al (2003) Type-1 interferon receptor deficiency reduces lupus-like disease in NZB mice. J Exp Med 197:777–788
3. Kono DH, Baccala R, Theofilopoulos AN (2003) Inhibition of lupus by genetic alteration of the interferon-alpha/beta receptor. Autoimmunity 36 (8):503–510
4. Agrawal H, Jacob N, Carreras E, Bajana S, Putterman C, Turner S et al (2009) Deficiency of type I IFN receptor in lupus-prone New Zealand mixed 2328 mice decreases dendritic cell numbers and activation and protects from disease. J Immunol 183:6021–6029
5. Baccala R, Gonzalez-Quintial R, Schreiber RD, Lawson BR, Kono DH, Theofilopoulos AN (2012) Anti-IFN-a/b receptor antibody treatment ameliorates disease in lupus-predisposed mice. J Immunol 189:5976–5984
6. Morand EF, Furie R, Tanaka Y, Bruce IN, Askanase AD, Richez C, et al (2020) TULIP-2 trial investigators. Trial of anifrolumab in active systemic lupus erythematosus. N Engl J Med 16:382 (3):211–221
7. Furie RA, Morand EF, Bruce IN, Manzi S, Kalunian KC, Vital EM et al (2019) Type I interferon inhibitor anifrolumab in active systemic lupus erythematosus (TULIP-1): a randomised, controlled, phase 3 trial. Lancet Rheumatol 1:e208–e219
8. Yee CS, Farewell V, Isenberg DA, Rahman A, Teh LS, Griffiths B et al (2007) British Isles Lupus Assessment Group 2004 index is valid for assessment of disease activity in systemic lupus erythematosus. Arthritis Rheum 56:4113–4119
9. Gladman DD, Ibañez D, Urowitz MB (2002) Systemic lupus erythematosus disease activity index 2000. J Rheumatol 29:288–291
10. Jolly M, Kazmi N, Mikolaitis RA, Sequeira W, Block JA (2013) Validation of the Cutaneous Lupus Disease Area and Severity Index (CLASI) using physician- and patient-assessed health outcome measures. J Am Acad Dermatol 68:618–623
11. Furie RA, Petri MA, Wallace DJ, Ginzler EM, Merrill JT, Stohl W et al (2009) Novel evidence based systemic lupus erythematosus responder index. Arthritis Rheum 61:1143–1151
12. Thanou A, Chakravarty E, James JA, Merrill JT (2014) Which outcome measures in SLE clinical trials best reflect medical judgement? Lupus Sci Med 1:e000005
13. Furie R, Khamashta M, Merrill JT, Werth VP, Kalunian K, Brohawn P, et al (2017) CD1013 study investigators. Anifrolumab, an anti-interferon-a receptor monoclonal antibody, in moderate-to-severe systemic lupus erythematosus. Arthritis Rheumatol 69(2):376–386
14. Tanaka Y, Takeuchi T, Okada M, Nakajima H, Kawai S, Nagashim T et al (2020) Safety and tolerability of anifrolumab, a monoclonal antibody targeting type I interferon receptor, in Japanese patients with systemic lupus erythematosus: a multicentre, phase 2, open-label study. Mod Rheumatol 30(1):101 108
15. Merrill JT, Furie R, Werth VP, Khamashta M, Drappa J, Wang L, et al (2018) Anifrolumab effects

on rash and arthritis: impact of the type I interferon gene signature in the phase IIb MUSE study in patients with systemic lupus erythematosus. Lupus Sci Med 5(1):e000284
16. Casey KA, Guo X, Smith MA, Wang S, Sinibaldi D, Sanjuan MA, et al (2018) Type I interferon receptor blockade with anifrolumab corrects innate and adaptive immune perturbations of SLE. Lupus Sci Med 5(1):e000286
17. Tummala R, Rouse T, Berglind A, Santiago L (2018) Safety, tolerability and pharmacokinetics of subcutaneous and intravenous anifrolumab in healthy volunteers. Lupus Sci Med 5(1):e000252
18. Bruce I, Nami A, Schwetje E, Pierson M, Chia Y, Kuruvilla D, et al (2019) PK/PD, safety and exploratory efficacy of subcutaneous anifrolumab in SLE: a phase-II study in interferon type I high patients with active skin disease [abstract]. Arthritis Rhematol 71(suppl 10). https://acrabstracts.org/abstract/pk-pd-safety-and-exploratory-efficacy-of-subcutaneous-anifrolumab-in-sle-a-phase-ii-study-in-interferon-type-i-high-patients-with-active-skin-disease/. Accessed March 21, 2020
19. Chatham W, Furie R, Saxena A, Brohawn P, Schwetje E, Abreu G, et al (2019) A phase 2, open-label extension study to evaluate long-term safety of anifrolumab in adults with systemic lupus erythematosus [abstract]. Arthritis Rheumatol 71(suppl 10). https://acrabstracts.org/abstract/a-phase-2-open-label-extension-study-to-evaluate-long-term-safety-of-anifrolumab-in-adults-with-systemic-lupus-erythematosus/. Accessed March 21, 2020
20. Khamashta M, Merrill JT, Werth VP, Furie R, Kalunian K, Illei GG, et al (2016) CD1067 study investigators. Sifalimumab, an anti-interferon-a monoclonal antibody, in moderate to severe systemic lupus erythematosus: a randomised, double-blind, placebo-controlled study. Ann Rheum Dis 75:1909–1916
21. Takeuchi T, Tanaka Y, Matsumura R, Saito K, Yoshimura M, Amano K et al (2020) Safety and tolerability of sifalimumab, an anti-interferon-a monoclonal antibody, in Japanese patients with systemic lupus erythematosus: a multicentre, phase 2, open-label study. Mod Rheumatol 30(1):93–100
22. Petri M, Wallace D, Spindler A, Chindalore V, Kalunian K, Mysler E et al (2013) Sifalimumab, a human anti-interferon-a monoclonal antibody, in systemic lupus erythematosus: a phase 1 randomized, controlled, dose-escalation study. Arthritis Rheum 65:1011–1021
23. Merrill JT, Wallace D, Petri M, Kirou K, Yao Y, White W et al (2011) Safety profile and clinical activity of sifalimumab, a fully human anti-interferon a monoclonal antibody, in systemic lupus erythematosus: a phase 1, multicentre, double bind randomised study. Ann Rheum Dis 70:1905–1913
24. Yao Y, Richman L, Higgs BW, Morehouse CA, de los Reyes M, Brohawn P, et al (2009) Neutralization of interferon-alpha/beta-inducible genes and downstream effect in a phase 1 trial of anti-intereferon-alpha monoclonal antibody in systemic lupus erythematosus. Arthritis Rheum 60:1785–1796
25. Kalunian KC, Merrill JT, Maciuca R, McBride J, Townsend M, Wei X et al (2016) A phase II study of the efficacy and safety of rontalizumab (rhuMAb interferon-a) in patients with systemic lupus erythematosus (ROSE). Ann Rheum Dis 75:196–202
26. McBride JM, Jiang J, Abbas A, Morimoto A, Li J, Maciuca R et al (2012) Safety and pharmacodynamics of rontalizumab in patients with systemic lupus erythematosus: results of a phase 1, placebo-controlled, double-blind, dose-escalation study. Arthritis Rheum 64:3666–3676
27. Tcherepanova M, Curtis M, Sale M, Miesowicz F, Nicolette C (2012) Results of a randomized placebo controlled phase Ia study of AGS-009, a humanized anti-interferon-alpha monoclonal antibody in subjects with systemic lupus erythematosus. Ann Rheum Dis 71(suppl 3):536–537
28. Morehouse C, Chang L, Wang L, Brohawn P, Ueda S, Ilei G, et al. Target modulation of a type I interferon (IFN) gene signature with sifalimumab or anifrolumab in systemic lupus erythematosus (SLE) patients in two open label phase 2 Japanese trials. In: Paper Presented at: 2014 American College of Rheumatology Annual Meeting, Boston, USA [abstract 719]. Available from http://acrabstracts.org/abstract/target-modulation-of-a-type-i-interferon-ifn-gene-signature-with-sifalimumab-or-anifrolumab-in-systemic-lupus-erythematosus-sle-patients-in-two-open-label-phase-2-japanese-trials/. Accessed 22 March, 2020
29. Lauwerys BR, Hachulla E, Spertini F, Lazaro E, Jorgensen C, Mariette X et al (2013) Down-regulation of interferon signature in systemic lupus erythematosus patients by active immunization with interferon a-kinoid. Arthritis Rheum 65:447–456
30. Houssiau FA, Thanou A, Mazur M, Ramiterre E, Gomez Mora DA, Misterska-Skora, et al (2020) IFN-a kinoid in systemic lupus erythematosus: results from a phase IIb, randomised, placebo-controlled study. Ann Rheum Dis 79(3):347–355
31. Mathian A, Amoura Z, Adam E, Colaone F, Hoekman M, Dhellin O et al (2011) Active immunisation of human interferon a transgenic mice with a human interferon a kinoid induces antibodies that neutralise interferon a in sera from patients with systemic lupus erythematosus. Ann Rheum Dis 70:1138–1143
32. Morimoto AM, Flesher DT, Yang J, Wolslegel K, Wang X, Brady A et al (2011) Association of endogeneous anti-interferon-a autoantibodies with decreased interferon-pathway and disease activity in patients with systemic lupus erythematosus. Arthritis Rheum 63:2407–2415
33. Zagury D, Buanec H, Mathian A, Larcier P, Burnett R, Amoura Z et al (2009). IFNa kinoid vaccine-induced neutralizing antibodies prevent

clinical manifestations in a lupus flare murine mode. Proc Natl Acad Sci USA 106: 5249–5294
34. Ducreaux J, Houssiau FA, Vandepapeliere P, Jorgensen C, Lazaro E, Spertini F et al (2016) Interferon a kinoid induces neutralizing anti-interferon a antibodies that decrease the expression of interferon-induced and B cell activation associated transcripts: analysis of extended follow-up data from the interferon a kinoid phase I/II study. Rheumatology (Oxford) 55(10):1901–1905
35. Franklyn K, Lau CS, Navarra SV, Louthrenoo W, Lateef A, Hamijoyo L et al (2016) Definition and initial validation of a Lupus Low Disease Activity State (LLDAS). Ann Rheum Dis 75:1615–1621
36. Petri M, Kim MY, Kalunian KC, Grossman J, Hahn BH, Sammaritano LR, et al. for the OCSELENA Trial (2005) Combined oral contraceptives in women with systemic lupus erythematosus. NEJM 353:2550–2558
37. Welcher AA, Boedigheimer M, Kivitz AJ, Amoura Z, Buyon J, Rudinskaya A et al (2015) Blockade of interferon-gamma normalizes interferon-regulated gene expression and serum CXCL10 levels in patients with systemic lupus erythematosus. Arthritis Rheumatol 67:2713–2722
38. Graninger WB, Hassfeld W, Pesau BB, Machold KP, Zielinski CC, Smolen JS (1991) Induction of systemic lupus erythematosus by intereferon gamma in a patient with rheumatoid arthritis. J Rheumatol 18:1621–1622
39. Robak E, Smolewski P, Wozniacka A, Sysa-Jedrzejowska A, Stepien H, Robak T (2004) Relationship between peripheral blood dendritic cells and cytokines involved in the pathogenesis of systemic lupus erythematosus. Eur Cytokine Networks 15:222–230
40. Rana A, Minz RW, Aggarwal R, Anand S, Pasricha N, Singh S (2012) Gene expression of cytokines (TNF-alpha, IFN-gamma), serum profiles of IL-17 and IL-23 in paediatric systemic lupus erythematosus. Lupus 21:1105–1112
41. Csiszar A, Nagy G, Gergely P, Pozsonyi T, Pocski E (2000) Increased interferon-gamma (IFN-gamma), IL-10 and decreased IL-4 mRNA expression in peripheral blood mononuclear cells (PBMC) from patients with systemic lupus erythematosus (SLE). Clin Exp Immunol 122:464–470
42. Hervier B, Beziat V, Haroche J, Mathian A, Lebon P, Ghillani-Dalbin P et al (2011) Phenotype and function of natural killer cells in systemic lupus erythematosus: excess interferon-gamma production in patients with active disease. Arthritis Rheum 63:1698–1706
43. Lemay S, Mao C, Singh AK (1996) Cytokine gene expression in the MRL/lpr model of lupus nephritis. Kidney Int 50:85–93
44. Enghard P, Langnickel D, Riemekasten G (2006) T cell cytokine imbalance towards production of IFN-gamma and IL-10 in NZB/W F1 lupus-prone mice is associated with autoantibody levels and nephritis. Scand J Rheumatol 35:209–216
45. Jacob CO, van der Meide PH, McDeviitt HO (1987) In vivo treatment of (NZB X NZW)F1 lupus-like nephritis with monoclonal antibody to gamma interferon. J Exp Med 166:798–803
46. Schmidt T, Paust HJ, Krebs CF, Turner JE, Kaffke A, Bennstein SB et al (2015) Function of the Th17/interleukin-17A immune response in murine lupus nephritis. Arthritis Rheumatol 67:475–487
47. Haas C, Ryffel B, Le Hir M (1998) IFN-gamma receptor deletion prevents autoantibody production and glomerulonephritis in lupus-prone (NZB x NZW)F1 mice. J Immunol 160:3713–3718
48. Werth VP, Fiorentino D, Sullivan BA, Boedigheimer MJ, Chiu K et al (2017) Brief report: Pharmacodynamics, safety and clinical efficacy of AMG 811, a human anti-intereferon-y antibody, in patients with discoid lupus erythematosus. Arthritis Rheumatol 69(5):1028–1034
49. Boedigheimer M, Martin D, Amoura Z, Sánchez-Guerrero J, Romero-Diaz J, Kivitz A et al (2017) Safety, pharmacokinetics and pharmacodynamics of AMG 811, an anti-interferon-γ monoclonal antibody, in SLE subjects without or with lupus nephritis. Lupus Sci Med 4:3000226
50. Van Vollenhoven RF, Hahn BH, Tsokos GC, Wagner CL, Lipsky P, Touma Z et al (2018) Efficacy and safety of ustekinumab, an IL-12 and IL-23 inhibitor, in patients with active systemic lupus erythematosus: results of a multicentre, double-blind, phase 2, randomised, controlled study. Lancet 392:1330–1339
51. Cesaroni M, Seridi L, Jordan J, Sweet, K, Keying MA, Franks C, et al (2019) Biomarker profiling reveals novel mechanistic insights into ustekinumab therapeutic responses in systemic lupus erythematosus [abstract]. Arthritis Rheumatol 71(suppl 10). https://acrabstracts.org/abstract/biomarker-profiling-reveals-novel-mechanistic-insights-into-ustekinumab-therapeutic-responses-in-systemic-lupus-erythematosus/. Accessed April 5, 2020
52. Chyuan IT, Tzeng HT, Chen JY (2019) Signaling pathways of type I and type III interferons and targeted therapies in systemic lupus erythematosus. Cells 8:E963
53. Jewell NA, Cline T, Mertz SE, Smirnov SV, Flaño E, Schindler C et al (2010) Lambda interferon is the predominant interferon induced by influenza A virus infection in vivo. J Virol 84:11515–11522
54. Amezcua-Guerra LM, Ferrusquía-Toriz D, Castillo-Martinez D, Márquez-Velasco R, Chávez-Rueda AK, Bojalil R (2015) Limited effectiveness for the therapeutic blockade of interferon alpha in systemic lupus erythematosus: a possible role for type III interferons. Rheumatology (Oxford) 54:203–205
55. Wang Y, Li T, Chen Y, Wei H, Sun R, Tian Z (2017) Involvement of NK cells in IL-28B-mediated immunity against influenza virus infection. J Immunol 199:1012–1020

56. Megjugorac NJ, Gallagher GE, Gallagher G (2009) Modulation of human plasmacytoid DC function by IFN-lambda1 (IL-29). J Leukocyte Biol 86:1359–1363
57. Koltsida O, Hausding M, Stavropoulos A, Koch S, Tzelepis G, Ubel C et al (2011) IL-28A (IFN-lambda2) modulates lung DC function to promote Th1 immune skewing and suppress allergic airway disease. EMBO Mol Med 3:348–361
58. Hagberg N, Rönnblom L (2015) Systemic lupus erythematosus - a disease with a dysregulated type I interferon system. Scand J Immunol 82:199–207
59. Wu Q, Yang Q, Lourenco E, Sun H, Zhang Y (2011) Interferon-lambda1 induces peripheral blood mononuclear cell-derived chemokines secretion in patients with systemic lupus erythematosus: its correlation with disease activity. Arthritis Res Ther 13:R88
60. Zahn S, Rehkämper C, Kümmerer BM, Ferring-Schmidt S, Bieber T, Tüting T et al (2011) Evidence for a pathophysiological role of keratinocyte-derived type III interferon (IFNλ) in cutaneous lupus erythematosus. J Invest Dermatol 131:133–140
61. Zickert A, Oke V, Parodis I, Svenungsson E, Sundström Y, Gunnarsson I (2016) Interferon (IFN-λ) is a potential mediator in lupus nephritis. Lupus Sci Med 3:e000170
62. Oon S, Huynh H, Tai TY, Ng M, Monaghan K, Biondo M, et al (2016) A cytotoxic anti-IL-3Ra antibody targets key cells and cytokines implicated in systemic lupus erythematosus. JCI Insight 1(6): e86131
63. Lauterbach H, Bathke B, Gilles S, Traidl-Hoffmann C, Luber CA, Fejer G et al (2010) Mouse CD8alpha + DCs and human BDCA3+ DCs are major producers of IFN-lambda in response to poly IC. J Exp Med 207:2703–2717
64. Ank N, West H, Bartholdy C, Eriksson K, Thomsen AR, Paludan SR (2006) Lambda interferon (IFN-lambda), a type III IFN, is induced by viruses and IFNs and displays potent antiviral activity against select virus infections in vivo. J Virol 80:4501–4509
65. Liu YJ (2005) IPC: professional type 1 interferon-producing cells and plasmacytoid dendritic cells precursors. Annu Rev Immunol 23:275–306
66. Banchereau J, Pascual V (2006) Type I interferon in systemic lupus erythematosus and other autoimmune diseases. Immunity 25(3):383–392
67. Means TK, Latz E, Hayashi F, Murali MR, Golenbock ET, Luster AD (2005) Human lupus autoantibody-DNA complexes activate DCs through cooperation of CD32 and TLR9. J Clin Invest 115:407–417
68. Lande R, Ganguly D, Facchinetti V, Frasca L, Conrad C, Gregorio J, et al (2011) Neutrophils activate plasmacytoid dendritic cells by releasing self-DNA-peptide complexes in systemic lupus erythematosus. Sci Transl Med 3:73ra19
69. Fiore N, Castellano G, Blasi A, Capobianco C, Loverre A, Montinaro V et al (2008) Immature myeloid and plasmacytoid dendritic cells infiltrate renal tubulointerstitium in patients with lupus nephritis. Mol Immunol 45:259–265
70. Tucci M, Quatraro C, Lomabardi L, Pelligrino C, Dammacco F, Silvestris F (2008) Glomerular accumulation of plasmacytoid dendritic cells in active lupus nephritis: role of interleukin-18. Arthritis Rheum 58:251–262
71. Migita K, Miyashita T, Maeda Y, Kimura H, Nakamura M, Yatsuhashi H et al (2005) Reduced blood BDCA2+ (lymphoid) and CD11c+ (myeloid) dendritic cells in systemic lupus erythematosus. Clin Exp Immunol 142:84–91
72. Cederblad B, Blomberg S, Vallin H, Perers A, Alm GV, Rönnblom L (1998) Paitents with systemic lupus erythematosus have reduced numbers of circulating natural interferon-a producing cells. J Autoimmun 11:465–470
73. Henriques A, Inês L, Carvalheiro T, Couto M, Andrade A, Pedreiro S et al (2012) Functional characterization of peripheral blood dendritic cells and monocytes in systemic lupus erythematosus. Rheumatol Int 32:863–869
74. Kwok SK, Lee JY, Park SH, Cho ML, Min SY, Park SH et al (2008) Dysfunctional interferon-alpha production by peripheral plasmacytoid dendritic cells upon Toll-like receptor-9 stimulation in patients with systemic lupus erythematosus. Arthritis Res Ther 10:R29
75. Jin O, Kavikondala S, Sun L, Fu RMok MY, Chan A, et al (2008) Systemic lupus erythematosus patients have increased number of circulating plasmacytoid dendritic cells, but decreased myeloid dendritic cells with deficient CD83 expression. Lupus 17:654–662
76. Farkas L, Beiske K, Lund-Johansen F, Brandtzaeg P, Jahsen F (2001) Plasmacytoid dendritic cells (natural interferon-a/b-producing cells) accumulate in cutaneous lupus erythematosus lesions. Am J Pathol 159:237–243
77. Blomberg S, Eloranta ML, Cederblad B, Nordlind K, Alm GV, Rönnblom L (2001) Presence of cutaneous interferon-alpha producing cells in patients with systemic lupus erythematosus. Lupus 10:484–490
78. Miyashita A, Fukushima S, Makino T, Yoshino Y, Yamashita J, Honda N et al (2014) Proportion of lymphocytic inflammation with CD123 positive cells in lupus erythematosus profundus predict a clinical response to treatment. Acta Derm Venereol 94:563–567
79. Jin O, Kavikondala S, Mok MY, Sun L, Gu J, Fu R et al (2010) Abnormalities in circulating plasmacytoid dendritic cells in patients with systemic lupus erythematosus. Arthritis Res Ther 12:R137
80. Morel L (2010) Genetics of SLE: evidence from mouse models. Nat Rev Rheumatol 6:348–357

81. Sisirak V, Ganguly D, Lewis KL, Couillault C, Tanaka L, Bolland S, et al (2014) Genetic evidence for the role of plasmacytoid dendritic cells in systemic lupus erythematosus. J Exp Med 1969–1976
82. Rowland SL, Riggs JM, Gilfillan S, Bugatti M, Vermi W, Kolbeck R et al (2014) Early, transient depletion of plasmacytoid dendritic cells ameliorates autoimmunity in a lupus model. J Exp Med 211:1977–1991
83. Davison LM, Jorgensen TN (2015) Sialic acid-binding immunoglobulin-type lectin H-positive plasmacytoid dendritic cells drive spontaneous lupus-like disease development in B6.Nba2 mice. Arthritis Rheumatol 67:1012–1022
84. Pellerin A, Otero K, Czerkowicz JM, Kerns HM, Shapiro RI, Ranger AM et al (2015) Anti-BDCA2 monoclonal antibody inhibits plasmacytoid dendritic cell activation through Fc-dependent and Fc-independent mechanisms. EMBO Mol Med 7:464–476
85. Furie R, Werth VP, Merola JF, Stevenson L, Reynolds TL, Naik H et al (2019) Monoclonal antibody targeting BDCA2 ameliorates skin lesions in systemic lupus erythematosus. J Clin Invest 129:1359–1371
86. Furie R, van Vollenhoven R, Kalunian K, Navarra S, Romero-Díaz J, Werth V, et al (2020) Efficacy and safety results from a phase 2, randomized, double-blind trial of BIIB059, an anti-blood dendritic cell antigen 2 antibody, in SLE [abstract]. Arthritis Rheumatol 72(suppl 10). http://acrabstracts.org/abstract/efficacy-and-safety-results-from-a-phase-2-randomized-double-blind-trial-of-biib059-an-anti-blood-dendritic-cell-antigen-2-antibody-in-sle/. Accessed April 11, 2021
87. Werth V, Furie R, Romero-Diaz J, Navarra S, Kalunian K, van Vollenhoven R, et al, and LILAC investigators (2020) OPO193 BIIB059, a humanized monoclonal antibody targeting BDCA2 on plasmacytoid dendritic cells (pDC), shows dose-related efficacy in the phase 2 LILAC study in patients (pts) with active cutaneous lupus erythematosus (LCE). Ann Rheum Dis 79:120–121
88. Henriquez JE, Crawford RB, Kaminski NE (2019) Suppression of CpG-ODN-mediated IFNa and TNFa response in human plasmacytoid dendritic cells (pDC) by cannabinoid receptor 2 (CB2)-specific agonists. Toxicol Appl Pharmacol 369:82–89
89. Dillmann C, Ringel C, Ringleb J, Mora J, Olesch C, Fink AF et al (2016) S1PR4 signaling attenuates ILT 7 internalization to limit IFN-a production by human plasmacytoid dendritic cells. J Immunol 196:1579–1590
90. Bardwell PD, Gu J, McCarthy D, Wallace C, Bryant S, Goess C et al (2009) The Bcl-2 family antagonist ABT-737 significantly inhibits multiple animal models of autoimmunity. J Immunol 182:7482–7489
91. Zhan Y, Carrington EM, Ko HJ, Vikstrom IB, Oon S, Zhang JG et al (2015) Bcl-2 antagonists kill plasmacytoid dendritic cells from lupus-prone mice and dampen interferon-a production. Arthritis Rheumatol 67:797–808
92. Hirai M, Kadowaki N, Kitawaki T, Fujita H, Takaori-Kondo A, Fukui R et al (2011) Bortezomib suppresses function and survival of plasmacytoid dendritic cells by targeting intracellular trafficking of Toll-like receptors and endoplasmic reticulum homeostasis. Blood 117(2):500–509
93. Ji J, Fan H, Li F, Li X, Dong G, Gong W et al (2015) A benzenediamine derivative fc-99 attenuates lupus-like syndrome in MRL/lpr mice related to suppression of pDC activation. Immunol Lett 168:355–365
94. Ishii T, Tanaka Y, Kawakami A, Saito K, Ichinose K, Fujii H et al (2018) Multicenter double-blind randomized controlled trial to evaluate the effectiveness and safety of bortezomib as a treatment for refractory systemic lupus erythematosus. Mod Rheumatol 28:982–992
95. Segarra A, Arredondo KV, Jaramillo J, Jatem E, Salcedo MT, Agraz I et al (2020) Efficacy and safety of bortezomib in refractory lupus nephritis: a single centre experience. Lupus 29:118–125
96. Alexander T, Sarfert R, Klotsche J, Kühl AA, Rubbert-Roth A, Lorenz HM et al (2015) The proteasome inhibitor bortezomib depletes plasma cells and ameliorates clinical manifestations of refractory systemic lupus erythematosus. Ann Rheum Dis 74:1474–1478
97. Lu P, Fleischmann R, Curtis C, Ignatenko S, Clarke SH, Desai M et al (2018) Safety and pharmacodynamics of venetoclax (ABT-199) in a randomized single and multiple ascending dose study in women with systemic lupus erythematosus. Lupus 27:290–302
98. Macanovic M, Sinicropi D, Shak S, Baughman S, Thiru S, Lachmann PJ (1996) The treatment of systemic lupus erythematosus (SLE) in NZB/W F_1 hybrid mice; studies with recombinant murine DNase and with dexamethasone. Clin Exp Immunol 106:243–252
99. Davis JC Jr, Manzi S, Yarboro C, Rairie J, McInnes I, Averthelyi D et al (1999) Recombinant human Dnase I (rhDNase) in patients with lupus nephritis. Lupus 8:68–76
100. Sun X, Widerman A, Agrawal N, Teal T, Tanaka L, Hudkins K et al (2013) Increased RNase expression reduces inflammation and prolongs survival in TLR7 transgenic mice. J Immunol 190:2536–2543
101. Burge DJ, Eisenman J, Byrnes-Blake K, Smolak P, Lau K, Cohen SB et al (2017) Safety, pharmacokinetics, and pharmacodynamics of RSLV-132, an RNase-Fc fusion protein in systemic lupus erythematosus: a randomized, double-blind, placebo-controlled study. Lupus 26:825–834
102. Resolve therapeutics. RSLV-132 demonstrates clinically meaningful improvement in patients with

systemic lupus erythematosus in phase 2a clinical trial. December 9 2020. Available at https://www.prnewswire.com/news-releases/rslv-132-demonstrates-clinically-meaningful-improvement-in-patients-with-systemic-lupus-erythematosus-in-phase-2a-clinical-trial-301189541.html. Accessed 11 April, 2021

103. Tillmanns S, Kolligs C, D'Cruz D, Doria A, Hachulla E, Voll R, et al. SM101, a novel recombinant, soluble, human FcyIIb receptor, in the treatment of systemic lupus erythematosus: results of a double-blind, placebo-controlled multicentre study. In: Paper Presented at: 2014 American College of Rheumatology Annual Meeting, Boston, USA [abstract 2833]. Available from https://acrabstracts.org/abstract/sm101-a-novel-recombinant-soluble-human-fcγiib-receptor-in-the-treatment-of-systemic-lupus-erythematosus-results-of-a-double-blind-placebo-controlled-multicenter-study/. Accessed 12 April 2020
104. Kono DH, Haraldsson MK, Lawson BR, Pollard KM, Koh YT, Du X et al (2009) Endosomal TLR signaling is required for anti-nucleic acid and rheumatoid factor autoantibodies in lupus. Proc Natl Acad Sci USA 106:12061–12066
105. Christensen SR, Shupe J, Nickerson K, Kashgarian M, Flavell RA, Schlomchik MJ (2006) Toll-like receptor 7 and TLR9 dictate autoantibody specificity and have opposing inflammatory and regulatory roles in a murine model of lupus. Immunity 25:417–428
106. Savarese E, Steinberg C, Pawar RD, Reindl W, Akira S, Anders HJ et al (2008) Requirement of Toll-like receptor 7 for pristane-induced production of autoantibodies and development of murine lupus nephritis. Arthritis Rheum 58:1107–1115
107. Savarese E, Chae OW, Trowitzsch S, Weber G, Kastner B, Akira S et al (2006) U1 small nuclear ribonucleoprotein immune complexes induce type I interferon in plasmacytoid dendritic cells through TLR7. Blood 107:3229–3234
108. Boulé MW, Broughton C, Mackay F, Akira S, Marshak-Rothstein A, Rifkin IR (2004) Toll-like receptor 9-dependent and -independent dendritic cell activation by chromatin-immunoglobulin G complexes. J Exp Med 199:1631–1640
109. Zorro S, Arias M, Riaño F, Paris S, Ramírez LA, Uribe O et al (2009) Response to ODN-CpG by B cells from patients with systemic lupus erythematosus correlates with disease activity. Lupus 18:718–726
110. Umiker BR, Andersson S, Fernandez L, Korgaokar P, Larbi A, Pilichowska M et al (2014) Dosage of X-linked toll-like receptor 8 determines gender differences in the development of systemic lupus erythematosus. Eur J Immunol 44:1503–1516
111. Tran NL, Manzin-Lorenzi C, Santiago-Raber ML (2015) Toll-like receptor 8 deletion accelerates autoimmunity in a mouse model of lupus through a Toll-like receptor 7-dependent mechanism. Immunology 145:60–70
112. The Canadian Hydroxychloroquine Study Group (1991). A randomized study of the effect of withdrawing hydroxychloroquine sulfate in systemic lupus erythematosus. N Engl J Med 324:150–154
113. Alarcón GS, McGwin G Jr, Bertoli AM, Fessler BJ, Calvo-Alén J, Bastian HM et al (2007) Effect of hydroxychloroquine on the survival of patients with systemic lupus erythematosus: data from LUMINA, a multiethnic US cohort (LUMINA L). Ann Rheum Dis 66:1168–1172
114. Cairoli E, Rebella M, Danese N, Garra V, Borba EF (2012) Hydroxychloroquine reduces low-density lipoprotein cholesterol levels in systemic lupus erythematosus: a longitudinal evaluation of the lipid-lowering effect. Lupus 21:1178–1182
115. Espinola RG, Pierangeli SS, Gharavi AE, Harris EN (2002) Hydroxychloroquine reverses platelet activation induced by human IgG antiphospholipid antibodies. Thromb Haemost 87:518–522
116. Wu YW, Tang W, Zuo JP (2015) Toll-like receptors: potential targets for lupus treatment. Acta Pharmacol Sin 36:1395–1407
117. Gardet A, Pellerin A, McCarl CA, Diwanji R, Wang W, Donaldson D et al (2019) Effect of in vivo hydroxychloroquine and ex vivo anti-BDCA2 mAb treatment on pDC IFNa production from patients affected with cutaneous lupus erythematosus. Front Immunol 10:275
118. Sacre K, Criswell LA, McCune JM (2012) Hydroxychloroquine is associated with impaired interferon-alpha and tumor necrosis factor-alpha production by plasmacytoid dendritic cells in systemic lupus erythematosus. Arthritis Res Ther 14:R155
119. Barrat FJ, Meeker T, Chan JH, Guiducci C, Coffman RL (2007) Treatment of lupus-prone mice with a dual inhibitor of TLR7 and TLR9 leads to reduction of autoantibody production and amelioration of disease symptoms. Eur J Immunol 37:3582–3586
120. Barrat FJ, Meeker T, Gregorio J, Chan JH, Uematsu S, Akira S et al (2005) Nucleic acids of mammalian origin can act as endogenous ligands for Toll-like receptors and may promote systemic lupus erythematosus. J Exp Med 202:113–119
121. Dynavax (DVAX) reports DV1179 did not meet endpoints in phase 1b/2a. August 7 2014. Available at https://www.streetinsider.com/Corporate+News/Dynavax+%28DVAX%29+Reports+DV1179+Did+Not+Meet+Endpoints+in+Phase+1b2a/9729788.html. Accessed 13 April, 2020
122. Kimball AB, Krueger J, Sullivan T, Arbeit RD. IMO-3100, an antagonist of Toll-like receptor (TLR) 7 and TLR9, demonstrates clinical activity in psoriasis patients with 4 weeks of treatment in a phase 2a trial. Available at https://www.iderapharma.com/wp-content/uploads/2015/11/iid_

2013_presentation_final.pdf. Accessed 13 April, 2020
123. Balak DM, van Doorn MB, Arbeit RD, Rijneveld R, Klaasen E, Sullivan T et al (2017) IMO-8400, a toll-like receptor 7, 8, and 9 antagonist, demonstrates clinical activity in a phase 2a, randomized, placebo-controlled trial in patients with moderate-to-severe plaque psoriasis. Clin Immunol 174:63–72
124. Zhu F, Yu D, Kandimalla E, La Monica N, Agrawal S. Treatment with IMO-3100, a novel TLR7 and TLR9 dual antagonist, inhibits disease development in lupus prone NZBW/F1 mice. Keystone Symposia: Dendritic Cells and the Initiation of Adaptive Immunity; Santa Fe 2011
125. Zhu F, Jiang W, Dong Y, Kandimalla E, La Monica N, Agrawal S (2012) IMO-8400, a novel TLR7, TLR8 and TLR9 antagonist, inhibits disease development in lupus-prone NZBW/F1 mice. J Immunol 188:119.12
126. Capolunghi F, Rosado MM, Cascioli S, Girolami E, Bordasco S, Vivarelli M et al (2010) Pharmacological inhibition of TLR9 activation blocks autoantibody production in human B cells from SLE patients. Rheumatology (Oxford) 49:2281–2289
127. Loiarro M, Capolunghi F, Fantò N, Gallo G, Campo S, Arseni B et al (2007) Pivotal advance: inhibition of MyD88 dimerization and recruitment of IRAK1 and IRAK4 by a novel peptidomimetic compound. J Leukoc Biol 82:801–810
128. Lipford G, Forsbach A, Zepp C, Nguyen T, Weeratna R, McCluskie M, et al. Selective toll-like receptor 7/8/9 antagonists for the oral treatment of autoimmune diseases. In: Paper Presented at: 2007 American College of Rheumatology Annual Meeting, Boston, USA [abstract 1596]. Available from https://acr.confex.com/acr/2007/webprogram/Paper8044.html. Accessed 13 April, 2020
129. AdisInsight, CpG 52364. 2010. Available at https://adisinsight.springer.com/drugs/800027264. Accessed 13 April, 2020
130. Mok CC (2019) The Jakinibs in systemic lupus erythematosus: progress and prospects. Expert Opin Investig Drugs 28:85–92
131. Markopoulou A, Kyttaris VC (2013) Small molecules in the treatment of systemic lupus erythematosus. Clin Immunol 148:359–368
132. Montealegre Sanchez GA, Reinhardt AL, Brogan P, Berkun Y, Brown D, Chira P et al (2013) Chronic atypical neutrophilic dermatosis with lipodystrophy and elevated temperatures (CANDLE): clinical characterization and initial response to Janus Kinase inhibition with baricitinib [abstract]. Arthritis Rheum 65(Suppl 10):1782
133. Liu Y, Jesus AA, Marrero B, Yang D, Ramsey SE, Montealegre Sanchez GA, Tenbrock K et al (2014) Activated STING in a vascular and pulmonary syndrome. NEJM 371:507–518
134. Wallace D, Furie R, Tanaka Y, Kalunian K, Mosca M, Petri M et al (2018) Baricitinib for systemic lupus erythematosus: a double-blind, randomised, placebo-controlled, phase 2 trial. Lancet 392:222–231
135. Keown A. Gilead and Galapagos' JAK inhibitor stumbles in mid-stage lupus trial. Available at https://www.biospace.com/article/gilead-and-galapagos-filgotinib-fails-in-mid-stage-lupus-trial/. Accessed April 19, 2020
136. Han P, Pohlmeyer C, Shang C, Cui Z, Lopez D, Clarke A, et al (2018) Monotherapy with filgotinib, a JAK1-selective inhibitor, reduces disease severity and alters immune cell subsets in the NZB/W F1 murine model of lupus [abstract]. Arthritis Rheumatol 70(suppl 10). https://acrabstracts.org/abstract/monotherapy-with-filgotinib-a-jak1-selective-inhibitor-reduces-disease-severity-and-alters-immune-cell-subsets-in-the-nzb-w-f1-murine-model-of-lupus/. Accessed April 19, 2020
137. van Vollenhoven RF, Layton M, Kahl L, Schifano L, Hachulla E, Machado D et al (2015) DRESS syndrome and reversible liver function abnormalities in patients with systemic lupus erythematosus treated with the highly selective JAK-1 inhibitor GSK2586184. Lupus 24:648–649
138. Presto JK, Okon LG, Feng R, Wallace DJ, Furie R, Fiorentino D et al (2018) Computerized planimetry to assess clinical responsiveness in a phase II randomized trial of topical R333 for discoid lupus erythematosus. Br J Derm 178:1308–1314
139. Hasni S, Gupta S, Davis M, Poncio E, Temesgen-Oyelakin Y, Biehl A, et al (2019). A phase 1B/2A trial of tofacitinib, an oral janus kinase inhibitor, in systemic lupus erythematosus. [abstract]. Lupus Sci Med. https://doi.org/10.1136/lupus-2019-lsm.183. Accessed April 19, 2020
140. Sarhan RA, Aboelenein HR, Sourour SK, Fawzy IO, Salah S, Abdelaziz AI (2015) Targeting E2F1 and c-Myc expression by microRNA-17-5p represses interferon-stimulated gene MxA in peripheral blood mononuclear cells of pediatric systemic lupus erythematosus patients. Discov Med 19:419–425
141. Han X, Wang Y, Zhang X, Qin Y, Qu B, Wu L et al (2016) MicroRNA-130b ameliorates murine lupus nephritis through targeting the type 1 interferon pathway on renal mesangial cells. Arthritis Rheumatol 68:2232–2243
142. Fabricius D, Neubauer M, Mandel B, Schütz C, Viardot A, Vollmer A et al (2010) Prostagladin E2 inhibits IFN-alpha secretion and Th1 costimulation by human plasmacytoid dendritic cells via E-prostanoid 2 and E-prostainoid 4 receptor engagement. J Immunol 184:677–684
143. Khoryati L, Augusto JF, Contin-Bordes C, Douchet I, Mitrovic S, Truchetet ME, et al (2016) Federation Hospitalo-Universitaire ACRONIM. IgE inhibits Toll-like receptor 7- and toll-like receptor-9-mediated expression of interferon-a by plasmacytoid dendritic cells in patients with systemic lupus erythematosus. Arthritis Rheumatol 68:2221–2231

144. Mina-Osorio P, LaStant J, Keirstead N, Whittard T, Ayala J, Stefanova S et al (2013) Suppression of glomerulonephritis in lupus-prone NZB x NZW mice by RN486, a selective inhibitor of Bruton's tyrosine kinase. Arthritis Rheum 65:2380–2391
145. Bender AT, Pereira A, Fu K, Samy E, Wu Y, Liu-Bujalski L et al (2016) Btk inhibition treats TLR7/IFN driven murine lupus. Clin Immunol 164:65–77
146. Katewa A, Wang Y, Hackney JA, Huang T, Suto E, Ramamoorthi N, et al (2017) Btk-specific inhibition blocks pathogenic plasma cell signatures and myeloid cell-associated damage in IFNa-driven lupus nephritis. JCI Insight 2(7):e90111
147. Wu L, Qin Y, Xia S, Dai M, Han X, Wu Y et al (2016) Identification of cyclin-dependent kinase 1 as a novel regulator of type I interferon signalling in systemic lupus erythematosus. Arthritis Rheumatol 68:1222–1232
148. Pauls E, Shapiro N, Peggie M, Young E, Sorcek R, Tan L et al (2012) Essential role for IKKb in production of type 1 interferons by plasmacytoid dendritic cells. J Biol Chem 287:19216–19228
149. Miyamoto R, Ito T, Nomura S, Amakawa R, Amuro H, Katashiba Y et al (2010) Inhibitor of IKB kinase activity, BAY 11–7082, interferes with interferon regulatory factor 7 nuclear translocation and type I interferon production by plasmacytoid dendritic cells. Arthritis Res Ther 12:R87
150. Wang H, Li T, Chen S, Gu Y, Ye S (2015) Neutrophil extracellular trap mitochondrial DNA and its autoantibody in systemic lupus erythematosus and a proof of concept trial of metformin. Arthritis Rheumatol 67:3190–3200
151. Oke V, Gunnarsson I, Dorschner J, Eketjäll S, Zickert A, Niewold T et al (2019) High levels of circulating interferons type I, type II and type III associate with distinct clinical features of active systemic lupus erythematosus. Arthritis Res Ther 21:107

6 The Concept of Co-Stimulatory Blockade in SLE

Alberta Hoi

Abstract

The beautifully orchestrated inflammatory cascade in response to a threat relies on an important mechanism called co-stimulation that occurs between the antigen presenting cells and T cells. Our understanding of how costimulatory molecules play a role in lupus pathogenesis will create novel ways to damper the inflammatory cascade and promote immune tolerance. The best-studied costimulatory molecules in SLE are the CD28/B7 system utilising CTLA4Ig to damper down the T cell response, but many more are of pertinent interest to downregulate the adaptive T-cell dependent humoral response. For example, the ICOS/ICOSL, CD40L/CD40, and PD1/PDL1 pathways are all relevant co-stimulatory molecules that are susceptible to manipulation, with the end-result on activation and maintenance of germinal centre response and production of high affinity autoantibodies. In this chapter, the biology of these costimulatory molecules will be summarised and the data supportive of them as emerging therapeutic targets in the SLE discussed.

Introduction

Autoreactive T cells are key players in autoimmune diseases, as they can act both as regulator and effectors cells, including cytotoxic ability and providing help to B cells. The activation of autoreactive T cells requires a critical process called co-stimulation, which can be modulated by endogenous or potentially exogenous ways. T cells, upon engagement of the T cell receptor by the antigen and MHC Class II complex on the antigen presenting cells, require a second signal before they can become fully activated [1]. Without this second signal provided by the costimulatory molecules, T cell receptor (TCR) mediated stimulation of T cell leads to an antigen specific unresponsiveness or so-called T cell anergy. This second signal allows fine tuning of the T cell response and provides feedback to the antigen presenting cells. In many instances, the antigen presenting cells are B cells or other haematopoietic cells. Increasingly it is also recognised that the ligands for some of these costimulatory molecules are present on endothelial or epithelial cells. The ability to interfere with

A. Hoi (✉)
Rheumatology Research Group, Centre for Inflammatory Diseases, School of Clinical Sciences at Monash Health, Monash University, Clayton, VIC 3168, Australia
e-mail: alberta.hoi@monash.edu

A. Hoi
Department of Rheumatology, Monash Health, Clayton, VIC 3168, Australia

A. Hoi
Department of Rheumatology, Austin Health, Heidelberg, VIC 3084, Australia

A. Hoi (ed.), *Pathogenesis of Systemic Lupus Erythematosus*,
https://doi.org/10.1007/978-3-030-85161-3_6

these interactions between costimulatory molecules may provide therapeutic interventions for patients with SLE.

CD28/B7 Pathway

The CD28/B2 pathway has been one of the best studied co-stimulatory molecules. The discovery began with the recognition that CD28 receptor is a transmembrane glycoprotein that serves as a stimulatory receptor for the activation of naive T cells [1]. Its ligands are the B7 family of proteins, namely B7.1 (or CD80) and B7.2 (or CD86), expressed constitutively on dendritic cells and macrophages, and can be rapidly upregulated on B cells following activation upon cross-linking of the immunoglobulin receptor. B7.1 and B7.2 share the common receptor, but has distinct biochemical features and roles in signalling transduction in different biological settings [2].

T cell activation requires at least two signals; the first driven by the interaction between T-cell receptor and MHC Class II molecule, and the second signal derives from the interaction between CD28 and CD80 or CD86. In the absence of co-stimulation through the CD28/B7 pathway T cells will undergo anergy, a tolerising mechanism in which T cells fail to react to the antigen presented. T cell anergy is an important aspect that can be manipulated for therapeutic use [3, 4] (Fig. 6.1).

The exact mechanism by which CD28 mediates the intracellular signalling is not yet entirely clear. CD28 shares overlapping structure with Cytotoxic T lymphocyte antigen 4 (CTLA4) and inducible costimulatory molecule (ICOS), but CD28 and CTLA4 mediates contrasting effect on T cell responses. The lipid kinase phosphatidylinositol 3-kinase (PI3K) has been reported to be a common signalling pathway for CD28-mediated effects [5]. Although CD28 ligation is critical in promoting proliferation and effector function of conventional T cells, it can also promote the anti-inflammatory function of regulatory T (Treg) cells. Therefore the effects of CD28 modulation may result in both pro- and anti-inflammatory roles depending on the cell type and context in which it is expressed [6].

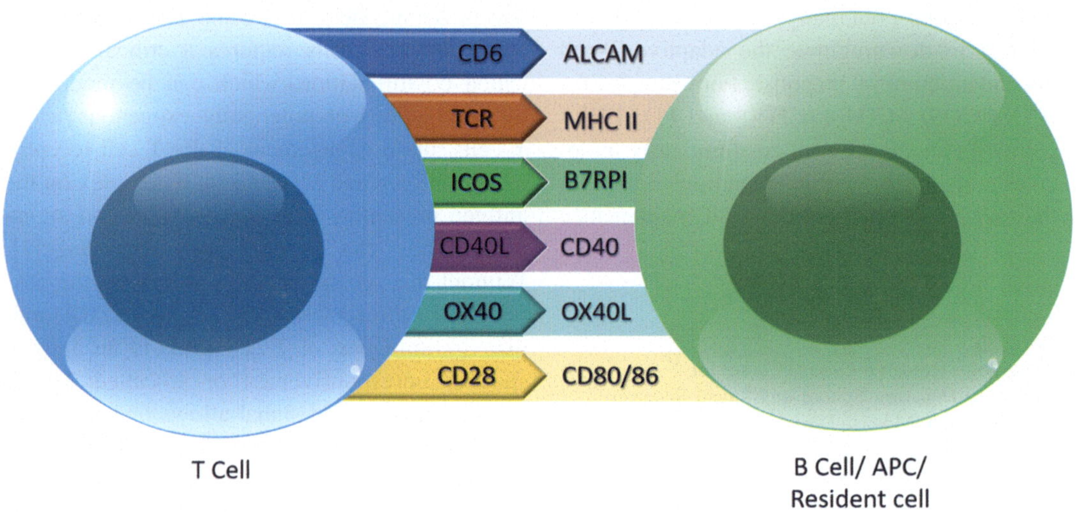

Fig. 6.1 Major co-stimulatory molecules in systemic lupus erythematosus. Expression of major co-stimulatory molecules as pairs on T cells and antigen presenting cells following initial interaction of MHC class II molecule with T cell receptor. Second signal from the interaction between these pairs provide a more sustained activation of both T cells and APC. B cells, platelets, monocytes, dendritic cells or local resident cells such as fibroblasts and epithelial cells can act as APC in some settings. Courtesy of Dr Bonnia Liu

Attempts to module CD28-mediated signals have revealed that by engineering CTLA4, a naturally occurring ligand of B7, CD28 signalling can be interrupted [1, 7]. By linking a soluble CTLA4 binding domain linked to an Ig constant region, the molecule was effective at binding CD80 and CD86, and therefore blocking access to these ligands. This fusion protein has been used in a number of animal models of allograft rejection and autoimmune diseases [3].

The first proof of concept that manipulation of costimulation may be an effective therapeutic strategy in autoimmune disease was an animal model of murine lupus where CTLA4Ig was used. CTLA4Ig prevented onset of disease in (NZB × NZW) F1 mice when administered consistently early in the model [8]. Treated mice have lower autoantibody production, less renal disease, and longer survival. Several subsequent murine studies have supported the notion of therapeutic potential of CTLA4Ig [4, 9, 10], but it appeared that the timing of administration was critical to the effectiveness of CTL4Ig in these models. In studies where co-stimulation blockade was initiated early in the disease, whether by long term parenteral administration or by gene therapy, there were consistent observation of a reduction in T cell dependent B cell hyperactivity and improvement in target organ inflammation [8, 9].

In established disease, monotherapy with CTLA4Ig was less effective but combination therapy with CTLA4Ig and cyclophosphamide could ameliorate disease and prolonged survival in a murine lupus nephritis model [10]. This synergistic action of cyclophosphamide and CTLA4 Ig appears to overcome the lag in clinical response mediated by co-stimulatory blockade alone.

Human Studies of Abatacept

Abatacept is the first generation commercially available CTLA4Ig approved for treatment of rheumatoid arthritis, and belatacept is another CTLA4Ig that has a higher affinity, is predominantly used in the transplant setting. To date, there have been three key studies examining the efficacy and safety of abatacept in SLE, one as phase IIb non-renal SLE and two on lupus nephritis [11–13]. The results unfortunately were conflicting and mostly did not meet the primary endpoints. When taken into account of the efficacy and safety data of abatacept outside SLE, costimulatory blockade using fusion protein CTLA4Ig remains a possible therapeutic option for SLE as an off-label therapy and remains amongst our therapeutic armamentarium for difficult cases.

In order to understand the utility of abatacept in SLE, it is important to examine these randomised placebo-controlled studies in detail, and therefore to understand its limitation. The first of these studies examined the use of abatacept in patients who has moderately active SLE, as defined by a disease activity instrument called British Isles Lupus Assessment Group (BILAG) index [11]. For each of eight organs or systems defined, the patient's presenting state can be classified as a BILAG score from A to E. A BILAG A score captures disease manifestations that when based on the physician's intention to treat the addition of an immunosuppressant or high dose corticosteroid would usually be required [14]. In comparison, a BILAG B score captures disease manifestations that are moderately active, but the physician usual treatment would be a gentler addition of an anti-inflammatory and medications such as hydroxychloroquine.

The 2010 study recruited 175 SLE patients with BILAG A or B score in either musculoskeletal, mucocutaneous, cardiovascular or respiratory system [11]. This phase IIb exploratory study compared the active drug arm (abatacept 10 mg/kg as per the "rheumatoid arthritis" protocol—intravenously on days 1, 15, 29 and then every 4 weeks) versus placebo while the patients were allowed to continue their background immunosuppressive medications. The primary endpoint of this study was a new flare according to the BILAG A or B definition, following corticosteroid tapering.

Following 12 months of treatment with abatacept or placebo, the proportion of patients having another flare by the BILAG A or B definition was almost the same between the abatacept and the placebo groups (79.7% vs. 82.5%) [11]. However, post hoc analyses revealed severe flares (BILAG A) were less frequent in the abatacept compared to the placebo group (40.7% versus 54.4%, with a treatment difference of −13.7 [95%CI −29.5, 2.1]). Furthermore, subgroup analysis revealed that those with musculoskeletal manifestation had the highest difference in flare rates, specifically 36.5% (abatacept group) versus 62.5% (placebo group), that is a treatment difference −26.1 [95% CI −47.4, −4.8]. Health-related quality of life, fatigue and sleep scores were all better in the abatacept treated group. Serious adverse events occurred more frequently in the abatacept group (19.8% versus 6.8%) but this was not statistically significant [11].

Two additional studies examined the role of abatacept in lupus nephritis, and both studies did not meet its primary or secondary efficacy endpoints, but there were lessons learnt from these randomised controlled studies on the way abatacept works and in how lupus nephritis trials can be conducted [12, 13]. The first of these was a 12-month phase II/III study which has recruited 298 patients with active lupus nephritis, defined by ISN/RPS Class III or IV proliferative changes on a biopsy done within 12 months of being screened, and the presence of active urinary sediment and urinary protein creatinine ratio of over 50 mg/mmol, as well as serologically activity with either an elevated anti-dsDNA or hypocomplementemia [12]. Patients with moderately severe renal impairment (serum creatinine at or greater than 3 mg/dL or 265umol/L or a recent increase of greater than 1 mg/dL or 88umol/L) were excluded.

Much discussion around how stringent the definition of the endpoints needs to be and their roles in influencing the outcome of the studies. In the study reported by Furie et al. the entire population had generally good response regardless of the treatment arms [12]. Only a third of patients had grumbling persistently active lupus nephritis, presumably related to the duration of pretreatment that the patients had received (over three months of mycophenolate and corticosteroid prior to entering the study). However, in the subset of patients with more severe disease at baseline, defined by having nephrotic range proteinuria, there was greater reduction in proteinuria in the treatment arm compared to placebo [12].

Exploratory analysis of the data set from this trial using different definitions of response has shown a difference between the abatacept-treated group compared to placebo, with an effect size as great as 18%, demonstrating that the choice of endpoints contributes to the outcome of the study [15]. In the case of the study by Furie et al., the choice was quite a stringent definition of complete renal response (CRR) (defined by normalisation of urinary protein creatinine ratio (UPCR) to < 0.03 g/mmol, normal urinary sediment, and estimated glomerular filtration rate could not drop by 10% or more of baseline values, or a six-month pre-flare value). These results had to be confirmed at two visits, one month apart [12].

In another attempt to evaluate efficacy of abatacept for treatment of lupus nephritis, the Abatacept and Cyclophosphamide Combination Efficacy and Safety Study (ACCESS) Trial Group studied the effect of abatacept vs placebo as an add-on therapy to patients receiving Euro-lupus cyclophosphamide regimen [13]. The main difference in the ACCESS trial in terms of inclusion criteria was a slightly higher threshold of proteinuria at 0.1 g/mmol. Biopsy was similarly performed within 12 months, with Class III or IV proliferative changes required as part of inclusion criteria, with or without concurrent Class V changes. The response criteria were also less stringent, with complete remission (CR) at week 24, defined as UPCR of less than 0.05 g/mmol, normalisation of serum creatinine or if abnormal no more than 25% above baseline, and prednisolone dose tapered to 10 mg per day at week 12. However, despite these adjustments, the ACCESS study also failed to reach their predefined endpoints.

Improvement in serological profile was observed in abatacept treated patients, and there did not seem to be an increase in adverse event rate when treatment was added to either background mycophenolate or Euro-lupus cyclophosphamide therapy. The potent background immunosuppression may explain the failure to demonstrate difference in efficacy between treatment arms.

The ACCESS trial had a second part to their study design in attempt to explore the potential of abatacept in restoring immune tolerance [16]. The study design was somewhat complicated, and numbers evaluated were small, but overall, it showed that 50% of those who had attained CR with the combination of immunosuppression with abatacept for 6 months, in combination of the Euro-lupus regimen of 3 months of low dose cyclophosphamide transitioning to azathioprine, were able to remain in remission despite withdrawn of abatacept.

The other finding of interest was that the response rates as measured immediately after the induction phase and the later maintenance phase may be different. It appeared that even though the addition of abatacept to conventional Euro-lupus regimen of cyclophosphamide and azathioprine did not show a difference in clinical response, over 40% of patients in each arm continue to improve beyond the 24 weeks of treatment, and eventually achieve CR by week 52. In this study, the addition of abatacept in the first 6 months did not show an improvement in clinical outcomes in the later maintenance phase of disease, but the study was said to not be powered to show a difference for this.

The findings of these human randomised controlled trials on the utility of abatacept in SLE were overall disappointing, given earlier speculation based on animal models that abatacept treatment should result in better disease control. The improvement in the arthritis group was probably not surprising, given the clinical efficacy seen in rheumatoid arthritis and in the psoriatic arthritis population [17, 18]. Abatacept has been used in SLE in refractory cases with some reports of benefit, particularly if the patient has predominantly musculoskeletal involvement [19].

Costimulatory Molecules That Regulate the Adaptive T Cell Dependent Humoral Response

Other than CD28/B7 system, there are a few costimulatory molecules that are critical in the augmentation of humoral immune response by T cells which form the basis of autoimmune diseases such as SLE. CD4 naïve T cells differentiate into specific subsets which distinct functions. One subset, called the follicular T helper cells (T_{FH}), are critical for optimal humoral immune response, which is dependent on the activation and maintenance of germinal centre, proliferation and selection of high affinity B cells that will then differentiate into long lived plasma cells.

The three key systems that have been best studied include ICOS/ICOSL, CD40L/CD40, and PD1/PDL1 signalling. While ICOS and CD28 share overlapping functions in immediate steps of T cell activation upon TCR ligation, ICOS can affect T cell effector functions.

ICOS/ICOSL System

ICOS is expressed on activated and memory T cells. Its engagement with ICOSL, which is expressed mainly on B cells and other antigen presenting cells, results in a second signal interaction that is critical for the development of follicular T helper cells (T_{FH}). This particular subset of CD4 + T cells are located in secondary lymphoid organs and play a critical role in humoral immunity in its part in providing B cell help [20–22]. ICOS/ICOSL signalling is essential for sustained T and B cell interaction.

In order to fulfil the function of the T_{FH} cells, a number of costimulatory molecules, not just ICOS, are upregulated in order to influence germinal centre interactions. ICOS can induce chemokine receptor 5 (CXCR5) on T_{FH} cells, and CD40L and PD-1 are also upregulated too upon antigen presentation [23].

The ligand for ICOS is ICOSL (also called the B7 Related Protein-1 (B7RP-1), or B7 Homolog 2 (abbreviated to B7H2)) which is expressed on CD19 B cells. ICOS/ICOSL signalling is critical for B cell survival and plasma cell differentiation in germinal centre. In animal models, mice lacking ICOSL or when treated with an anti-ICOSL antibody had a blunted Th2 immunity and an impaired T cell dependent antibody response [24].

ICOS/ICOSL signalling also has an antibody-independent role by its effect on CD11c myeloid dendritic cells. ICOSL has also been shown to have regulatory roles in inflammatory cytokine production such as IL-4, IL-10 and Il-13 [25, 26]. In a study where ICOSL was specifically deleted in CD11c myeloid cells in a murine lupus model, disease was ameliorated as shown by reduction in proteinuria, reduction of inflammatory infiltrate and improved histopathology [27]. In contrast to the model where ICOS was completely deleted, we observed relative preservation of autoantibody titre in this specific model, suggesting that there are differential roles for B cells and myeloid cells when acting as antigen presenting cells, and there is an autoantibody independent role for ICOS in self-reactive T cells at target organs of SLE [27].

One of the key cytokines produced by T_{FH} cells is IL-21 which sustain the expansion of germinal centre. Both IL-21 and IL-6 contribute to T_{FH} differentiation and upregulation of a range of co-stimulatory molecules not just ICOS, but CD40L and Programmed Cell Death Protein-1 (PD-1) which are essential for B cell activation and evolution of long-term memory B cell and plasma cells [28–30]. In human, mutation in ICOS results in impaired T effector memory cell haemostasis and defective humoral immune responses.

Interestingly, intestinal inflammation has been shown to be relatively resistant to B7 blockade, which may be due to unique features of commensal intestinal microbes expressing accessory ICOSL or OX40L. Activation of a subset of T cells by different these accessory costimulatory molecules can lead to susceptibility to inflammation despite B7 blockade [31]. B7 blockade can result in a paradoxical suppression of regulatory T cells which further promotes intestinal inflammation by Th17 polarisation, which is attenuated by neutralising ICOSL [31].

Novel way of targeting ICOS/ICOSL pathway is under development, and in particular AMG557 has been developed as a fully human IgG2 bispecific antibody against ICOSL and BAFF. Interruption of this pathway will hopefully be effective in conditions such as lupus or Sjogren. There is some evidence that dual ICOSL/BAFF inhibition may be more efficacious than a single target based on animal studies [32]. A small, dose-finding human SLE study has shown promise for AMG557 to be efficacious in controlling lupus arthritis [33].

CD40L/CD40 Pathway

The CD40L/CD40 pathway is another accessory costimulatory pair, that belongs to the TNFR family, which has been quite well-studied. CD40L, also called CD154, is expressed predominantly on activated CD4 + T cells and upregulated upon CD28 co-stimulation. Its ligand, CD40, is constitutively found on a broad range of immune cells including B cells, CD8 + cytotoxic T cells, dendritic cells, monocytes, platelets and macrophages, as well as a range of non-haematopoietic cells such as fibroblasts, epithelial and endothelial cells [34–36].

The function of CD40L/CD40 interaction serves as a positive costimulatory signal to induce proliferation of CD4 + T cells, particularly the subset of follicular T helper cells (T_{FH}). It has dominant effects via the CD40 signalling on B cells, playing a major part in an effective humoral T-cell dependent adaptive immune response. CD40 signalling is critical for B cells to produce high titre isotype-switched high affinity antibody. It also triggers B cell to proliferate, adhere and differentiate, and in vivo it has been shown to be critical for germinal cell formation and progression [37–39]. Interruption of the CD40 signalling can interfere with the generation of memory B cell and long-lived plasma cells.

Superimposed on the critical role in germinal cell formation and isotype class switching, CD40 signalling on non-B cells also play an important role in promoting an inflammatory response. It is the constitutive expression of CD40 on monocytes and dendritic cells that promote cytokine production, cell survival and leucocyte recruitment. A range of non-haematopoietic cell lines such as endothelial and epithelial cells, fibroblasts and neuronal cells, express CD40 and can promote upregulation of other costimulatory and adhesion molecules, increased cytokine production and cross-presentation of antigen [40]. In addition, CD40L also occurs in soluble form, and is also upregulated on activated B cells, platelets, monocytes, natural killer cells, mast cells, and basophils, in addition to activated T cells [41]. Other ligands recognised for CD40L include fibronectin receptor Very Late Activation Antigen-5 (VLA-5) and Macrophage-1 leukocyte integrin (Mac-1), which play a role in leucocyte recruitment in venules in angiotensin II-mediated thrombosis and inflammation [42, 43].

Its broad cellular effects can be understood from the structure of CD40, which belongs to the TNF receptor family of molecules. The engagement of CD40 by CD40L results in conformational change and clustering to recruit specific adapter proteins known as TNFR-associated factors (TRAFs) to the cytoplasmic domain of CD40 [44]. TRAF proteins can activate a number of key intracellular signalling pathways such as NFκB, MAPK, PI3K and phospholipase C, which play a critical role for the activation of B cells and dendritic cells. CD40 engagement can also have a number of non-TRAF medicated effects including activation of Janus Family kinase 3 (JAK3) [44].

Animal studies have confirmed the essential role of CD40L/CD40 interaction in autoimmune diseases particularly SLE, with anti-CD40L antibody treatment of mice with lupus nephritis showing significant amelioration of disease [45]. The first human study of BG9588 in human lupus nephritis in an open-label study demonstrated some clinical efficacy but raised some concerns regarding potential thromboembolic events relating to platelet activation caused by cross-linking of anti-CD40L antibodies to platelet CD40L and FCγRIIa on adjacent platelets [46, 47]. A different anti-CD40L monoclonal antibody, ABI793, also reported similar thromboembolic events [48].

Recently a humanised antagonistic anti-CD40 antibody (BI 655,064) has reported some benefits in patients with rheumatoid arthritis in a Phase 2a study. There was some improvement in biological parameters but overall did not demonstrate significant clinical efficacy endpoint at week 12 [49]. The efficacy of BI 655,064 in lupus nephritis is currently investigated in a phase 2 study. Other anti-CD40 monoclonal antibody (CFZ533) is also under development for SLE. A novel CD40L binding protein VIB4920, which lacks an Fc domain, may avoid the platelet related adverse effects [50].

OX40 and OX40L

OX40L is coded by *TNFSF4*, one of the fifty susceptibility genes discovered in genome-wide association studies (GWAS) in SLE [51–53]. OX40L is a type II transmembrane protein expressed on a variety of APC such as dendritic cells, B cells and macrophages, but also found on activated T cells, mast cells and vascular endothelial cells [54]. It binds OX40 (or CD134), which is expressed on activated CD4 + T cells, including T regulatory cells. This pathway appears to be critical for regulating adaptive T cell immune responses, promoting effector T cell phenotype and both CD4 and CD8 T cell memory. Upon OX40L ligation, there are a range of effects depending on the cell types. In B cells, immunoglobulin production is increased, and CD40L signalling is facilitated, and B cells are encouraged to differentiate into plasma cells. Upon OX40L ligation on the endothelial cells, there is increased expression of RANTES/CCL5, promoting leucocyte trafficking via trans-endothelial migration. OX40L signalling on dendritic cells promotes increased cytokine production such as TNF, IL-12, IL-1 and IL-6, and increased expression of other costimulatory molecules such as CD80 and CD86 and CD40. OX40L expression is promoted by interferon-gamma signalling [55].

The overall biological effect of OX40L/OX40 signalling is to promote an effector T cell phenotype, with increased effector cytokine production and enhanced cell mobility. OX40 has a broadly stimulatory effect on a range of T cells, but it has a distinct inhibitory action on the suppressive effect of Treg. OX40 signal may act as a negative regulator of Treg functions [56]. In the presence of OX40, it is less likely for CD4 + T cells to differentiate into FOXP3 + Treg [57]. OX40 agonism however appears to have differential effects on Treg in some reports, with maintenance and homeostasis of the CD4 + CD25 + suppressor T cells which increased tolerance and anergy [58].

The effect of OX40/OX40L interaction is demonstrated in a transgenic mouse model that constitutively expressed OX40L on T cells which revealed an autoimmune phenotype [59]. Using an in vivo agonist and antagonist, the biological effects of this axis was confirmed showing an increased number of follicular helper T cells and plasmablasts in the spleen and kidney of diseased mice, and the use of an antagonist OX40-Fc fusion protein significantly delayed the onset of severe proteinuria and improved survival [60]. In human SLE, the frequency of OX40 expressing circulating T cells, both CD4 + and IL17 producing, has been shown to be increased and correlate with SLE activity particularly renal disease [61, 62]. In another study, OX40L expressing myeloid APC correlated with disease activity and the frequency of T_{FH} cells [63]. The beneficial effects of OX40/OX40L blockade have been explored in various animal models. One study examined the function of OX40L on T cells and B cells using a conditional knockout model, revealing that even though OX40L have distinct role in T cell activation, OX40L deletion on B cells can have a T cell independent reduction in the secondary humoral response observed in this model [64]. OX40L blockade has also been explored in murine model of systemic sclerosis, demonstrating a potential protection in inflammation-driven skin, lung and vessel fibrosis [65]. Understanding of OX40/OX40L interactions have progressed at a steady pace, and the development of therapies that utilise either the augmentative effect on OX40/OX40L in immuno-oncology or the antagonistic effect in autoimmune diseases will hopefully not be in too-distant future.

PD-1/PD-L Pathway

The recent success of checkpoint inhibition in oncology has shed light on potential ways to manipulate the immunosuppressive signalling from this pathway. Programmed cell death protein 1 (PD-1) is an important negative regulator of co-stimulation, that is expressed on a variety of immune cells. It competes with CD28 for binding to CD80 and C86, but also serves as a receptor that bind with PD-L to resist positive signals through T cell receptors and CD28 (Fig. 6.2) [66]. PD-L is broadly expressed on a variety of immune cells, such as T cells, B cells, dendritic cells, monocytes and neutrophils, and can be induced by. There is evidence of

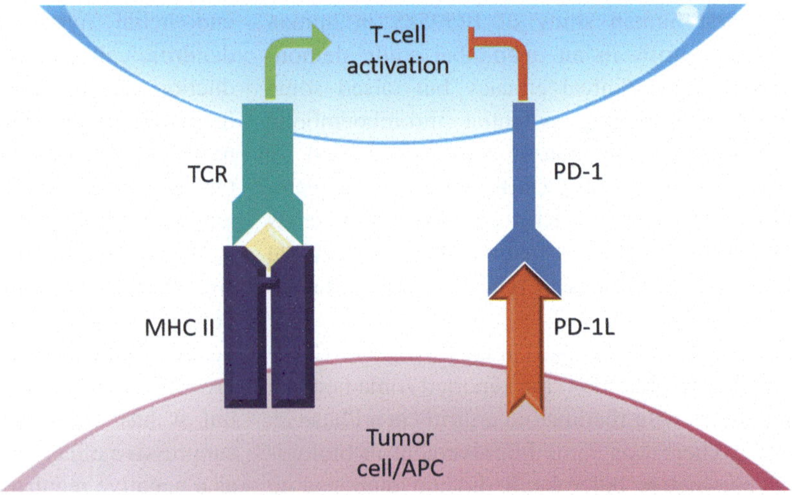

Fig. 6.2 Schematic interaction of PD-1 on T cells and PD-1L on tumour cell or APC. PD-1 is an important negative regulator of co-stimulation and upon interaction of with PD-1L provide inhibitory signal to the T cell to resist activation by other costimulatory signals. Courtesy of Dr Bonnia Liu

increased expression of PD-L on neutrophils of human SLE patients and some correlation with disease activity and severity.

To date there are five antibody-based inhibitors targeting PD-1/PD-L pathway, including two anti-PD-1 antibodies (nivolumab and pembrolizumab) and three anti-PD-L1 antibodies (avelumab, atezolizumab and durvalumab) that has been approved by the Food and Drug Administration (FDA) for treatment of malignancy. Ways to modulate the axis for autoimmune diseases is still to be elucidated, but lessons have been learnt from the occurrence of autoimmune inflammatory disorder as an adverse effect from treatment using these anti-PD-1 or PD-L therapies [67].

CD6/ALCAM

CD6 lymphocyte receptor has been relatively difficult to study until recently even though it is one of the first lymphocyte surface markers identified [68]. It is a cell surface glycoprotein expressed by most T cells and a small subset of mature B cells, and functions to stabilise adhesive contacts between T cell and APC. This stabilisation step leads to an optimisation for subsequent proliferation and differentiation [69, 70]. ALCAM, or CD166, and CD318 are two know ligands for CD6, and are widely expressed by both immune cells, and epithelial, keratinocytes, fibroblasts, and mesenchymal cells. Soluble forms of CD6 isoforms are also expressed by T and B cells.

It is currently unclear whether CD6-dependent signalling results in a stimulatory or inhibitor effects on T cell activation as the effects were conflicting depending on the experimental system used. In the CD6 knockout mice, we have observed an autoimmune phenotype but there was lower proliferative responses of splenocytes with a higher induction of regulatory T cells that have a defective suppressive activity on T and B cell proliferation [68]. The mice showed a reduced Th17 differentiation.

CD6 targeting may be relevant for a number of autoimmune diseases. The recent discovery of a humanised anti-CD6 monoclonal antibody (itolizumab) has led to renewed interest to study this co-stimulatory pathway. The use of anti-CD6 antibody in animal models of SLE has showed promising results, published recently in abstract form. Preliminary results of evidence of upregulation of ALCAM on tubular cells and macrophages in patients with active lupus nephritis. Correlation between urinary ALCAM with disease activity measures in SLE patients added to the basis of continued exploration of anti-CD6 strategy for the treatment of SLE.

Conclusions

Our improved understanding of the immunopathogenesis of SLE has led us to several therapeutic discoveries that aim to modulate co-stimulatory signalling, in order to downregulate the interaction between antigen presenting cells and T cells in SLE. Abatacept is the first costimulatory blocking molecule that has now been available for the treatment of rheumatoid arthritis for more than a decade, and despite the lack of supportive evidence from randomised controlled studies in lupus, it has been used successfully in some refractory cases of SLE on an off-label basis. Many other emerging therapeutics are on the horizon. Specifically, the ability to downregulate both expansion of autoreactive T cells as well as T cell dependent humoral responses by several other costimulatory pathways, such as ICOS/ICOSL, CD40L/CD40, OX40L/OX40, PD-1/PD-1L and CD6/ALCAM, continue to provide strong scientific rationale to modulate these costimulatory pathways. The possibility of restoring tolerance by targeting these pathways will make it an ideal strategy in the treatment of autoimmune diseases.

References

1. Linsley PS, Brady W, Urnes M, Grosmaire LS, Damle NK, Ledbetter JA (1991) CTLA-4 is a second receptor for the B cell activation antigen B7. J Exp Med 174(3):561–569

2. Bluestone JA (1995) New perspectives of CD28-B7-mediated T cell costimulation. Immunity 2(6):555–559
3. Sayegh MH, Akalin E, Hancock WW, Russell ME, Carpenter CB, Linsley PS et al (1995) CD28-B7 blockade after alloantigenic challenge in vivo inhibits Th1 cytokines but spares Th2. J Exp Med 181 (5):1869–1874
4. Chu EB, Hobbs MV, Wilson CB, Romball CG, Linsley PS, Weigle WO (1996) Intervention of CD4+ cell subset shifts and autoimmunity in the BXSB mouse by murine CTLA4Ig. J Immunol 156 (3):1262–8
5. Rudd CE, Schneider H (2003) Unifying concepts in CD28, ICOS and CTLA4 co-receptor signalling. Nat Rev Immunol 3(7):544–556
6. Lyddane C, Gajewska BU, Santos E, King PD, Furtado GC, Sadelain M (2006) Cutting edge: CD28 controls dominant regulatory T cell activity during active immunization. J Immunol 176(6):3306–3310
7. Linsley PS, Wallace PM, Johnson J, Gibson MG, Greene JL, Ledbetter JA et al (1992) Immunosuppression in vivo by a soluble form of the CTLA-4 T cell activation molecule. Science (New York, NY) 257(5071):792–795
8. Finck BK, Linsley PS, Wofsy D (1994) Treatment of murine lupus with CTLA4Ig. Science (New York, NY) 265(5176):1225–1227
9. Mihara M, Tan I, Chuzhin Y, Reddy B, Budhai L, Holzer A et al (2000) CTLA4Ig inhibits T cell-dependent B-cell maturation in murine systemic lupus erythematosus. J Clin Investig 106(1):91–101
10. Daikh DI, Wofsy D (2001) Cutting edge: reversal of murine lupus nephritis with CTLA4Ig and cyclophosphamide. J Immunol 166(5):2913–6
11. Merrill JT, Burgos-Vargas R, Westhovens R, Chalmers A, D'Cruz D, Wallace DJ, et al (2010) The efficacy and safety of abatacept in patients with non-life-threatening manifestations of systemic lupus erythematosus: results of a twelve-month, multicenter, exploratory, phase IIb, randomized, double-blind, placebo-controlled trial. Arthritis Rheum 62(10):3077–87. https://doi.org/10.1002/art.27601. PMID: 20533545
12. Furie R, Nicholls K, Cheng T-T, Houssiau F, Burgos-Vargas R, Chen S-L et al (2014) Efficacy and safety of abatacept in lupus nephritis: a twelve-month, randomized, double-blind study. Arthritis Rheumatol 66(2):379–389
13. Group TAT (2014) Treatment of lupus nephritis with abatacept: the abatacept and cyclophosphamide combination efficacy and safety Sstudy. Arthritis Rheumatol 66(11):3096–3104
14. Isenberg DA, Rahman A, Allen E, Farewell V, Akil M, Bruce IN et al (2005) BILAG (2004). Development and initial validation of an updated version of the british isles lupus assessment group's disease activity index for patients with systemic lupus erythematosus. Rheumatology (Oxford) 44(7):902–6
15. Wofsy D, Hillson JL, Diamond B (2012) Abatacept for lupus nephritis: Alternative definitions of complete response support conflicting conclusions. Arthritis Rheum 64(11):3660–3665
16. Group AT (2014) Treatment of lupus nephritis with abatacept: the abatacept and cyclophosphamide combination efficacy and safety study. Arthritis Rheumatol 66(11):3096–3104
17. Kremer JM, Russell AS, Emery P, Abud-Mendoza C, Szechinski J, Westhovens R et al (2011) Long-term safety, efficacy and inhibition of radiographic progression with abatacept treatment in patients with rheumatoid arthritis and an inadequate response to methotrexate: 3-year results from the AIM trial. Ann Rheum Dis 70(10):1826–1830
18. Mease P, Genovese MC, Gladstein G, Kivitz AJ, Ritchlin C, Tak PP et al (2011) Abatacept in the treatment of patients with psoriatic arthritis: results of a six-month, multicenter, randomized, double-blind, placebo-controlled, phase II trial. Arthritis Rheum 63 (4):939–948
19. Danion F, Rosine N, Belkhir R, Gottenberg JE, Hachulla E, Chatelus E et al (2016) Efficacy of abatacept in systemic lupus erythematosus: a retrospective analysis of 11 patients with refractory disease. Lupus 25(13):1440–1447
20. McAdam AJ, Greenwald RJ, Levin MA, Chernova T, Malenkovich N, Ling V et al (2001) ICOS is critical for CD40-mediated antibody class switching. Nature 409(6816):102–105
21. Bossaller L, Burger J, Draeger R, Grimbacher B, Knoth R, Plebani A et al (2006) ICOS deficiency is associated with a severe reduction of CXCR5+CD4 germinal center Th cells. J Immunol 177(7):4927–4932
22. Hu YL, Metz DP, Chung J, Siu G, Zhang M (2009) B7RP-1 blockade ameliorates autoimmunity through regulation of follicular helper T cells. J Immunol 182 (3):1421–1428
23. Jogdand GM, Mohanty S, Devadas S (2016) Regulators of Tfh Cell Differentiation. Front Immunol 7:520
24. Wikenheiser DJ, Stumhofer JS (2016) ICOS co-stimulation: friend or foe? Front Immunol 7:304
25. Nurieva RI, Mai XM, Forbush K, Bevan MJ, Dong C (2003) B7h is required for T cell activation, differentiation, and effector function. Proc Natl Acad Sci U S A 100(24):14163–14168
26. Khayyamian S, Hutloff A, Büchner K, Gräfe M, Henn V, Kroczek RA et al (2002) ICOS-ligand, expressed on human endothelial cells, costimulates Th1 and Th2 cytokine secretion by memory CD4+ T cells. Proc Natl Acad Sci USA 99(9):6198–6203
27. Teichmann LL, Cullen JL, Kashgarian M, Dong C, Craft J, Shlomchik MJ (2015) Local triggering of the ICOS coreceptor by CD11c(+) myeloid cells drives organ inflammation in lupus. Immunity 42(3):552–565
28. Higuchi T, Aiba Y, Nomura T, Matsuda J, Mochida K, Suzuki M et al (2002) Cutting edge: ectopic expression

of CD40 ligand on B cells induces lupus-like autoimmune disease. J Immunol 168(1):9–12
29. Good-Jacobson KL, Szumilas CG, Chen L, Sharpe AH, Tomayko MM, Shlomchik MJ (2010) PD-1 regulates germinal center B cell survival and the formation and affinity of long-lived plasma cells. Nat Immunol 11(6):535–542
30. Liu D, Xu H, Shih C, Wan Z, Ma X, Ma W et al (2015) T-B-cell entanglement and ICOSL-driven feed-forward regulation of germinal centre reaction. Nature 517(7533):214–218
31. Xin L, Jiang TT, Chaturvedi V, Kinder JM, Ertelt JM, Rowe JH et al (2014) Commensal microbes drive intestinal inflammation by IL-17-producing CD4+ T cells through ICOSL and OX40L costimulation in the absence of B7–1 and B7–2. Proc Natl Acad Sci USA 111(29):10672–10677
32. Zhang M, Lee F, Knize A, Jacobsen F, Yu S, Ishida K et al (2019) Development of an ICOSL and BAFF bispecific inhibitor AMG 570 for systemic lupus erythematosus treatment. Clin Exp Rheumatol 37(6):906–914
33. Cheng LE, Amoura Z, Cheah B, Hiepe F, Sullivan BA, Zhou L et al (2018) Brief report: a randomized, double-blind, parallel-group, placebo-controlled, multiple-dose study to evaluate AMG 557 in patients with systemic lupus erythematosus and active lupus arthritis. Arthritis Rheumatol 70(7):1071–1076
34. Banchereau J, Dubois B, Fayette J, Burdin N, Brière F, Miossec P et al (1995) Functional CD40 antigen on B cells, dendritic cells and fibroblasts. Adv Exp Med Biol 378:79–83
35. Bourgeois C, Rocha B, Tanchot C (2002) A role for CD40 expression on CD8+ T cells in the generation of CD8+ T cell memory. Science 297(5589):2060–2063
36. van Kooten C (2000) Immune regulation by CD40-CD40-l interactions—2; Y2K update. Front Biosci 5: D880–D693
37. Hollenbaugh D, Ochs HD, Noelle RJ, Ledbetter JA, Aruffo A (1994) The role of CD40 and its ligand in the regulation of the immune response. Immunol Rev 138:23–37
38. Foy TM, Laman JD, Ledbetter JA, Aruffo A, Claassen E, Noelle RJ (1994) gp39-CD40 interactions are essential for germinal center formation and the development of B cell memory. J Exp Med 180 (1):157–163
39. Kawabe T, Naka T, Yoshida K, Tanaka T, Fujiwara H, Suematsu S et al (1994) The immune responses in CD40-deficient mice: impaired immunoglobulin class switching and germinal center formation. Immunity 1(3):167–178
40. Quezada SA, Jarvinen LZ, Lind EF, Noelle RJ (2004) CD40/CD154 interactions at the interface of tolerance and immunity. Annu Rev Immunol 22:307–328
41. Belkhir R, Gestermann N, Koutero M, Seror R, Tost J, Mariette X et al (2014) Upregulation of membrane-bound CD40L on CD4+ T cells in women with primary Sjögren's syndrome. Scand J Immunol 79(1):37–42
42. Michel NA, Zirlik A, Wolf D (2017) CD40L and Its Receptors in Atherothrombosis-An Update. Front Cardiovasc Med 4:40
43. Senchenkova EY, Russell J, Vital SA, Yildirim A, Orr AW, Granger DN et al (2018) A critical role for both CD40 and VLA5 in angiotensin II-mediated thrombosis and inflammation. FASEB J 32(6):3448–3456
44. Bishop GA, Xie P (2007) Multiple roles of TRAF3 signaling in lymphocyte function. Immunol Res 39 (1–3):22–32
45. Wang X, Huang W, Schiffer LE, Mihara M, Akkerman A, Hiromatsu K et al (2003) Effects of anti-CD154 treatment on B cells in murine systemic lupus erythematosus. Arthritis Rheum 48(2):495–506
46. Boumpas DT, Furie R, Manzi S, Illei GG, Wallace DJ, Balow JE et al (2003) A short course of BG9588 (anti-CD40 ligand antibody) improves serologic activity and decreases hematuria in patients with proliferative lupus glomerulonephritis. Arthritis Rheum 48(3):719–727
47. Robles-Carrillo L, Meyer T, Hatfield M, Desai H, Dávila M, Langer F et al (2010) Anti-CD40L immune complexes potently activate platelets in vitro and cause thrombosis in FCGR2A transgenic mice. J Immunol 185(3):1577–1583
48. Schuler W, Bigaud M, Brinkmann V, Di Padova F, Geisse S, Gram H et al (2004) Efficacy and safety of ABI793, a novel human anti-human CD154 monoclonal antibody, in cynomolgus monkey renal allotransplantation. Transplantation 77(5):717–726
49. Visvanathan S, Daniluk S, Ptaszyński R, Müller-Ladner U, Ramanujam M, Rosenstock B et al (2019) Effects of BI 655064, an antagonistic anti-CD40 antibody, on clinical and biomarker variables in patients with active rheumatoid arthritis: a randomised, double-blind, placebo-controlled, phase IIa study. Ann Rheum Dis 78(6):754–760
50. Karnell JL, Albulescu M, Drabic S, Wang L, Moate R, Baca M et al (2019) A CD40L-targeting protein reduces autoantibodies and improves disease activity in patients with autoimmunity. Sci Transl Med. 11(489):eaar6584. https://doi.org/10.1126/scitranslmed.aar6584. PMID: 31019027
51. Cunninghame Graham DS, Graham RR, Manku H, Wong AK, Whittaker JC, Gaffney PM et al (2008) Polymorphism at the TNF superfamily gene TNFSF4 confers susceptibility to systemic lupus erythematosus. Nat Genet 40(1):83–89
52. Manku H, Graham DS, Vyse TJ (2009) Association of the co-stimulator OX40L with systemic lupus erythematosus. J Mol Med (Berl) 87(3):229–234
53. Harley JB, Alarcón-Riquelme ME, Criswell LA, Jacob CO, Kimberly RP, Moser KL et al (2008) Genome-wide association scan in women with systemic lupus erythematosus identifies susceptibility variants in ITGAM, PXK, KIAA1542 and other loci. Nat Genet 40(2):204–210

54. Webb GJ, Hirschfield GM, Lane PJ (2016) OX40, OX40L and Autoimmunity: a Comprehensive Review. Clin Rev Allergy Immunol 50(3):312–332
55. Kurche JS, Haluszczak C, McWilliams JA, Sanchez PJ, Kedl RM (2012) Type I IFN-dependent T cell activation is mediated by IFN-dependent dendritic cell OX40 ligand expression and is independent of T cell IFNR expression. J Immunol 188(2):585–593
56. Takeda I, Ine S, Killeen N, Ndhlovu LC, Murata K, Satomi S et al (2004) Distinct roles for the OX40-OX40 ligand interaction in regulatory and nonregulatory T cells. J Immunol 172(6):3580–3589
57. So T, Croft M (2007) Cutting edge: OX40 inhibits TGF-beta- and antigen-driven conversion of naive CD4 T cells into CD25+Foxp3+ T cells. J Immunol 179(3):1427–1430
58. Gavin MA, Clarke SR, Negrou E, Gallegos A, Rudensky A (2002) Homeostasis and anergy of CD4 (+)CD25(+) suppressor T cells in vivo. Nat Immunol 3(1):33–41
59. Murata K, Nose M, Ndhlovu LC, Sato T, Sugamura K, Ishii N (2002) Constitutive OX40/OX40 ligand interaction induces autoimmune-like diseases. J Immunol 169(8):4628–4636
60. Sitrin J, Suto E, Wuster A, Eastham-Anderson J, Kim JM, Austin CD et al (2017) The Ox40/Ox40 ligand pathway promotes pathogenic Th cell responses, plasmablast accumulation, and lupus nephritis in NZB/W F1 Mice. J Immunol 199 (4):1238–1249
61. Patschan S, Dolff S, Kribben A, Dürig J, Patschan D, Wilde B et al (2006) CD134 expression on CD4+ T cells is associated with nephritis and disease activity in patients with systemic lupus erythematosus. Clin Exp Immunol 145(2):235–242
62. Dolff S, Quandt D, Wilde B, Feldkamp T, Hua F, Cai X et al (2010) Increased expression of costimulatory markers CD134 and CD80 on interleukin-17 producing T cells in patients with systemic lupus erythematosus. Arthritis Res Ther 12(4):R150
63. Jacquemin C, Schmitt N, Contin-Bordes C, Liu Y, Narayanan P, Seneschal J et al (2015) OX40 ligand contributes to human lupus pathogenesis by promoting T follicular helper response. Immunity 42 (6):1159–1170
64. Cortini A, Ellinghaus U, Malik TH, Cunninghame Graham DS, Botto M, Vyse TJ (2017) B cell OX40L supports T follicular helper cell development and contributes to SLE pathogenesis. Ann Rheum Dis 76 (12):2095–2103
65. Elhai M, Avouac J, Hoffmann-Vold AM, Ruzehaji N, Amiar O, Ruiz B et al (2016) OX40L blockade protects against inflammation-driven fibrosis. Proc Natl Acad Sci USA 113(27):E3901–E3910
66. Keir ME, Butte MJ, Freeman GJ, Sharpe AH (2008) PD-1 and its ligands in tolerance and immunity. Annu Rev Immunol 26:677–704
67. Sakurai Y, Usui Y, Hattori T, Takeuchi M, Takayama K, Karasawa Y et al (2021) Programmed cell death-1 (PD-1) pathway deficiency enhances autoimmunity leading to dacryoadenitis of mice. Am J Pathol 191(6):1077–1093. https://doi.org/10.1016/j.ajpath.2021.02.014. Epub 2021 Mar 8. PMID: 33705751
68. Consuegra-Fernández M, Martínez-Florensa M, Aranda F, de Salort J, Armiger-Borràs N, Lozano T et al (2017) Relevance of CD6-mediated interactions in the regulation of peripheral T-cell responses and tolerance. Front Immunol 8:594
69. Gimferrer I, Calvo M, Mittelbrunn M, Farnós M, Sarrias MR, Enrich C et al (2004) Relevance of CD6-mediated interactions in T cell activation and proliferation. J Immunol 173(4):2262–2270
70. Zimmerman AW, Joosten B, Torensma R, Parnes JR, van Leeuwen FN, Figdor CG (2006) Long-term engagement of CD6 and ALCAM is essential for T-cell proliferation induced by dendritic cells. Blood 107(8):3212–3220

Cytokines: Their Role in Amplifying SLE Pathogenesis

7

Bonnia Liu and Alberta Hoi

Abstract

Cytokines are small soluble mediators released by immune cells and tissue that play a pivotal part in the pathogenesis of SLE. They are proteins that are important in cell signalling, and able to modulate immune responses rapidly. We review the pleiotropic actions of cytokines and their network in health and disease state such as SLE. Cytokine dysregulation contributes to each phase of disease development in SLE, from failure of restoration of self-tolerance, aberrant production of pathogenic autoantibodies, perpetuation of systemic or end organ inflammation, or promotion of local tissue damage. We present the key cytokines that have emerged as important players in our current understanding of lupus pathogenesis and summarise the translational research to date that could bring us closer to potential therapeutics in SLE.

Introduction

Systemic lupus erythematosus (SLE), a systemic autoimmune disease of considerable heterogeneity, is characterised by the dysfunction of both the innate and adaptive immune systems. Cytokines are small soluble mediators released by many immune cell subsets and tissue and play pivotal roles in modulating both the innate and adaptive immune responses. The intricate cascade of cytokine in molecular signalling highlights the constant requirement for regulation and balance to maintain immune homeostasis. Both deficient and excessive of production of cytokines, in combination with abnormal responsiveness of immune cells, can favour the development of immune-mediated diseases.

Cytokines play an active part to SLE pathogenesis as the key soluble factors that serve to promote and amplify an immune response, after a loss of tolerance is implicated. The cytokine network can act on multiple cell types and target organs with considerable redundancy. The literature on the roles that cytokines play in SLE pathogenesis has evolved. Some earlier studies have shown some conflicting results, but this may be explained by knowledge of the different

B. Liu · A. Hoi (✉)
Department of Rheumatology, Austin Health, Heidelberg, Victoria 3084, Australia
e-mail: alberta.hoi@monash.edu

B. Liu
e-mail: bonnia.liu@austin.org.au

A. Hoi
Rheumatology Research Group, Centre for Inflammatory Diseases, School of Clinical Sciences at Monash Health, Monash University, Clayton, Victoria 3168, Australia

Department of Rheumatology, Monash Health, Clayton, Victoria 3168, Australia

A. Hoi (ed.), *Pathogenesis of Systemic Lupus Erythematosus*,
https://doi.org/10.1007/978-3-030-85161-3_7

phases of autoimmunity and disease processes in murine and human models [1–5].

The pleiotropic actions of cytokines and the interplay of different cytokines can result in cytokine dysregulation that manifest differently whether it was a failure of restoration of self-tolerance, aberrant production of pathogenic autoantibodies, perpetuation of systemic or end organ inflammation, or promotion of local tissue damage [6–8]. While numerous cytokines have been implicated for pathological processes in SLE, a useful paradigm to understand the pleiotropic nature of specific cytokines is by functional grouping of T helper 1 (Th1) and T helper 2 (Th2) cytokines, based on initial observations made on murine T helper cells, but it is now recognised that these cytokines can be produced by a variety of other cell types [9].

Autoimmune phenotype has traditionally been considered to be an imbalance between Th1 and Th2 cytokines, but this dichotomy is now replaced by the discovery of a new Th subsets, called the Th17 cytokines and T helper regulatory (Treg) cytokines. Cytokines produced via these pathways may act on different stages of disease development in SLE. The classification into Th1, Th2, Th17 and Treg cytokines could be an over-simplification of the complex cytokine loops, but it is a useful paradigm to simplify our understanding based on their ability to enhance cellular immune responses in particular directions (Fig. 7.1). There have been emerging data on particular cytokines such as IL-9 and IL-22 that are produced by a number of different T helper cell subtypes and may play unique roles in disease pathogenesis.

Other cytokines are produced by B cells, such as B-lymphocyte stimulator (BlyS), that is well studied in its ability to promote differentiation and survival of B cells. The discussion of BlyS in SLE will be included in a different chapter.

Insights gained regarding the characteristics of cytokine dysregulation in SLE have been pivotal to understanding disease aetiology and the development of effective therapeutics. The role of cytokines in the loss of tolerance, development of autoimmunity, and subsequently the propagation of inflammation and end-organ damage has been the subject of intense investigation in the last decades [10]. This review will summarise our current understanding of the role of cytokines played in the pathogenesis of both murine and human models of systemic lupus erythematosus. Specific chapters will be dedicated to the studies of BlyS and interferons particularly Type I interferons. We will also review the current data in relation to any therapeutic agent has been developed for the particular cytokine.

Upon certain stimulating conditions, naïve CD4 + T cells differentiate into different subtypes of T helper cells known as Th1, Th2, Th17 and Treg. Activation of transcription factors ultimately regulates the differentiation of Th lineages which result in the production of signature cytokines. IL: interleukins; Th: T-helper; Treg: regulatory T cells; T-bet: transcription factor T-bet; IRF: interferon regulatory factor; STAT: signal transducer and activator of transcription; GATA: transcription factor GATA; IFN: interferon; ROR: RAR-related orphan receptors; Foxp3: forkhead box P3; CCL20: chemokine (C–C motif) ligand 20; TGF-β: transforming growth factor beta.

Type 1 T Helper Cell Cytokines

Th1-type cytokines tend to produce the proinflammatory responses responsible for host defence against intracellular pathogens, induction of delayed-type hypersensitivity responses, and support cell-mediated functions such as macrophage activation (Fig. 7.2) [11, 12]. The signature cytokines include interferon (IFN)-γ, interleukin (IL)-2, and tumour necrosis factor-α (TNF-α) [13]. The key transcription factors involved in the response of Th1 pathway included T-bet, IRF-1 and the STAT4-signalling cascade [14].

Th1 differentiation is initiated by the recognition of peptide antigen-MHC class II complex presented by APC. A number of cytokines such as IL-12 and IFNγ are produced to induce this differentiation. Once differentiated, its signature cytokines such as IFNγ and TNF can promote a range of downstream effects such as increased

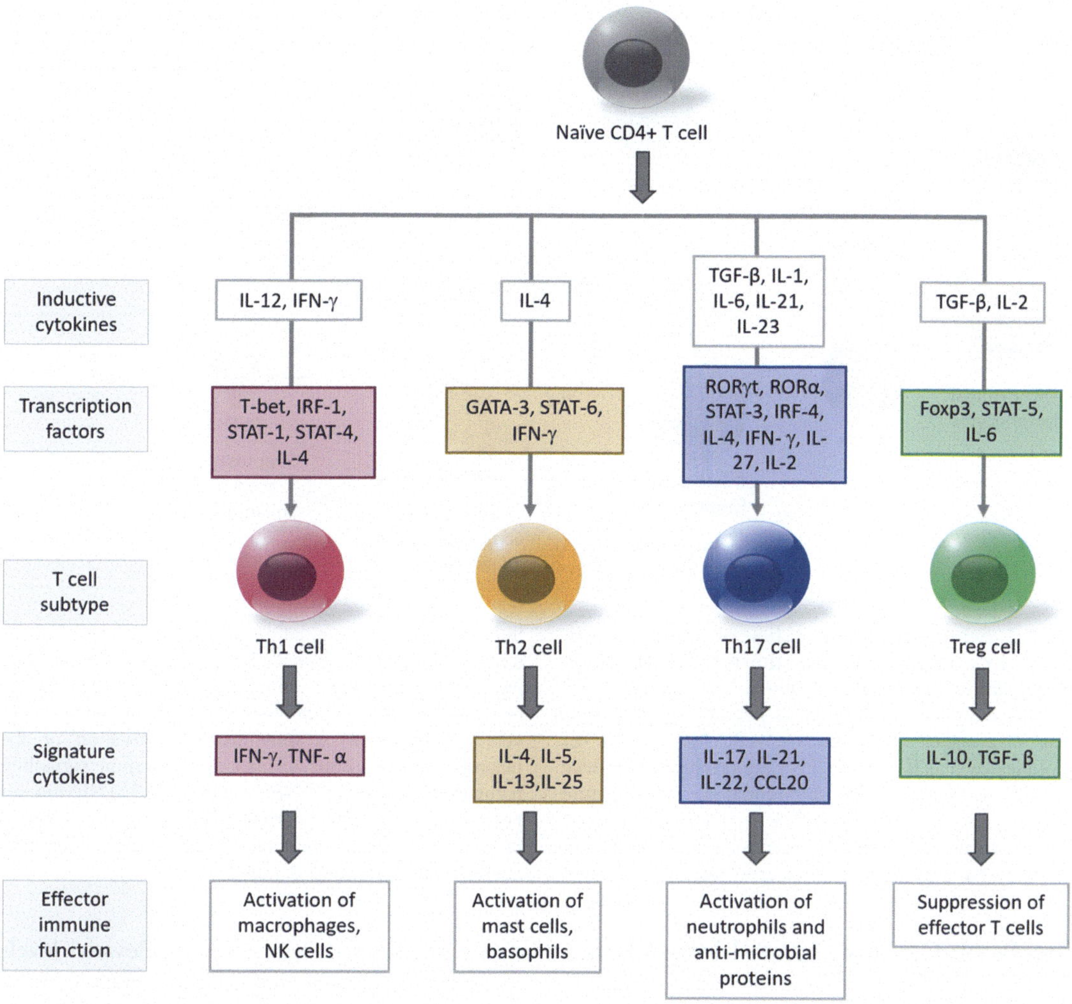

Fig. 7.1 **Differentiation of naïve CD4 + T cells.** Upon certain stimulating conditions, naïve CD4 + T cells differentiate into different subtypes of T helper cells known as Th1, Th2, Th17 and Treg. Activation of transcription factors ultimately regulates the differentiation of Th lineages which result in the production of signature cytokines. IL: interleukins; Th: T-helper; Treg: regulatory T cells; T-bet: transcription factor T-bet; IRF: interferon regulatory factor; STAT: signal transducer and activator of transcription; GATA: transcription factor GATA; IFN: interferon; ROR: RAR-related orphan receptors; Foxp3: forkhead box P3; CCL20: chemokine (C-C motif) ligand 20; TGF-β: transforming growth factor beta

cytotoxic T cell activity and activation of macrophages at site of inflammation. The regulation of TNFα mediated inflammation by TNFR1 and TNFR2, expressed by macrophages, is important to control the extent of inflammatory response by a balance between immunogenic death versus persistence of inflammatory activation of the macrophages via both distinct and common signalling pathways mediated by TNFR1 and TNFR2. TNF: tumour necrosis factor; IFN: interferon; FADD: Fas associated protein with death domain; TNFR1 and 2: Tumour necrosis factor receptor 1 and 2.

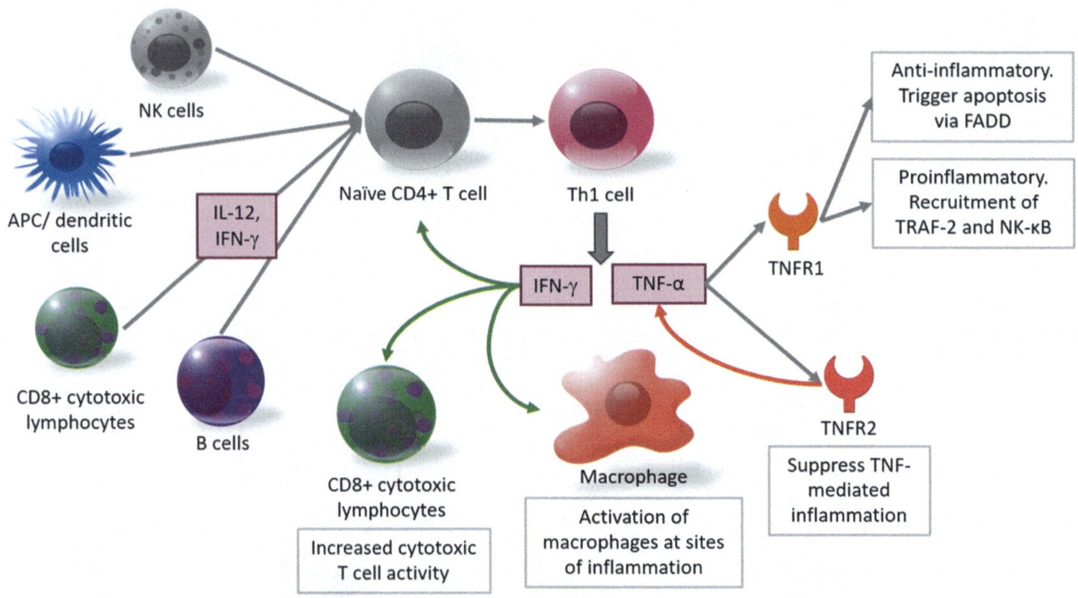

Fig. 7.2 Schematic representation of Th1 differentiation, Th1 cytokines and their downstream effects on the regulation of immune response. Th1 differentiation is initiated by the recognition of peptide antigen-MHC class II complex presented by APC. A number of cytokines such as IL-12 and IFNγ are produced to induce this differentiation. Once differentiated, its signature cytokines such as IFNγ and TNF can promote a range of downstream effects such as increased cytotoxic T cell activity and activation of macrophages at site of inflammation. The regulation of TNFα mediated inflammation by TNFR1 and TNFR2, expressed by macrophages, is important to control the extent of inflammatory response by a balance between immunogenic death versus persistence of inflammatory activation of the macrophages via both distinct and common signalling pathways mediated by TNFR1 and TNFR2. TNF: tumour necrosis factor; IFN: interferon; FADD: Fas associated protein with death domain; TNFR1 and 2: Tumour necrosis factor receptor 1 and 2

Interferon-Gamma (Type II interferon)

Interferon-gamma (IFN-γ) is the sole Type II interferon. It is structurally unrelated to type I IFNs, binds to a different receptor, and is encoded by a separate chromosomal locus. IFN-γ can be produced by antigen-presenting cells (APCs), NK cells, Th1 lymphocytes, CD8 + cytotoxic lymphocytes and B-cells in response to IL-12 and IL-18 [15, 16]. Interferon-gamma (IFN-γ) activates macrophages at the site of inflammation, contributes to cytotoxic T-cell activity, and is strongly associated with Th1 responses. It induces differentiation of naïve T cells into Th1 cells and triggers Th1 differentiation in an autocrine manner [17].

Gene polymorphisms of IFN-γ have been associated with susceptibility to SLE in patients [18, 19]. IFN-γ appears to be involved in the early stages of SLE pathogenesis with elevated levels appearing in serum before the appearance of autoantibodies and SLE diagnosis [20]. Transgenic mice models with overexpression of IFN-γ developed lupus nephritis and the production of antinuclear autoantibodies [21]. Although there are some conflicting human studies regarding the correlation of serum IFN-γ levels and development of lupus nephritis [22, 23].

Animal models have revealed that the chronological variation in serum levels may reflect the complex role of this cytokine in the regulation of systemic autoimmunity. Treatment of SLE-prone IFN-γ-deficient MRL/*lpr* mice with recombinant IFN-γ appears to be protective against inflammatory infiltration of the kidneys and the lung if administered to younger animals, but treatment later in life accelerates disease manifestations [24]. In addition, high levels of

IFN-γ were found in lymphoid organs and in disease-affected tissues of mice models, particularly at the late stages of the disease [25]. Patients with SLE were found to have increased levels of IFN-γ which correlated with disease activity [26].

As a therapeutic target, two phase one clinical trials investigating the efficacy and safety of anti-IFN-γ antibody, AMG811, have demonstrable some IFN-g associated biomarkers change with the agent, but in the small Phase 1b studies in patients with fairly quiescent non-renal SLE, discoid lupus and lupus nephritis, there was no clinical effect observed [27–29].

Tumour Necrosis Factor-α

Tumour Necrosis Factor-α (TNF-α) is a cytokine belonging to the group of TNF family and is produced by a variety of cells, including activated macrophages and monocytes as well as B cells, T cells, keratinocytes, renal cells, and fibroblasts [30]. It is expressed in two forms, a membrane-bound trimer and as a secreted soluble form [31]. TNF-α has both immunoregulatory and proinflammatory functions. The mains roles of TNF-α includes the activation of macrophages and dendritic cells, promoting lymphoid germinal centres development, and the activation and proliferation of B and T lymphocytes [32]. The differential bioactivities are achieved through binding to main two distinct receptors, TNF receptor 1 (TNFR1) and TNFR2. Activation of TNFR1 result in dual function. Binding may trigger apoptotic and anti-inflammatory signals through Fas-associated death domain (FADD) and activation of the caspase cascade, while alternatively exerting a proinflammatory response via the recruitment of TNF receptor-associated factor 2 (TRAF-2) and activation of NF-κB [33, 34]. TNFR2 activation appears to autoregulate and suppress TNF-mediated inflammatory response [35].

Several studies point to a pathogenic role for TNF-α in murine SLE. However, studies often highlight the conflicting effects of TNF-α in different murine SLE models which further underscore this cytokine's dual proinflammatory and immunoregulatory roles [36–38]. In human studies, similar conflicting findings have been demonstrated. Some studies found a correlation between elevated TNF-α and SLE disease activity, while others show that TNF-α levels were higher in patients with inactive disease, suggesting a protective role of TNF-α [39, 40].

Anti-TNF therapies have been investigated extensively in other areas of rheumatology and have been shown to be highly effective in conditions such as rheumatoid arthritis, psoriatic arthritis and ankylosing spondylitis [37, 41, 42]. It has been used in a few small open labelled studies in SLE which showed some improvement in clinical activity in patients who were moderately active with inflammatory arthritis and renal disease [32, 43]. However, in these small open labelled short-term studies, there were suggestions that infliximab, the anti-TNF therapy used in these studies, was associated with increased serological activity with higher level of autoantibodies detected [32, 43]. A report of the follow-up study of these patients, a total of 13, who have received a course of infliximab treatment following by azathioprine maintenance immunosuppression, suggested that despite a mix of clinical manifestations and severity, there were some long-term benefit up to 5 years, in certain clinical manifestations [44].

There was another slightly larger study with a controlled arm published in 2009 which explored the efficacy of anti-TNF therapy in SLE [45]. Twenty-seven SLE patients received either infliximab at 3 mg/kg (n = 9) for 24 weeks in a protocol similar to that in rheumatoid arthritis or were allocated to the controlled arm (n = 18). There were some clinical improvements in terms of glucocorticoid dose, disease activity measurement by SLEDAI change at 6 months, and physical global assessment scores at 6 months [45].

Another open labelled study of 43 SLE patients with predominantly articular and serositis were treated with etanercept in addition to standard immunosuppression. Outcomes were measured in terms of clinical remission in terms of symptoms or SLEDAI score of less than 4.

The study was published in an abstract form, indicating a rate of clinical improvement but some required switching to adalimumab. Some had a rise in their serology but were not associated with clinically significant [46].

Despite these observations, there has not been much interest in developing anti-TNF therapy in SLE. In part, this was due to various case reports of TNF-induced lupus observed in patients with rheumatoid arthritis or Crohn's disease who have developed SLE-like autoimmune disease and resulted in the production of autoantibodies after treatment with TNF blockade [47–52]. The randomised controlled phase 2 study of etanercept in lupus nephritis was proposed but was terminated due to these concerns [53].

In reality, based on experience of these small studies, anti-TNF blockade has been used anecdotally by some clinicians [54]. The basis that Th1 cytokines such as TNF-α and IFN-γ are important mediators of inflammation in SLE is robust, but the careful balance of other aspect of immune responses that may be tipped by anti-TNF therapy. Unfortunately, given the overall role of TNF-α in human SLE is controversial, this probably means that the likelihood of a properly conducted randomised controlled clinical trial examining the utility of anti-TNF therapy in SLE may not happen in the near future.

Type 2 T Helper Cell Cytokines

In contrast to Th1 cytokines which promote inflammation and cellular immune response, Th2 cytokines (IL-4, IL-5, IL-6, IL-10, IL-13) are responsible for the development of B-cell mediated humoral immunity, IgE production, activation of eosinophils and deactivation of monocytes [55]. Differentiation of naïve T cells into Th2 cells is driven by the activation of the T-cell receptor and IL-4 receptor via the STAT6 signally cascade and selective expression of the c-maf-proto-oncogene (Fig. 7.3) [56–59]. Th1 and Th2 differentiation and cytokine production counter-regulate the expansion and function of the other subtype through reciprocal activity of IL-4 and IFN-γ [60].

In SLE the understanding of the role of Th2 cytokines has evolved. Many of the conventional Th2 cytokines such as IL-6 and IL-10 are now regarded as key inductive and signature cytokines of the Th17 and T regulator cell pathways, we will therefore discuss them in their respective pathways. Cytokines play a critical role in the promotion of B cell mediated humoral immunity, but it is far more complex of the conventional paradigm of the list of Th2 cytokines.

Th2 differentiation mediates the humoral immune response, which is promoted by a range of Th2 cytokines such as IL-4, IL-5, IL-23 and

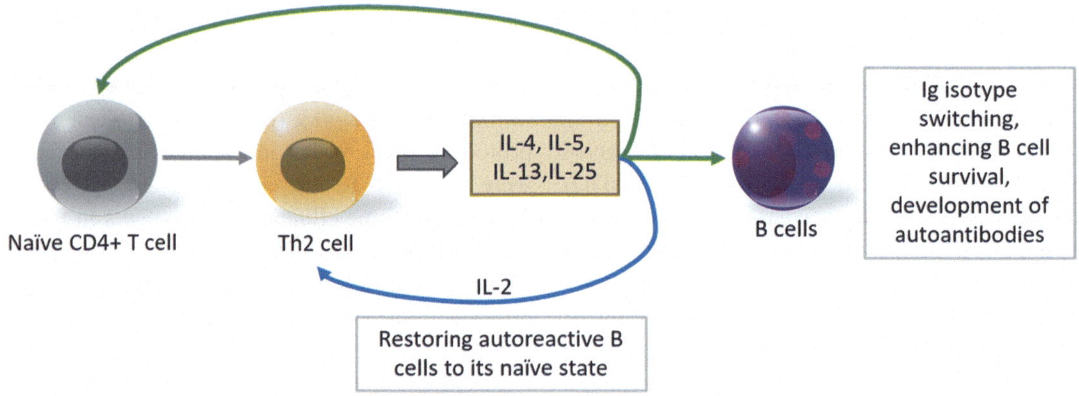

Fig. 7.3 Schematic representation of Th2 differentiation and the role of IL-4 in promoting B cell proliferation and isotype switching. Th2 differentiation mediates the humoral immune response, which is promoted by a range of Th2 cytokines such as IL-4, IL-5, IL-23 and IL-25. The differentiated Th2 cells can in turn activate B cells that recognise the same antigen. This so-called linked recognition of antigen constitutes a range of T cell dependent antibody responses, such as maintenance of B cell survival, antibody production and immunoglobulin isotype switching

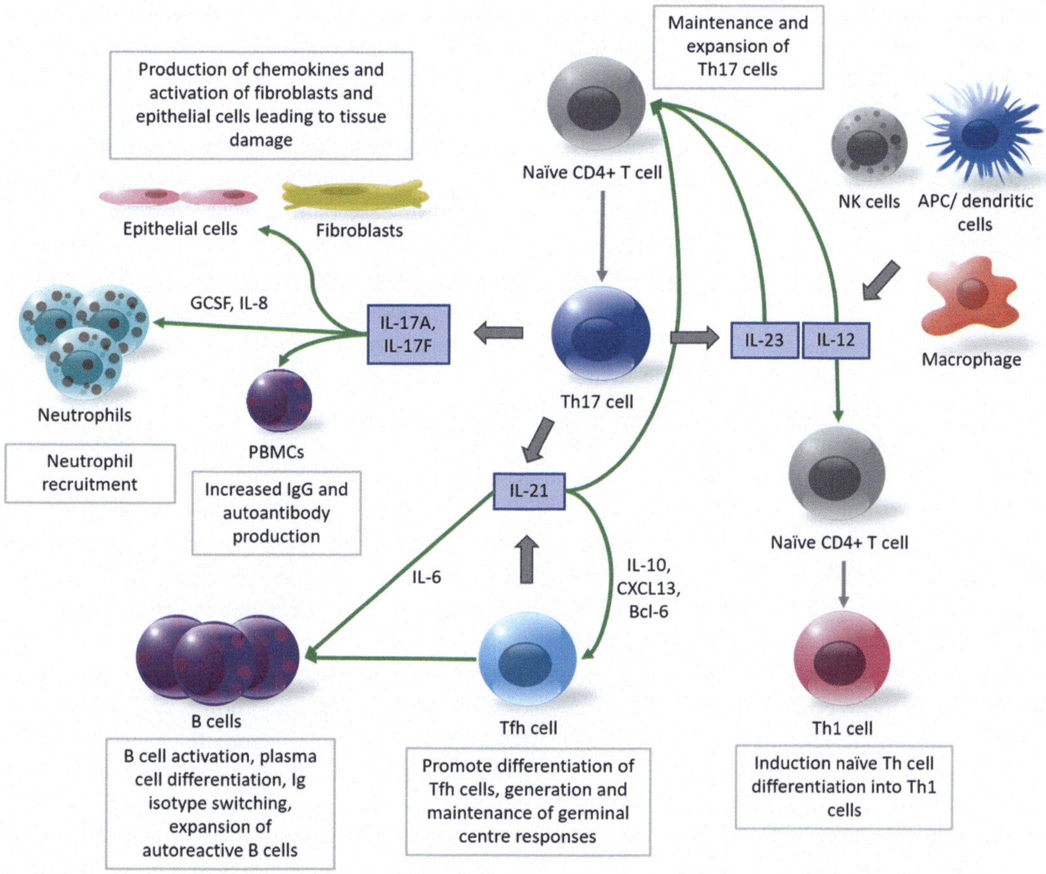

Fig. 7.4 Th17 and Th17 cytokines interactions on the innate and adaptive immune system. Th17 T cells are distinct lineage of effector T cells, that are characterised by their production of IL-17, and other cytokines such as IL-21 and IL-23. Its actions on epithelial cells and fibroblasts, as well as neutrophils account for its role in the protective response to pathogens, but is also important in mediating disease activity and associated tissue damage in autoimmune diseases. A robust inflammatory response is also mediated by B cells and Tfh cells. PMBC: peripheral blood mononuclear cells; GCSF: granulocyte colony-stimulating factor; Tfh: T follicular helper cells

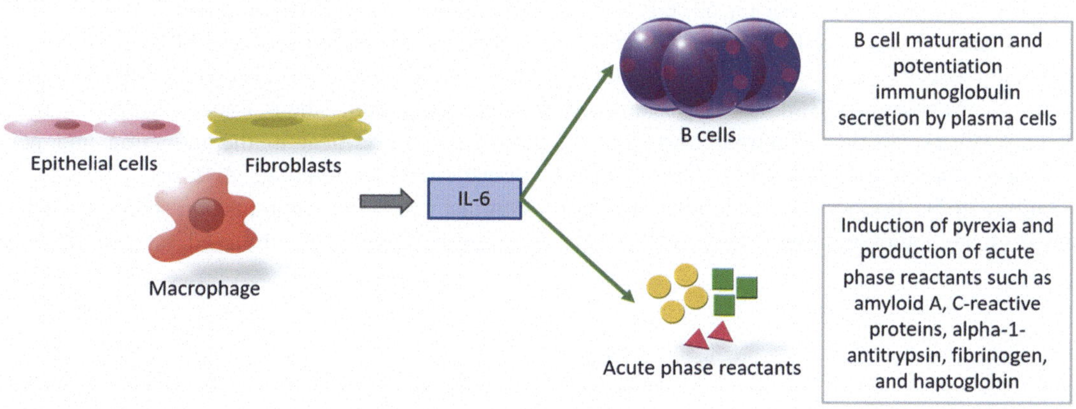

Fig. 7.5 The role of IL-6. IL-6 has pleotropic effects on a variety of cell types and plays a key role in the regulation of inflammation. Among many of its effects, its ability to induce B cell maturation and acute phase response are most relevant in the SLE pathogenesis

IL-25. The differentiated Th2 cells can in turn activate B cells that recognise the same antigen. This so-called linked recognition of antigen constitutes a range of T cell dependent antibody responses, such as maintenance of B cell survival, antibody production and immunoglobulin isotype switching.

Interleukin-4

Interleukin-4 is the prototypal cytokine which promotes the differentiation Th2 pathway and plays a powerful role in suppressing the Th1 response [59]. IL-4 acts via the signalling pathway through tyrosine phosphorylation of the Signal Transducer and Activator of Transcription 6 (STAT6), and GATA binding protein-3 was the key transcription protein, essential for Th2 differentiation, [56]. It has a number of different roles relevant to the development of lupus, some of which may appear conflicting and therefore the development of IL-4 as a potential target for modulation has been challenging [61].

As the key cytokine that promotes immunoglobulin (Ig) isotype switching, IL-4 is known to drive switching to a predominantly IgG1 subclass, which is consider as a more Th2 phenotype (Fig. 7.3). Typically, IgG2a and IgG3 subclasses are more pathogenic in mice and are promoted by Th1 cytokines such as IFN-γ [3, 62, 63]. IL-4 overexpression appears to increase IgG1 while decreasing IgG2a and IgG3 expression in a murine lupus nephritis model, although this effect can be inconsistent [3, 62, 64].

IL-4 has a direct B cell stimulatory effect by enhancing B cell survival. In transgenic murine model over-expressing IL-4, there was a polyclonal B cell expansion and activity observed and developed autoimmune lupus-like disease manifestations [65]. Furthermore, administration of anti-IL-4 antibodies inhibited the production of autoantibodies and prevented the onset of lupus nephritis in a few lupus-prone mouse models [59, 66]. IL-4 promotes autoimmunity and production of autoantibodies through inhibition of apoptosis of autoreactive B cells and thus contribute to the development or aggravation of autoantibody-mediated disease pathogenesis [67].

The conflicting roles of IL-4 is also seen with its direct tissue effects promoted by IL-4. In two separate models where IL-4 expression was ameliorated by an anti-IL-4 approach or in the STAT6 knockout model, end organ inflammation in the kidney appears to be reduced, whereas in an IL-4 transgenic mice developed glomerulosclerosis [59, 68].

The pursuit of further understanding of the complex role of IL-4 continues, particularly around its T cell suppressor effect. IL-4 is a key cytokine that assist T regulatory cells to modulate their immune response and to take on a key tolerogenic role. In the IL-4 knockout mice, Treg mediated immune suppression was compromised [69]. In SLE patients, serum IL-4 also appeared to be lower than controls but IL-4 level did not appear to correlate with disease activity [70, 71]. There was also a downregulation of IL-4 receptor in SLE patients. Emerging data suggest that IL-4 may play a synergistic role with low dose IL-2 in restoring autoreactive B cells to its naïve state [72].

Type 17 Helper Cell Response

In addition to the classical paradigm of Th1/Th2 balance in the pathogenesis of SLE, the subset of Type 17 T helper (Th17) cells and the cytokines association with their expansion have been shown to be critical in the induction and perpetuation of autoimmune diseases. [73, 74]. In contrast to Th1 immune responses, Th17 responses have been observed in host defence against extracellular pathogens. The signature cytokine produced by the Th17 cells is interleukin-17 (IL-17) but other cytokines such as IL-21 and IL-22 are also detected in high concentration in Th17 cells (Fig. 7.4) [75]. The key transcription factors are Retinoic acid-related orphan nuclear receptors (ROR)α and ROR-γt, which have overlapping role in amplifying the Th17 responses [76]. Th17 differentiation is also dependent on the STAT3 intracellular transcription factor [77].

Th17 cells develop from naïve T helper cells under the influence of inflammatory cytokines such as IL-6, which superimposed on the presence of transforming growth factor-β (TGF-β), to promote Th17 differentiation. Other cytokines such as IL-21 and IL-23 further amplify the process [78]. The differentiation into Th17 lineage cells is amenable to *in vitro* manipulation with, anti-IFN-γ, anti-IL-4, or both which can steer differentiation away from the Th1 or Th2 profiles [74]. Th17 differentiation requires presence of TGF-β, which also has links to regulatory T cell development and function [79, 80]. Th17 cytokines can also derived from certain CD8 + cytotoxic T cells as well as some innate lymphoid cells, which are tissue resident cells that contribute to the Th17 responses [81–83].

Th17 cells and Th17 cytokines play an important role in a wide range of immune responses including recruitment of neutrophils and macrophages to affected tissue, tissue remodelling and repair, and production of antimicrobial proteins [84]. Deficiency in Th17 cell differentiation predispose the host to bacterial and fungal infections at mucosal surfaces [85]. Aberrant activity of Th17 cytokines have been implicated in a wide range of autoimmune manifestations including rheumatoid arthritis and psoriasis [86, 87]. In this section, we will explore the role of Th17 cytokines in the pathogenesis of SLE.

Th17 T cells are distinct lineage of effector T cells, that are characterised by their production of IL-17, and other cytokines such as IL-21 and IL-23. Its actions on epithelial cells and fibroblasts, as well as neutrophils account for its role in the protective response to pathogens, but is also important in mediating disease activity and associated tissue damage in autoimmune diseases. A robust inflammatory response is also mediated by B cells and Tfh cells. PMBC: peripheral blood mononuclear cells; GCSF: granulocyte colony-stimulating factor; Tfh: T follicular helper cells.

Interleukin-17

The interleukin-17 (IL-17) family consists of IL-17A to IL-17F and each cytokine can be composed of either homodimers or heterodimers, with IL-17A being the most intensely studied cytokine [88]. While the IL-17A cytokine expression is generally thought to be predominantly observed in T-cells, it has also been found to be produced by cells of the innate immune system such as neutrophils, eosinophils and natural killer cells [89].

The effects of IL-17 are widespread and IL-17 receptor (IL-17R) is distributed ubiquitously in various tissues with the highest levels being found on hematopoietic tissues. Of interest connective tissue such as fibroblasts, epithelial cells, and endothelial cells are the most responsive to IL-17 (Fig. 7.4) [90]. In vitro studies have indicated the proinflammatory function of IL-17 and have highlighted their role in neutrophil recruitment through the induction of granulocyte colony-stimulating factor (GCSF) and IL-8 [74].

A number of murine models that showed an important role for IL-17 [91, 92]. Expansion of Th17 cells and high expression of IL-17 in lymphocytes of MRL/*lpr* and B6/*lpr* mice. Inhibition of IL-17 signalling prevented formation of germinal centre and decreased humoral response in IL-17 receptor deficient mice (93 Nat Immunol 2008).

There has been emerging evidence to suggest a role for IL-17 in the pathogenesis of human SLE. Serum levels of IL-17 have been demonstrated to be elevated in SLE patients and correlated with disease activity [94–96]. Serum levels did not appear to correlate with specific lupus manifestations but IL-17 producing T cells have also been found in kidneys of patients with lupus nephritis [95, 97]. Increased local levels of IL-17 have also been found in end-organs commonly affected by SLE including skin and central nervous system [98, 99]. In lupus nephritis, urinary expression of Th17-related genes, including IL-17 and IL-23, is increased and is associated with disease activity and predict poor histopathological outcomes after immunosuppressive therapy [97, 100]. The induction of IL-17 in human SLE is not entirely clear, but a hypothesis based on the over-production of IFN-α by activated plasmacytoid dendritic cells, producing an abundance of IL-6 and IL-23 which are potent inducers of Th17 differentiation [101].

In the pathogenesis of SLE, IL-17 appears to potentiate localised inflammatory response by inducing local production of chemokines and activation of fibroblasts and epithelial cells leading to tissue damage [79]. IL-17 increases IgG production and autoantibody production from peripheral blood mononuclear cells (PBMCs) from lupus nephritis patients in a dose-dependent manner, not observed in healthy controls [102]. This overproduction could be suppressed through IL-17 blockade.

As a therapeutic target, several antibodies targeting IL-17 have been approved for the treatment of inflammatory conditions such as ankylosing spondylitis, psoriasis and psoriatic arthritis [103]. To date, none have been approved for use in patients with SLE. Indirect effects of IL-17 blockade through inhibition of IL-12 and IL-23 are discussed in the relevant sections of this chapter. One case report exists of a patient with refractory lupus nephritis complicated by psoriasis vulgaris found to have elevated Th17 cells in peripheral blood and in renal tissue. Treatment with an anti-IL-17 antibody (secukinumab) resulted in improvement in disease activity of SLE both clinically and biochemically with reduction in autoantibody levels [104].

Interleukin-21

IL-21 plays a major role in the regulation of humoral responses and promotes B cell activation and plasma cell differentiation and class-switching of immunoglobulin isotypes [105]. IL-21 is predominantly produced by Th17 cells under the influence of IL-6 [106, 107]. However, IL-21 is also an important cytokine produced by follicular helper T (Tfh) cells and acts in an autocrine manner, together with other cytokines such as IL-10 and CXC-chemokine ligand 13 (CXCL13), to promote differentiation of Tfh cells [108, 109]. Bcl-6 transcription factor was the key transcription factor that is critical for Tfh generation [110].

In SLE emerging data supports that IL-21 mediates its pathogenic role by enhancing Tfh functions [105]. Tfh play an essential role in the generation and maintenance of germinal centre responses in secondary lymphoid organs [108]. Tfh cells provide critical co-stimulation to B cells to eventually differentiate into antibody secreting plasma cells and memory B cells, and this is promoted by IL-21 [111]. The expansion of autoreactive B cells has been shown to be present in diseased kidney in SLE patients, and IL-21 can potently induce differentiation of these cells into Ig-secretary autoreactive plasma cells [112].

Serum IL-21 are elevated in SLE and single-nucleotide polymorphisms in both IL-21 and the IL-21 receptor have also been associated with susceptibility to SLE [113–115]. In addition, higher levels of IL-21 have been associated with expansion of specific autoreactive B cell subsets which correlated with disease manifestation [112]. In lupus-prone mice models, inhibition of IL-21R resulted in reduced proteinuria, histological glomerulonephritis, reduced levels of circulating dsDNA autoantibodies and reduced skin lesions and lymphadenopathy [116].

It will be important to take in account of other potential effects of IL-21 in SLE. There is a correlation between IL-21 expressing CD4 T cells observed in SLE patients compared to controls and the level of Th17 cells [106]. In addition, IL-21 plays an important role in amplification of Th17 cells via positive feedback loop which may have implications in autoimmune disorders [117]. IL-21 is also a powerful inducer of cytotoxic T cells and NK cells [118].

The current development of IL-21 blockade in human is limited due to the mixed results seen in experimental models. While there has been interests in developing an anti-IL-21 strategy in autoimmune diseases such as SLE, rheumatoid arthritis and Crohn's, further work needs to be done to elucidate the way to interfere with this cytokine action.

Interleukin-12

Interleukin-12 structurally belongs to a family of cytokines containing IL-12, IL-23, IL-27 and IL-35. IL-12 has a heterodimeric structure which includes two covalently linked subunits, p40 and p35 [119]. IL-12 and IL-23 both contain the subunit p40. Despite their structural similarity, IL-12 and IL-23 have differing roles in the regulation of Th1 and Th17 responses respectively [120]. IL-23 will be discussed in more detail in the following section.

IL-12 is produced by cells of the innate immune system such as dendritic cells and macrophages in response to microbial products, particularly in the production against intracellular parasites, and therefore can considered as an inductive cytokine for the Th1 immune response [119, 120]. Subsequent binding to IL-12 receptors (IL-12R) stimulates the activity of janus kinase 2 (JAK2) and tyrosine kinase 2 (TYK2), leading to phosphorylation of signal transducer and activator of transcription (STAT) family member, STAT4 [121]. Activation of STAT4 via IL-12 leads to the production of interferon-gamma (IFN-γ) and induces naïve Th cell differentiation into Th1 cells [120, 122]. In essence, IL-12 represents a link between the innate and adaptive proinflammatory responses.

The importance of the IL-12 and STAT4 activation pathway in the pathogenesis of SLE have been demonstrated in literature and offer a potential therapeutic target. Single nucleotide polymorphism (SNP) in the STAT4 gene which results in increased IL-12-induced phosphorylation of STAT4 and IFN-γ production is associated with increased susceptibility to developing SLE and associated with more severe disease phenotype with lupus nephritis [123]. Although interestingly, polymorphisms within the IL-12p40 subunit genes was not associated with susceptibility or severity of SLE in a Spanish population [29]. Studies on lupus-prone MRL/*lpr* mice characterized by fatal nephritis and autoantibodies, demonstrated that intrarenal IL-12 could elicit nephritis by promoting accumulation of IFN-γ-secreting T cells [124]. IL-12p40 deficiency in MRL/*lpr* resulted in delayed nephritis and protected against systemic pathologies of SLE. The protective effect could however be mediated by the absence of IL-23 activity, given IL-23 also shares the p40 subunit.

Interleukin-23

Interleukin-23 (IL-23) is a member of the heterodimeric family of IL-12 cytokines consisting of members including IL-12, IL-23 and IL-35 [125]. It is produced by macrophages, dendritic cells, keratinocytes, and other antigen-presenting cells as well as T and NK lymphocytes [126]. IL-23 is a heterodimer molecule consisting of the p19 and p40 subunit [127]. The p40 subunit is shared with IL-12, while the p19 subunit is specific to IL-23. Two receptors bind IL-23, the IL-12Rβ1 which is also shared with IL-12, and the IL-23-specific receptor (IL-23R) [128]. Signal transduction occurs through the activation of TYK2 via IL-12Rβ1, and JAK2 via IL-23R, which in turn phosphorylates STAT3[128].

The primary role of IL-23 appears to be the maintenance and expansion of IL-17 secreting phenotype of Th17 cells. [129]. Blockade of IL-23 action may be an effective way to reduce proinflammatory drive of the Th17 populations. While IL-17 can be produced by other non-Th17 cellular source, such as NK cells and neutrophils, the dominant production of IL-17 would be attenuated by IL-23 blockade, due to its effect on the downregulation of the Th17 cell lineage. The IL-17/ IL23 axis has been shown to be pivotal in the pathophysiology of psoriasis, spondyloarthritis and inflammatory bowel disease, and has a sound basis in SLE [130, 131].

Murine models have supported this observation; mice deficient in IL-23p19 have diminished ability to mediate inflammation and thus emphasising the critical role of IL-23 in sustaining Th17-driven immune response [132]. IL-23R deficiency resulted in decreased T-cell dependent autoantibody production, demonstrating that IL-23R signalling is important for T cell-dependent humoral responses in lupus prone mice [133, 134]. In the lupus-prone MRL/*lpr* mice, extracted T-cells lacking both CD4 and CD8 surface markers (double-negative T cells, DNT) was found to express high amounts of IL-17 cytokines [91]. In addition, as disease severity increased, as did the expression of IL-17 and IL-23 [91]. These IL-17 secreting DNT cells were also found in the kidney biopsies of patients with active lupus nephritis [135]. Furthermore, IL-23 treatment of extracted DNT contributed to the preferential enhancement of the Th17 cell response over the Th1 pathway by increasing expression of IL-23R and further production of IL-17A, while simultaneously decreasing production of both IL-2 and IFN-γ [91, 133]. In addition, mouse models injected with IL-23-treated lymphocytes developed manifestations of lupus nephritis while injection of untreated lymphocytes did not [91]. Conversely, the use of neutralising anti-IL-23 antibodies in lupus prone MRL/*lpr* mice resulted in less severe nephritis [136].

In human studies, patients with active SLE were found to have elevated production of p19 and p40 mRNA levels compared to inactive SLE [137]. Localised elevation of IL-23 in the kidneys positively correlated with severity of lupus nephritis in patients [126]. Increased serum IL-23 levels were shown to be associated with other SLE manifestations in humans, including cutaneous involvement and serositis [94].

Ustekinumab, is the first approved biologic targeting IL-23. It binds to the p40 subunit and was originally developed to neutralize effects of IL-12 but also inhibits IL-23 due to the shared subunits [138]. In its phase 2 study ustekinumab initially showed some promising data in patients with moderately severe SLE, but the phase III study was suspended as a result of a lack of efficacy from the pre-planned interim efficacy analysis [139, 140]. There is ongoing interest in developing other anti-IL-23 strategies particularly in the treatment for psoriasis. Three other monoclonal antibodies, namely guselkumab, tildrakizumab and risankizumab, which selectively inhibit the IL-23p19 subunit have been developed but currently no specific study planned in SLE [138, 141, 142].

Interleukin-6

IL-6 is synthesized predominantly by monocytes, macrophages, fibroblasts and endothelial cells and to a lesser extent T and B cell lymphocytes [143]. It has pleotropic effects on a variety of cell types and plays a key role in the regulation of inflammation (***Fig. 7.5***). It is well known for its effect on the induction of pyrexia and production of acute phase reactants such as amyloid A, C-reactive proteins, alpha-1-antitrypsin, fibrinogen, and haptoglobin [144].

One of the most important actions of IL-6 is its ability to stimulate the final stages of B lymphocyte maturation and potentiates immunoglobulin secretion by plasma cells [145, 146]. IL-6 is also critical in the generation of Th17 cells, from naïve T cells [80]. The balance between Th17 and Treg differentiation appears to be related to the cytokine milieu of IL-6 and TGF-β [147].

Elevated levels of IL-6 in serum, urine and renal glomeruli in murine models of SLE suggest that IL-6 plays a role in the production of autoantibodies may be correlation with specific end-organ manifestations such as lupus nephritis, serositis and anaemia [104, 148–151]. In murine models, elevated levels have been observed before the appearance of autoantibodies in preclinical SLE, suggesting the role of IL-6 may be more important in the early stages of tolerance dysregulation [20, 152].

To date there have been a number of early phase clinical trials studying the efficacy of IL-6 inhibition in human SLE. Sirukumab, a monoclonal antibody to IL-6 was studied in a randomised controlled trial of Class III or IV lupus nephritis. The treatment did not result in a reduction in proteinuria at week 24 [153]. Another anti-IL-6 monoclonal antibody was studied in a phase 2 randomised controlled study in non-renal SLE but did not meet its primary endpoint of SRI-4 at 24 weeks. The trial however reported a BILAG-based combined Lupus Assessment (BICLA) response and reduction in flare rate [154]. Vobarilizumab, a novel anti-IL6R drug that was developed, also failed to meet its primary endpoints in a dose finding phase 2 study. There were 2 deaths in the treatment arms [155].

Tocilizumab, a widely available anti-IL6R had showed promising preliminary activity in a phase 1 study with reduction in disease activity measured by SLE Disease Activity Index (SLEDAI) and Systemic Lupus Activity Measure (SLAM). However, to date, there has not been any further study planned with this agent [156].

IL-6 has pleotropic effects on a variety of cell types and plays a key role in the regulation of inflammation. Among many of its effects, its ability to induce B cell maturation and acute phase response are most relevant in the SLE pathogenesis.

T Regulatory Cell Cytokines

Interleukin-2

IL-2 and related members of the superfamily IL-5, and IL-21 are multifunctional autocrine and paracrine growth factors produced predominantly by activated CD4 + T cells. IL-2 plays a pivotal role in both T and B cell activation and proliferation [157]. While it has been considered in some old literature as a Th1 cytokine, we have classified it under a number of key cytokines for the regulation of Tregs in autoimmune diseases.

IL-2 plays a central role in determining the fate of T lymphocytes by promoting their cell cycle progression (Fig. 7.6). It controls the growth and apoptosis of T lymphocytes and termination of T cell immune response [158, 159]. It has a complex role, as an essential factor for growth and survival in regulatory T cells (Tregs), and therefore maintenance of self-tolerance, and on the other hand, IL-2 can suppress the differentiation of Tfh and Th17 cells, via its function as the activator of T cells apoptosis, promoting activation induced cell death [158]. It can also augment the NK cell and cytotoxic T cell activity, and immunoglobulin production in plasma cells [160].

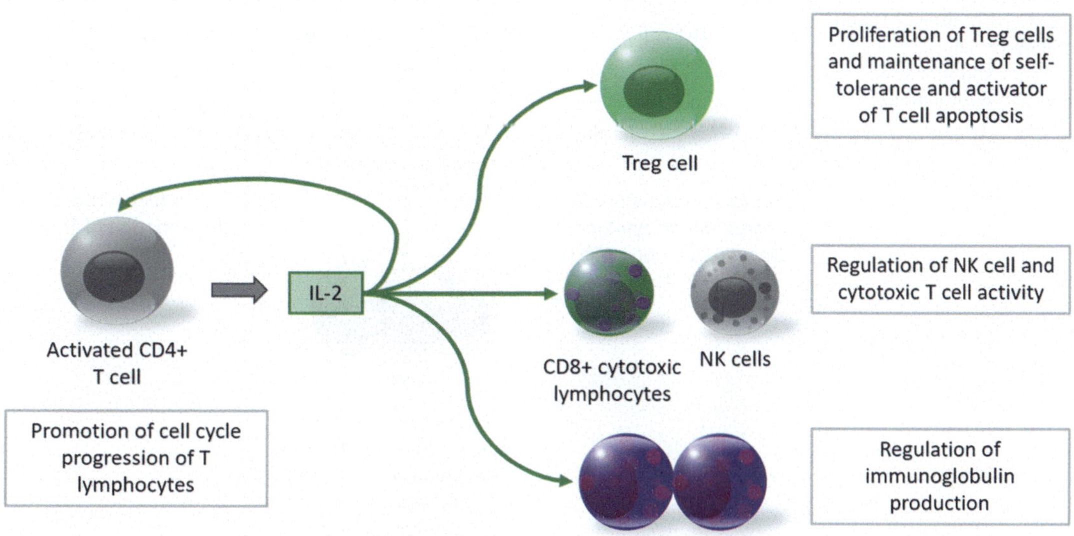

Fig. 7.6 The role of IL-2 regulation of lymphocyte proliferation and differentiation through regulation of growth and apoptosis. IL-2 is a potent cytokine that plays a critical role in T cell growth and differentiation into effector cells such as Treg and CD8+ cytotoxic cells. It has both immune stimulatory and regulatory roles depending on the level and timing of its production

In its immunostimulatory role, IL-2 is effective in priming CD8 + T cells and promotes T cell activity. IL-2 signals received during the priming period of CD8 + T cell s are essential for the programming of proliferation-competent memory CD8 + T cells. T cell activation and proliferation stimulated by cognate antigenic interaction is enhanced by IL-2. This is the basis of calcineurin inhibitors act on reducing T cell activity and its IL-2 production.

In patients with SLE, we have observed a blunted IL-2 production and decreased T cell response to IL-2 at the receptor level, which correlated with defects in Treg homeostasis [161–163] [164–166]. In vitro experiments have demonstrated that administration of IL-2 to CD4 + T cells isolated from SLE patients can restore the Treg defects. The lower doses of IL-2 administration had the most sustained expansion of Treg [167]. The first pilot studies in human have shown that low dose IL-2 treatment in SLE was associated with a reduction in disease activity, along with an upregulation of Treg cells, and a downregulation of Tfh and Th17 cells [167–169]. The first randomised controlled phase 2 study examining the efficacy of intermittent dosing of IL-2 over 12 weeks has shown us that even though it did not meet its primary endpoint of SRI4 response at 12 weeks, treatment arm had a much higher complete remission rate in patients with lupus nephritis by 24 weeks [170].

IL-2 is a potent cytokine that plays a critical role in T cell growth and differentiation into effector cells such as Treg and CD8 + cytotoxic cells. It has both immune stimulatory and regulatory roles depending on the level and timing of its production.

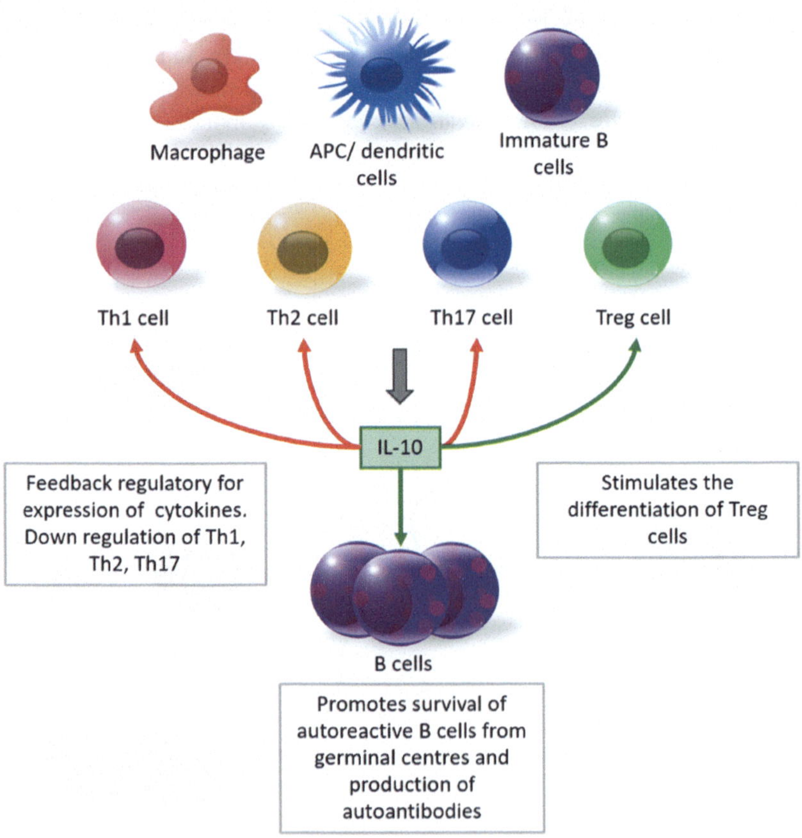

Fig. 7.7 The role of IL-10 as a negative feedback regulator of immune activation. IL-10 has a role in prevention of tissue damage caused by excessive inflammatory effector responses during infection or in autoinflammatory conditions. Due to the pleotropic nature of IL-10 in SLE, IL-10 promoted B cell proliferation and Ig class switching, and increased autoantibody production

Interleukin-10

IL-10 is produced by numerous cells of the innate and adaptive immune system, including macrophages, dendritic cells, B cells and all subsets of helper T cells including Th1, Th2 and Th17 (Fig. 7.7) [171–174]. Its anti-inflammatory effect, as reflected by its other name human cytokine synthesis inhibitory factor. It is sometimes listed as "on-effect", IL-10 produced by different Th cell subsets play an important role as feedback regulator for different Th2 cytokines, as it can downregulate expression of Th1 cytokines and enhances B cell survival, proliferation and antibody production. Immature B cell production of IL-10 suppresses Th1 and Th17 cell differentiation, and stimulates the differentiation of Treg cells [175]. IL-10 have also been shown to recruit CD8 + T cells and induce cytotoxicity activity in both mouse and human studies [176, 177]. IL-10 signalling acts predominantly though the Jak/STAT pathway with activation of transcription factor STAT3 [178]. IL-10 functions primarily as a negative feedback regulator to reduce tissue damage caused by excessive inflammatory effector responses during infection or in autoinflammatory conditions [179].

Due to the predominant regulatory effects of IL-10, it follows that IL-10 deficiency would develop an exaggerated immune response as demonstrated by mice models of IL-10 deficiency which develop inflammatory bowel disease and appear to potentiate clearance of intracellular pathogens [180, 181]. However, in addition to the inhibitory effects, in vitro and animal models have suggested IL-10 acts as a growth factor for B cells and promotes survival of autoreactive B cells from germinal centres, even though these effects can be circumvented by IL-21 [182]. In SLE, serum IL-10 was found to be higher than controls, and IL-10 promoted B cell proliferation and Ig class switching, and increased autoantibody production [183–189]. Early studies blocking IL-10 activity in lupus-prone mice resulted in reduced levels of autoantibodies and increased survival, while administration of IL-10 accelerated autoimmunity [135, 190]. In human, a small pilot open labelled study using a murine anti-IL-10 monoclonal antibodies administered to six patient with active SLE resulted in decreased prednisolone use and improved disease activity [191]. A small prospective randomised controlled proof of concept study with a novel anti-IL-10 antibody (BT063) was started but no preliminary data has been published to date [192]. On the other hand, IL-10 response can be augmented by administration of exogenous IL-10. This has been trialled in a small open labelled study of patients with rheumatoid arthritis [193]. It is presently unclear that among the many pleiotropic and regulatory actions of IL-10, which of these pathways is the most important in the pathophysiology of SLE. While many believe that the capacity of Il-10 to amplify the humoral immune response supports an inhibitory strategy to IL-10, other cell or organ specific approaches are being considered. There is some emerging data on specific actions of IL-10 in the central nervous system (CNS), with ability to limit glial inflammatory response and provide protection against CNS insult (194).

IL-10 also has a role in prevention of tissue damage caused by excessive inflammatory effector responses during infection or in autoinflammatory conditions. As part of the pleotropic nature of IL-10 in SLE, IL-10 is known to promote B cell proliferation and Ig class switching, and increase autoantibody production.

Conclusions

SLE is a complex and clinically heterogenous disease and represents a major challenge for the development of targeted therapy. In this chapter, we have explored the pathogenic roles of distinct T help cell subsets Th1, Th2, Th17 and Treg cells, and their associated cytokines. *In vitro* and *in vivo* studies highlight the intricate balance required for the normal immune function and the maintenance of immune tolerance. Advances in our understanding of the complex molecular interaction resulting in cytokine dysregulation in SLE has provided the window of opportunity to intervene in disease pathogenesis. In recent years, clinical trials have explored several molecular targets including IL-2, IL-6, IL-12 and IL-23, although none has yet to receive approval for the treatment of SLE. With time, utilisation of translational medical research to identify novel molecular targets in conjunction with rigorous clinical trial design will hopefully result in progress towards a more effective therapeutic paradigm.

References

1. Wong CK, Ho CY, Li EK, Lam CW (2000) Elevation of proinflammatory cytokine (IL-18, IL-17, IL-12) and Th2 cytokine (IL-4) concentrations in patients with systemic lupus erythematosus. Lupus 9(8):589–593
2. Horwitz DA, Gray JD, Behrendsen SC, Kubin M, Rengaraju M, Ohtsuka K et al (1998) Decreased production of interleukin-12 and other Th1-type cytokines in patients with recent-onset systemic lupus erythematosus. Arthritis Rheum 41(5):838–844
3. Takahashi S, Fossati L, Iwamoto M, Merino R, Motta R, Kobayakawa T et al (1996) Imbalance towards Th1 predominance is associated with acceleration of lupus-like autoimmune syndrome in MRL mice. J Clin Invest 97(7):1597–1604
4. Akahoshi M, Nakashima H, Tanaka Y, Kohsaka T, Nagano S, Ohgami E et al (1999) Th1/Th2 balance of peripheral T helper cells in systemic lupus erythematosus. Arthritis Rheum 42(8):1644–1648
5. Guimarães PM, Scavuzzi BM, Stadtlober NP, Franchi Santos LFdR, Lozovoy MAB, Iriyoda TMV, et al. (2017) Cytokines in systemic lupus erythematosus: far beyond Th1/Th2 dualism lupus: cytokine profiles. Immunol Cell Biol 95 (9):824–831
6. Reynolds JA, McCarthy EM, Haque S, Ngamjanyaporn P, Sergeant JC, Lee E et al (2018) Cytokine profiling in active and quiescent SLE reveals distinct patient subpopulations. Arthritis Res Ther 20(1):173
7. Kaul A, Gordon C, Crow MK, Touma Z, Urowitz MB, van Vollenhoven R et al (2016) Systemic lupus erythematosus. Nat Rev Dis Primers 2 (1):16039
8. Tsokos GC, Lo MS, Reis PC, Sullivan KE (2016) New insights into the immunopathogenesis of systemic lupus erythematosus. Nat Rev Rheumatol 12(12):716–730
9. Mosmann TR, Cherwinski H, Bond MW, Giedlin MA, Coffman RL. (1986) Two types of murine helper T cell clone. I. Definition according to profiles of lymphokine activities and secreted proteins. J Immunol 136(7):2348–2357
10. Muhammad Yusoff F, Wong KK, Mohd Redzwan N (2020) Th1, Th2, and Th17 cytokines in systemic lupus erythematosus. Autoimmun 53 (1):8–20. https://doi.org/10.1080/08916934.2019.1693545
11. Berger A (2000) Th1 and Th2 responses: what are they? BMJ 321(7258):424
12. Lourenço EV, La Cava A (2009) Cytokines in systemic lupus erythematosus. Curr Mol Med 9 (3):242–254
13. Liblau RS, Singer SM, McDevitt HO (1995) Th1 and Th2 CD4+ T cells in the pathogenesis of organ-specific autoimmune diseases. Immunol Today 16 (1):34–38
14. Mullen AC, High FA, Hutchins AS, Lee HW, Villarino AV, Livingston DM et al (2001) Role of T-bet in commitment of TH1 cells before IL-12-dependent selection. Science 292(5523):1907–1910
15. Yoshimoto T, Takeda K, Tanaka T, Ohkusu K, Kashiwamura S, Okamura H et al (1998) IL-12 up-regulates IL-18 receptor expression on T cells, Th1 cells, and B cells: synergism with IL-18 for IFN-gamma production. J Immunol 161(7):3400–3407
16. Frucht DM, Fukao T, Bogdan C, Schindler H, O'Shea JJ, Koyasu S (2001) IFN-gamma production by antigen-presenting cells: mechanisms emerge. Trends Immunol 22(10):556–560
17. Yang J, Murphy TL, Ouyang W, Murphy KM (1999) Induction of interferon-γ production in Th1 CD4+ T cells: evidence for two distinct pathways for promoter activation. Eur J Immunol 29(2):548–555
18. Kim K, Cho SK, Sestak A, Namjou B, Kang C, Bae SC (2010) Interferon-gamma gene polymorphisms associated with susceptibility to systemic lupus erythematosus. Ann Rheum Dis 69(6):1247–1250
19. Leng R-X, Pan H-F, Liu J, Yang X-K, Zhang C, Tao S-S et al (2016) Evidence for genetic association of TBX21 and IFNG with systemic lupus erythematosus in a Chinese Han population. Sci Rep 6(1):22081
20. Munroe ME, Lu R, Zhao YD, Fife DA, Robertson JM, Guthridge JM et al (2016) Altered type II interferon precedes autoantibody accrual and elevated type I interferon activity prior to systemic lupus erythematosus classification. Ann Rheum Dis 75(11):2014–2021
21. Seery JP, Carroll JM, Cattell V, Watt FM (1997) Antinuclear autoantibodies and lupus nephritis in transgenic mice expressing interferon gamma in the epidermis. J Exp Med 186(9):1451–1459
22. Uhm WS, Na K, Song GW, Jung SS, Lee T, Park MH et al (2003) Cytokine balance in kidney tissue from lupus nephritis patients. Rheumatology (Oxford) 42(8):935–938

23. Min DJ, Cho ML, Cho CS, Min SY, Kim WU, Yang SY et al (2001) Decreased production of interleukin-12 and interferon-gamma is associated with renal involvement in systemic lupus erythematosus. Scand J Rheumatol 30(3):159–163
24. Nicoletti F, Di Marco R, Zaccone P, Xiang M, Magro G, Grasso S et al (2000) Dichotomic effects of IFN-gamma on the development of systemic lupus erythematosus-like syndrome in MRL-lpr / lpr mice. Eur J Immunol 30(2):438–447
25. Prud'homme GJ, Kono DH, Theofilopoulos AN (1995) Quantitative polymerase chain reaction analysis reveals marked overexpression of interleukin-1 beta, interleukin-1 and interferon-gamma mRNA in the lymph nodes of lupus-prone mice. Mol Immunol 32(7):495–503
26. Lit LC, Wong CK, Li EK, Tam LS, Lam CW, Lo YM (2007) Elevated gene expression of Th1/Th2 associated transcription factors is correlated with disease activity in patients with systemic lupus erythematosus. J Rheumatol 34(1):89–96
27. Boedigheimer MJ, Martin DA, Amoura Z, Sánchez-Guerrero J, Romero-Diaz J, Kivitz A, et al. (2017) Safety, pharmacokinetics and pharmacodynamics of AMG 811, an anti-interferon-γ monoclonal antibody, in SLE subjects without or with lupus nephritis. Lupus Sci Med 4(1):e000226. https://doi.org/10.1136/lupus-2017-000226
28. Werth VP, Fiorentino D, Sullivan BA, Boedigheimer MJ, Chiu K, Wang C et al (2017) Brief Report: Pharmacodynamics, Safety, and Clinical Efficacy of AMG 811, a Human Anti-Interferon-γ Antibody, in Patients With Discoid Lupus Erythematosus. Arthritis & rheumatology (Hoboken, NJ). 69(5):1028–1034
29. Martin DA, Amoura Z, Romero-Diaz J, Chong YB, Sanchez-Guerrero J, Chan T et al (2015) A multiple dose study of AMG 811 (Anti-IFN-Gamma) in subjects with systemic lupus erythematosus and active nephritis. Ann Rheumatic Dis 74:337
30. Postal M, Appenzeller S (2011) The role of Tumor Necrosis Factor alpha (TNF α) in the pathogenesis of systemic lupus erythematosus. Cytokine 56 (3):537–543
31. Grell M, Douni E, Wajant H, Löhden M, Clauss M, Maxeiner B et al (1995) The transmembrane form of tumor necrosis factor is the prime activating ligand of the 80 kDa tumor necrosis factor receptor. Cell 83(5):793–802
32. Aringer M, Smolen JS (2018) The role of tumor necrosis factor-alpha in systemic lupus erythematosus. Arthritis Res Ther 10(1):202
33. MacEwan DJ (2002) TNF ligands and receptors–a matter of life and death. Br J Pharmacol 135 (4):855–875
34. Vandenabeele P, Declercq W, Beyaert R, Fiers W (1995) Two tumour necrosis factor receptors: structure and function. Trends Cell Biol 5 (10):392–399
35. McDevitt H, Munson S, Ettinger R, Wu A (2002) Multiple roles for tumor necrosis factor-alpha and lymphotoxin alpha/beta in immunity and autoimmunity. Arthritis Res 4(Suppl 3):S141–S52. https://doi.org/10.1186/ar570
36. Gordon C, Ranges GE, Greenspan JS, Wofsy D (1989) Chronic therapy with recombinant tumor necrosis factor-alpha in autoimmune NZB/NZW F1 mice. Clin Immunol Immunopathol 52(3):421–434
37. Kollias G, Kontoyiannis D (2002) Role of TNF/TNFR in autoimmunity: specific TNF receptor blockade may be advantageous to anti-TNF treatments. Cytokine Growth Factor Rev 13(4–5):315–321
38. Zhou T, Edwards CK 3rd, Yang P, Wang Z, Bluethmann H, Mountz JD (1996) Greatly accelerated lymphadenopathy and autoimmune disease in lpr mice lacking tumor necrosis factor receptor I. J Immunol 156(8):2661–2665
39. Gómez D, Correa PA, Gómez LM, Cadena J, Molina JF, Anaya JM (2004) Th1/Th2 cytokines in patients with systemic lupus erythematosus: is tumor necrosis factor alpha protective? Semin Arthritis Rheum 33(6):404–413
40. Studnicka-Benke A, Steiner G, Petera P, Smolen JS (1996) Tumour necrosis factor alpha and its soluble receptors parallel clinical disease and autoimmune activity in systemic lupus erythematosus. Br J Rheumatol 35(11):1067–1074
41. Coates LC, Marzo-Ortega H, Bennett AN, Emery P (2010) Anti-TNF Therapy in Ankylosing Spondylitis: Insights for the Clinician. Ther Adv Musculoskelet Dis. 2(1):37–43
42. Mease PJ (2002) Tumour necrosis factor (TNF) in psoriatic arthritis: pathophysiology and treatment with TNF inhibitors. Ann Rheum Dis 61(4):298–304
43. Aringer M, Graninger WB, Steiner G, Smolen JS (2004) Safety and efficacy of tumor necrosis factor alpha blockade in systemic lupus erythematosus: an open-label study. Arthritis Rheum 50(10):3161–3169
44. Aringer M, Houssiau F, Gordon C, Graninger WB, Voll RE, Rath E et al (2009) Adverse events and efficacy of TNF-alpha blockade with infliximab in patients with systemic lupus erythematosus: long-term follow-up of 13 patients. Rheumatology (Oxford) 48(11):1451–1454
45. Uppal SS, Hayat SJ, Raghupathy R (2009) Efficacy and safety of infliximab in active SLE: a pilot study. Lupus 18(8):690–697
46. Cortés-Hernández J, Egri N, Vilardell-Tarrés M, Ordi-Ros J (2015) Etanercept in refractory lupus arthritis: An observational study. Semin Arthritis Rheum 44(6):672–679
47. Katz U, Zandman-Goddard G (2010) Drug-induced lupus: an update. Autoimmun Rev 10(1):46–50
48. Feldmann M, Brennan FM, Maini RN (1996) Role of cytokines in rheumatoid arthritis. Annu Rev Immunol 14:397–440

49. Soforo E, Baumgartner M, Francis L, Allam F, Phillips PE, Perl A (2010) Induction of systemic lupus erythematosus with tumor necrosis factor blockers. J Rheumatol 37(1):204–205
50. Swale VJ, Perrett CM, Denton CP, Black CM, Rustin MHA (2003) Etanercept-induced systemic lupus erythematosus. Clin Exp Dermatol 28 (6):604–607
51. Cairns AP, Duncan MKJ, Hinder AE, Taggart AJ (2002) New onset systemic lupus erythematosus in a patient receiving etanercept for rheumatoid arthritis. Ann Rheum Dis 61(11):1031
52. Mor A, Bingham C, Barisoni L, Lydon E, Belmont HM (2005) Proliferative lupus nephritis and leukocytoclastic vasculitis during treatment with etanercept. J Rheumatol 32(4):740
53. Dall'Era M, Wofsy D (2013) Etanercept for the treatment of lupus nephritis. Identifier NCT00447265: national library of Medicine [updated February 12. ClinicalTrials.gov identifier (NCT number): NCT00447265]. Available from: https://ClinicalTrials.gov/show/NCT00447265 .
54. Aringer M, Smolen JS (2012) Therapeutic blockade of TNF in patients with SLE-promising or crazy? Autoimmun Rev 11(5):321–325
55. Mosmann TR, Sad S (1996) The expanding universe of T-cell subsets: Th1, Th2 and more. Immunol Today 17(3):138–146
56. Takeda K, Tanaka T, Shi W, Matsumoto M, Minami M, Kashiwamura S et al (1996) Essential role of Stat6 in IL-4 signalling. Nature 380 (6575):627–630
57. Zheng W-p, Flavell RA (1997) The transcription factor GATA-3 Is necessary and sufficient for Th2 Cytokine Gene Expression in CD4 T Cells. Cell 89 (4):587–596
58. Ho IC, Hodge MR, Rooney JW, Glimcher LH (1996) The Proto-Oncogene c-<em>maf</em> Is Responsible for Tissue-Specific Expression of Interleukin-4. Cell 85(7):973–983
59. Singh RR, Saxena V, Zang S, Li L, Finkelman FD, Witte DP, et al. (2003)nDifferential contribution of IL-4 and STAT6 vs STAT4 to the development of lupus nephritis. J Immunol 170(9):4818–4825
60. Jäger A, Kuchroo VK (2010) Effector and regulatory T-cell subsets in autoimmunity and tissue inflammation. Scand J Immunol 72(3):173–184
61. Singh RR (2003) IL-4 and many roads to lupuslike autoimmunity. Clin Immunol 108(2):73–79
62. Ohnishi K, Ebling FM, Mitchell B, Singh RR, Hahn BH, Tsao BP (1994) Comparison of pathogenic and non-pathogenic murine antibodies to DNA: antigen binding and structural characteristics. Int Immunol 6(6):817–830
63. Santiago ML, Fossati L, Jacquet C, Müller W, Izui S, Reininger L (1997) Interleukin-4 protects against a genetically linked lupus-like autoimmune syndrome. J Exp Med 185(1):65–70
64. Deocharan B, Marambio P, Edelman M, Putterman C (2003) Differential effects of interleukin-4 in peptide induced autoimmunity. Clin Immunol 108 (2):80–88
65. Erb KJ, Rüger B, von Brevern M, Ryffel B, Schimpl A, Rivett K (1997) Constitutive expression of interleukin (IL)-4 in vivo causes autoimmune-type disorders in mice. J Exp Med 185(2):329–339
66. Nakajima A, Hirose S, Yagita H, Okumura K (1997) Roles of IL-4 and IL-12 in the development of lupus in NZB/W F1 mice. J Immunol 158 (3):1466
67. Mori M, Morris SC, Orekhova T, Marinaro M, Giannini E, Finkelman FD (2000) IL-4 Promotes the Migration of Circulating B Cells to the Spleen and Increases Splenic B Cell Survival. J Immunol 164(11):5704
68. Rüger BM, Erb KJ, He Y, Lane JM, Davis PF, Hasan Q (2000) Interleukin-4 transgenic mice develop glomerulosclerosis independent of immunoglobulin deposition. Eur J Immunol 30 (9):2698–2703
69. Yang W-C, Hwang Y-S, Chen Y-Y, Liu C-L, Shen C-N, Hong W-H, et al (2017) Interleukin-4 supports the suppressive immune responses elicited by regulatory T cells. Front Immunol 8:1508. https://doi.org/10.3389/fimmu.2017.01508
70. Elewa EA, Zakaria O, Mohamed EI, Boghdadi G (2014) The role of interleukins 4, 17 and interferon gamma as biomarkers in patients with Systemic Lupus Erythematosus and their correlation with disease activity. The Egyptian Rheumatologist. 36 (1):21–27
71. Sugimoto K, Morimoto S, Kaneko H, Nozawa K, Tokano Y, Takasaki Y et al (2002) Decreased IL-4 Producing CD4 + T Cells in Patients with Active Systemic Lupus Erythematosus-relation to IL-12R Expression. Autoimmunity 35(6):381–387
72. Hsu H-C, Liu S, Gao M, Wu Q, Yang P, Chatham WW, et al (2020) IL-4 synergizes with low-dose IL-2 to restore systemic lupus erythematosus B cells at the resting naive status. J Immunol 204(1 Supplement):218.8
73. Harrington LE, Hatton RD, Mangan PR, Turner H, Murphy TL, Murphy KM et al (2005) Interleukin 17-producing CD4+ effector T cells develop via a lineage distinct from the T helper type 1 and 2 lineages. Nat Immunol 6(11):1123–1132
74. Park H, Li Z, Yang XO, Chang SH, Nurieva R, Wang YH et al (2005) A distinct lineage of CD4 T cells regulates tissue inflammation by producing interleukin 17. Nat Immunol 6(11):1133–1141
75. Qu N, Xu M, Mizoguchi I, Furusawa J, Kaneko K, Watanabe K, et al (2013) Pivotal roles of T-helper 17-related cytokines, IL-17, IL-22, and IL-23, in inflammatory diseases. Clin Dev Immunol 2013:968549. https://doi.org/10.1155/2013/968549 . Epub 2013 Jul 14

76. Yang XO, Pappu BP, Nurieva R, Akimzhanov A, Kang HS, Chung Y et al (2008) T helper 17 lineage differentiation is programmed by orphan nuclear receptors ROR alpha and ROR gamma. Immunity 28(1):29–39
77. Yang XO, Panopoulos AD, Nurieva R, Chang SH, Wang D, Watowich SS et al (2007) STAT3 regulates cytokine-mediated generation of inflammatory helper T cells. J Biol Chem 282(13):9358–9363
78. Dong C (2008) TH17 cells in development: an updated view of their molecular identity and genetic programming. Nat Rev Immunol 8(5):337–348
79. Weaver CT, Harrington LE, Mangan PR, Gavrieli M, Murphy KM (2006) Th17: an effector CD4 T cell lineage with regulatory T cell ties. Immunity 24(6):677–688
80. Mangan PR, Harrington LE, O'Quinn DB, Helms WS, Bullard DC, Elson CO et al (2006) Transforming growth factor-beta induces development of the T(H)17 lineage. Nature 441(7090):231–234
81. Langrish CL, Chen Y, Blumenschein WM, Mattson J, Basham B, Sedgwick JD et al (2005) IL-23 drives a pathogenic T cell population that induces autoimmune inflammation. J Exp Med 201 (2):233–240
82. Korn T, Oukka M, Kuchroo V, Bettelli E (2007) Th17 cells: effector T cells with inflammatory properties. Semin Immunol 19(6):362–371
83. Louten J, Boniface K, de Waal MR (2009) Development and function of TH17 cells in health and disease. J Allergy Clin Immunol 123(5):1004–1011
84. Littman DR, Rudensky AY (2010) Th17 and regulatory T cells in mediating and restraining inflammation. Cell 140(6):845–858
85. Ouyang W, O'Garra A (2019) IL-10 Family Cytokines IL-10 and IL-22: from Basic Science to Clinical Translation. Immunity 50(4):871–891
86. Furuzawa-Carballeda J, Vargas-Rojas MI, Cabral AR (2007) Autoimmune inflammation from the Th17 perspective. Autoimmun Rev 6(3):169–175
87. Lowes MA, Bowcock AM, Krueger JG (2007) Pathogenesis and therapy of psoriasis. Nature 445 (7130):866–873
88. Ouyang W, Kolls JK, Zheng Y (2008) The biological functions of T helper 17 cell effector cytokines in inflammation. Immunity 28(4):454–467
89. Koga T, Ichinose K, Kawakami A, Tsokos GC (2019) The role of IL-17 in systemic lupus erythematosus and its potential as a therapeutic target. Expert Rev Clin Immunol 15(6):629–637
90. Moseley TA, Haudenschild DR, Rose L, Reddi AH (2003) Interleukin-17 family and IL-17 receptors. Cytokine Growth Factor Rev 14(2):155–174
91. Zhang Z, Kyttaris VC, Tsokos GC (2009) The role of IL-23/IL-17 axis in lupus nephritis. J Immunol 183(5):3160–3169
92. Kang H-K, Liu M, Datta SK (2007) Low-Dose Peptide Tolerance Therapy of Lupus Generates Plasmacytoid Dendritic Cells That Cause Expansion of Autoantigen-Specific Regulatory T Cells and Contraction of Inflammatory Th17 Cells. J Immunol 178(12):7849
93. Hsu HC, Yang P, Wang J, Wu Q, Myers R, Chen J et al (2008) Interleukin 17-producing T helper cells and interleukin 17 orchestrate autoreactive germinal center development in autoimmune BXD2 mice. Nat Immunol 9(2):166–175
94. Wong CK, Lit LC, Tam LS, Li EK, Wong PT, Lam CW (2008) Hyperproduction of IL-23 and IL-17 in patients with systemic lupus erythematosus: implications for Th17-mediated inflammation in auto-immunity. Clin Immunol 127(3):385–393
95. Vincent FB, Northcott M, Hoi A, Mackay F, Morand EF (2013) Clinical associations of serum interleukin-17 in systemic lupus erythematosus. Arthritis Res Ther 15(4):R97
96. Yu B, Guan M, Peng Y, Shao Y, Zhang C, Yue X et al (2011) Copy number variations of interleukin-17F, interleukin-21, and interleukin-22 are associated with systemic lupus erythematosus. Arthritis Rheum 63(11):3487–3492
97. Zickert A, Amoudruz P, Sundström Y, Rönnelid J, Malmström V, Gunnarsson I (2015) IL-17 and IL-23 in lupus nephritis - association to histopathology and response to treatment. BMC Immunol 16(1):7
98. Tanasescu C, Balanescu E, Balanescu P, Olteanu R, Badea C, Grancea C et al (2010) IL-17 in cutaneous lupus erythematosus. Eur J Intern Med 21(3):202–207
99. Lu XY, Zhu CQ, Qian J, Chen XX, Ye S, Gu YY (2010) Intrathecal cytokine and chemokine profiling in neuropsychiatric lupus or lupus complicated with central nervous system infection. Lupus 19(6):689–695
100. Kwan BC, Tam LS, Lai KB, Lai FM, Li EK, Wang G et al (2009) The gene expression of type 17 T-helper cell-related cytokines in the urinary sediment of patients with systemic lupus erythematosus. Rheumatology (Oxford) 48(12):1491–1497
101. Yu C-F, Peng W-M, Oldenburg J, Hoch J, Bieber T, Limmer A et al (2010) Human Plasmacytoid Dendritic Cells Support Th17 Cell Effector Function in Response to TLR7 Ligation. J Immunol 184 (3):1159
102. Dong G, Ye R, Shi W, Liu S, Wang T, Yang X et al (2003) IL-17 induces autoantibody overproduction and peripheral blood mononuclear cell overexpression of IL-6 in lupus nephritis patients. Chin Med J (Engl) 116(4):543–548
103. Beringer A, Miossec P (2019) Systemic effects of IL-17 in inflammatory arthritis. Nat Rev Rheumatol 15(8):491–501
104. Richards HB, Satoh M, Shaw M, Libert C, Poli V, Reeves WH (1998) Interleukin 6 dependence of anti-DNA antibody production: evidence for two

pathways of autoantibody formation in pristane-induced lupus. J Exp Med 188(5):985–990
105. Gensous N, Schmitt N, Richez C, Ueno H, Blanco P (2016) T follicular helper cells, interleukin-21 and systemic lupus erythematosus. Rheumatology 56 (4):516–523
106. Korn T, Bettelli E, Gao W, Awasthi A, Jäger A, Strom TB et al (2007) IL-21 initiates an alternative pathway to induce proinflammatory T(H)17 cells. Nature 448(7152):484–487
107. Wurster AL, Rodgers VL, Satoskar AR, Whitters MJ, Young DA, Collins M et al (2002) Interleukin 21 is a T helper (Th) cell 2 cytokine that specifically inhibits the differentiation of naive Th cells into interferon gamma-producing Th1 cells. J Exp Med 196(7):969–977
108. Vinuesa CG, Tangye SG, Moser B, Mackay CR (2005) Follicular B helper T cells in antibody responses and autoimmunity. Nat Rev Immunol 5 (11):853–865
109. Nurieva RI, Chung Y, Hwang D, Yang XO, Kang HS, Ma L et al (2008) Generation of T follicular helper cells is mediated by interleukin-21 but independent of T helper 1, 2, or 17 cell lineages. Immunity 29(1):138–149
110. Nurieva RI, Chung Y, Martinez GJ, Yang XO, Tanaka S, Matskevitch TD et al (2009) Bcl6 mediates the development of T follicular helper cells. Science 325(5943):1001–1005
111. Deng X-M, Yan S-X, Wei W (2015) IL-21 acts as a promising therapeutic target in systemic lupus erythematosus by regulating plasma cell differentiation. Cell Mol Immunol 12(1):31–39
112. Wang S, Wang J, Kumar V, Karnell JL, Naiman B, Gross PS et al (2018) IL-21 drives expansion and plasma cell differentiation of autoreactive CD11chiT-bet+ B cells in SLE. Nat Commun 9(1):1758
113. Sawalha AH, Kaufman KM, Kelly JA, Adler AJ, Aberle T, Kilpatrick J et al (2008) Genetic association of interleukin-21 polymorphisms with systemic lupus erythematosus. Ann Rheum Dis 67 (4):458–461
114. Webb R, Merrill JT, Kelly JA, Sestak A, Kaufman KM, Langefeld CD et al (2009) A polymorphism within IL21R confers risk for systemic lupus erythematosus. Arthritis Rheum 60(8):2402–2407
115. Dolff S, Abdulahad WH, Westra J, Doornbos-van der Meer B, Limburg PC, Kallenberg CG et al (2011) Increase in IL-21 producing T-cells in patients with systemic lupus erythematosus. Arthritis Res Ther 13:R157. https://doi.org/10.1186/ar3474
116. Herber D, Brown TP, Liang S, Young DA, Collins M, Dunussi-Joannopoulos K (2007) IL-21 has a pathogenic role in a lupus-prone mouse model and its blockade with IL-21R.Fc reduces disease progression. J Immunol 178(6):3822–3830
117. Bettelli E, Carrier Y, Gao W, Korn T, Strom TB, Oukka M et al (2006) Reciprocal developmental pathways for the generation of pathogenic effector TH17 and regulatory T cells. Nature 441 (7090):235–238
118. Allard EL, Hardy MP, Leignadier J, Marquis M, Rooney J, Lehoux D et al (2007) Overexpression of IL-21 promotes massive CD8+ memory T cell accumulation. Eur J Immunol 37(11):3069–3077
119. Langrish CL, McKenzie BS, Wilson NJ, de Waal MR, Kastelein RA, Cua DJ (2004) IL-12 and IL-23: master regulators of innate and adaptive immunity. Immunol Rev 202:96–105
120. Trinchieri G (2003) Interleukin-12 and the regulation of innate resistance and adaptive immunity. Nat Rev Immunol 3(2):133–146
121. Watford WT, Hissong BD, Bream JH, Kanno Y, Muul L, O'Shea JJ (2004) Signaling by IL-12 and IL-23 and the immunoregulatory roles of STAT4. Immunol Rev 202:139–156
122. Manetti R, Parronchi P, Giudizi MG, Piccinni MP, Maggi E, Trinchieri G et al (1993) Natural killer cell stimulatory factor (interleukin 12 [IL-12]) induces T helper type 1 (Th1)-specific immune responses and inhibits the development of IL-4-producing Th cells. J Exp Med 177(4):1199–1204
123. Hagberg N, Joelsson M, Leonard D, Reid S, Eloranta ML, Mo J et al (2018) The STAT4 SLE risk allele rs7574865[T] is associated with increased IL-12-induced IFN-γ production in T cells from patients with SLE. Ann Rheum Dis 77(7):1070–1077
124. Kikawada E, Lenda DM, Kelley VR (2003) IL-12 deficiency in MRL-Fas(lpr) mice delays nephritis and intrarenal IFN-gamma expression, and diminishes systemic pathology. J Immunol 170(7):3915–3925
125. Floss DM, Schröder J, Franke M, Scheller J (2015) Insights into IL-23 biology: From structure to function. Cytokine Growth Factor Rev 26(5):569–578
126. Larosa M, Zen M, Gatto M, Jesus D, Zanatta E, Iaccarino L et al (2019) IL-12 and IL-23/Th17 axis in systemic lupus erythematosus. Exp Biol Med 244 (1):42–51
127. Teng MW, Bowman EP, McElwee JJ, Smyth MJ, Casanova JL, Cooper AM et al (2015) IL-12 and IL-23 cytokines: from discovery to targeted therapies for immune-mediated inflammatory diseases. Nat Med 21(7):719–729
128. Parham C, Chirica M, Timans J, Vaisberg E, Travis M, Cheung J et al (2002) A Receptor for the Heterodimeric Cytokine IL-23 Is Composed of IL-12Rβ1 and a Novel Cytokine Receptor Subunit, IL-23R. J Immunol 168(11):5699–5708
129. Stritesky GL, Yeh N, Kaplan MH (2008) IL-23 promotes maintenance but not commitment to the Th17 lineage. J Immunol 181(9):5948–5955
130. Murphy CA, Langrish CL, Chen Y, Blumenschein W, McClanahan T, Kastelein RA et al (2003) Divergent pro- and antiinflammatory roles for IL-23 and IL-12 in joint autoimmune inflammation. J Exp Med 198(12):1951–1957

131. Tang C, Chen S, Qian H, Huang W (2012) Interleukin-23: as a drug target for autoimmune inflammatory diseases. Immunology 135(2):112–124
132. Ghilardi N, Kljavin N, Chen Q, Lucas S, Gurney AL, De Sauvage FJ (2004) Compromised humoral and delayed-type hypersensitivity responses in IL-23-deficient mice. J Immunol 172(5):2827–2833
133. Dai H, He F, Tsokos GC, Kyttaris VC (2017) IL-23 limits the production of IL-2 and promotes autoimmunity in lupus. J Immunol 199(3):903–910
134. Kyttaris VC, Zhang Z, Kuchroo VK, Oukka M, Tsokos GC (2010) Cutting edge: IL-23 receptor deficiency prevents the development of lupus nephritis in C57BL/6-lpr/lpr mice. J Immunol 184(9):4605–4609
135. Yin Z, Bahtiyar G, Zhang N, Liu L, Zhu P, Robert ME et al (2002) IL-10 regulates murine lupus. J Immunol 169(4):2148–2155
136. Kyttaris VC, Kampagianni O, Tsokos GC (2013) Treatment with anti-interleukin 23 antibody ameliorates disease in lupus-prone mice. Biomed Res Int 2013:861028. https://doi.org/10.1155/2013/861028
137. Huang X, Hua J, Shen N, Chen S (2007) Dysregulated expression of interleukin-23 and interleukin-12 subunits in systemic lupus erythematosus patients. Mod Rheumatol 17(3):220–223
138. Krueger JG, Ferris LK, Menter A, Wagner F, White A, Visvanathan S et al (2015) Anti-IL-23A mAb BI 655066 for treatment of moderate-to-severe psoriasis: Safety, efficacy, pharmacokinetics, and biomarker results of a single-rising-dose, randomized, double-blind, placebo-controlled trial. J Allergy Clin Immunol 136(1):116–24.e7
139. van Vollenhoven RF, Hahn BH, Tsokos GC, Wagner CL, Lipsky P, Touma Z et al (2018) Efficacy and safety of ustekinumab, an IL-12 and IL-23 inhibitor, in patients with active systemic lupus erythematosus: results of a multicentre, double-blind, phase 2, randomised, controlled study. Lancet 392(10155):1330–1339
140. van Vollenhoven RF, Hahn BH, Tsokos GC, Lipsky P, Fei K, Gordon RM et al (2020) Maintenance of Efficacy and Safety of Ustekinumab Through One Year in a Phase II Multicenter, Prospective, Randomized, Double-Blind, Placebo-Controlled Crossover Trial of Patients With Active Systemic Lupus Erythematosus. Arthritis Rheumatol 72(5):761–768
141. Reich K, Papp KA, Blauvelt A, Tyring SK, Sinclair R, Thaçi D et al (2017) Tildrakizumab versus placebo or etanercept for chronic plaque psoriasis (reSURFACE 1 and reSURFACE 2): results from two randomised controlled, phase 3 trials. Lancet 390(10091):276–288
142. Langley RG, Tsai TF, Flavin S, Song M, Randazzo B, Wasfi Y et al (2018) Efficacy and safety of guselkumab in patients with psoriasis who have an inadequate response to ustekinumab: results of the randomized, double-blind, phase III NAVIGATE trial. Br J Dermatol 178(1):114–123
143. Tackey E, Lipsky PE, Illei GG (2004) Rationale for interleukin-6 blockade in systemic lupus erythematosus. Lupus 13(5):339–343
144. Naka T, Nishimoto N, Kishimoto T (2002) The paradigm of IL-6: from basic science to medicine. Arthritis Res 4(Suppl 3):S233–S242
145. Nishimoto N, Kishimoto T (2006) Interleukin 6: from bench to bedside. Nat Clin Pract Rheumatol 2(11):619–626
146. Kitani A, Hara M, Hirose T, Harigai M, Suzuki K, Kawakami M et al (1992) Autostimulatory effects of IL-6 on excessive B cell differentiation in patients with systemic lupus erythematosus: analysis of IL-6 production and IL-6R expression. Clin Exp Immunol 88(1):75–83
147. Xu L, Kitani A, Fuss I, Strober W (2007) Cutting edge: regulatory T cells induce CD4+CD25-Foxp3-T cells or are self-induced to become Th17 cells in the absence of exogenous TGF-beta. J Immunol 178(11):6725–6729
148. Mihara M, Ohsugi Y (1990) Possible role of IL-6 in pathogenesis of immune complex-mediated glomerulonephritis in NZB/W F1 mice: induction of IgG class anti-DNA autoantibody production. Int Arch Allergy Appl Immunol 93(1):89–92
149. Spronk PE, ter Borg EJ, Limburg PC, Kallenberg CG (1992) Plasma concentration of IL-6 in systemic lupus erythematosus; an indicator of disease activity? Clin Exp Immunol 90(1):106–110
150. Ripley BJM, Goncalves B, Isenberg DA, Latchman DS, Rahman A (2005) Raised levels of interleukin 6 in systemic lupus erythematosus correlate with anaemia. Ann Rheum Dis 64(6):849–853
151. Iwano M, Dohi K, Hirata E, Kurumatani N, Horii Y, Shiiki H et al (1993) Urinary levels of IL-6 in patients with active lupus nephritis. Clin Nephrol 40(1):16–21
152. Lu R, Munroe ME, Guthridge JM, Bean KM, Fife DA, Chen H et al (2016) Dysregulation of innate and adaptive serum mediators precedes systemic lupus erythematosus classification and improves prognostic accuracy of autoantibodies. J Autoimmun 74:182–193
153. Rovin BH, van Vollenhoven RF, Aranow C, Wagner C, Gordon R, Zhuang Y et al (2016) A Multicenter, Randomized, Double-Blind, Placebo-Controlled Study to Evaluate the Efficacy and Safety of Treatment With Sirukumab (CNTO 136) in Patients With Active Lupus Nephritis. Arthritis Rheumatol 68(9):2174–2183
154. Wallace DJ, Strand V, Merrill JT, Popa S, Spindler AJ, Eimon A et al (2017) Efficacy and safety of an interleukin 6 monoclonal antibody for the treatment of systemic lupus erythematosus: a phase II dose-ranging randomised controlled trial. Ann Rheum Dis 76(3):534

155. Ablynx A (2019) Phase II study to evaluate safety and efficacy of ALX-0061 in subjects with systemic lupus erythematosus. Identifier NCT02437890: national library of medicine [updated February 26 Available from: https://clinicaltrials.gov/ct2/show/NCT02437890
156. Illei GG, Shirota Y, Yarboro CH, Daruwalla J, Tackey E, Takada K et al (2010) Tocilizumab in systemic lupus erythematosus: data on safety, preliminary efficacy, and impact on circulating plasma cells from an open-label phase I dosage-escalation study. Arthritis Rheum 62(2):542–552
157. Lieberman LA, Tsokos GC (2010) The IL-2 defect in systemic lupus erythematosus disease has an expansive effect on host immunity. J Biomed Biotechnol 2010:740619
158. Abbas AK (2003) The control of T cell activation vs. tolerance. Autoimmun Rev 2(3):115–118
159. Fontenot JD, Rasmussen JP, Gavin MA, Rudensky AY (2005) A function for interleukin 2 in Foxp3-expressing regulatory T cells. Nat Immunol 6(11):1142–1151
160. Mitra S, Leonard WJ (2018) Biology of IL-2 and its therapeutic modulation: Mechanisms and strategies. J Leukoc Biol 103(4):643–655
161. Alcocer-Varela J, Alarcón-Segovia D (1982) Decreased production of and response to interleukin-2 by cultured lymphocytes from patients with systemic lupus erythematosus. J Clin Invest 69 (6):1388–1392
162. de Faucal P, Godard A, Peyrat MA, Moreau JF, Soulillou JP. (1984) Impaired IL2 production by lymphocytes of patients with systemic lupus erythematosus. Ann Immunol (Paris) 135d(2):161–172
163. Ishida H, Kumagai S, Umehara H, Sano H, Tagaya Y, Yodoi J, et al (1987) Impaired expression of high affinity interleukin 2 receptor on activated lymphocytes from patients with systemic lupus erythematosus. J Immunol139(4):1070–1074
164. Schorle H, Holtschke T, Hünig T, Schimpl A, Horak I (1991) Development and function of T cells in mice rendered interleukin-2 deficient by gene targeting. Nature 352(6336):621–624
165. Suzuki H, Kündig TM, Furlonger C, Wakeham A, Timms E, Matsuyama T et al (1995) Deregulated T cell activation and autoimmunity in mice lacking interleukin-2 receptor beta. Science 268 (5216):1472–1476
166. Malek TR, Yu A, Vincek V, Scibelli P, Kong L (2002) CD4 regulatory T cells prevent lethal autoimmunity in IL-2Rbeta-deficient mice. Implications for the nonredundant function of IL-2. Immunity 17(2):167–178
167. von Spee-Mayer C, Siegert E, Abdirama D, Rose A, Klaus A, Alexander T et al (2016) Low-dose interleukin-2 selectively corrects regulatory T cell defects in patients with systemic lupus erythematosus. Ann Rheum Dis 75(7):1407–1415
168. He J, Zhang X, Wei Y, Sun X, Chen Y, Deng J et al (2016) Low-dose interleukin-2 treatment selectively modulates CD4(+) T cell subsets in patients with systemic lupus erythematosus. Nat Med 22(9):991–993
169. Humrich JY, von Spee-Mayer C, Siegert E, Bertolo M, Rose A, Abdirama D et al (2019) Low-dose interleukin-2 therapy in refractory systemic lupus erythematosus: an investigator-initiated, single-centre phase 1 and 2a clinical trial. The Lancet Rheumatology. 1(1):e44–e54
170. He J, Zhang R, Shao M, Zhao X, Miao M, Chen J et al (2020) Efficacy and safety of low-dose IL-2 in the treatment of systemic lupus erythematosus: a randomised, double-blind, placebo-controlled trial. Ann Rheum Dis 79(1):141
171. Boonstra A, Rajsbaum R, Holman M, Marques R, Asselin-Paturel C, Pereira JP et al (2006) Macrophages and myeloid dendritic cells, but not plasmacytoid dendritic cells, produce IL-10 in response to MyD88- and TRIF-dependent TLR signals, and TLR-independent signals. J Immunol 177 (11):7551–7558
172. Gabryšová L, Howes A, Saraiva M, O'Garra A. The regulation of IL-10 expression (2014) In: Fillatreau S, O'Garra A, editors. Interleukin-10 in health and disease. Berlin, Heidelberg: Springer Berlin Heidelberg p 157–190
173. Fillatreau S, Sweenie CH, McGeachy MJ, Gray D, Anderton SM (2002) B cells regulate autoimmunity by provision of IL-10. Nat Immunol 3(10):944–950
174. Saraiva M, O'Garra A (2010) The regulation of IL-10 production by immune cells. Nat Rev Immunol 10(3):170–181
175. Flores-Borja F, Bosma A, Ng D, Reddy V, Ehrenstein MR, Isenberg DA, et al. (2013) CD19 +CD24hiCD38hi B cells maintain regulatory T cells while limiting TH1 and TH17 differentiation. Science Translational Medicine. 5(173):173ra23. doi: 10.1126/scitranslmed.3005407.
176. Jinquan T, Larsen CG, Gesser B, Matsushima K, Thestrup-Pedersen K (1993) Human IL-10 is a chemoattractant for CD8+ T lymphocytes and an inhibitor of IL-8-induced CD4+ T lymphocyte migration. J Immunol 151(9):4545–4551
177. Groux H, Bigler M, de Vries JE, Roncarolo MG (1998) Inhibitory and stimulatory effects of IL-10 on human CD8+ T cells. J Immunol 160(7):3188–3193
178. Weber-Nordt RM, Riley JK, Greenlund AC, Moore KW, Darnell JE, Schreiber RD (1996) Stat3 recruitment by two distinct ligand-induced, tyrosine-phosphorylated docking sites in the interleukin-10 receptor intracellular domain. J Biol Chem 271(44):27954–27961
179. O'Garra A, Barrat FJ, Castro AG, Vicari A, Hawrylowicz C (2008) Strategies for use of IL-10 or its antagonists in human disease. Immunol Rev 223(1):114–131
180. Kühn R, Löhler J, Rennick D, Rajewsky K, Müller W (1993) Interleukin-10-deficient mice develop chronic enterocolitis. Cell 75(2):263–274

181. Gazzinelli RT, Wysocka M, Hieny S, Scharton-Kersten T, Cheever A, Kühn R et al (1996) In the absence of endogenous IL-10, mice acutely infected with Toxoplasma gondii succumb to a lethal immune response dependent on CD4+ T cells and accompanied by overproduction of IL-12. IFN-gamma and TNF-alpha. J Immunol. 157(2):798–805
182. Go NF, Castle BE, Barrett R, Kastelein R, Dang W, Mosmann TR et al (1990) Interleukin 10, a novel B cell stimulatory factor: unresponsiveness of X chromosome-linked immunodeficiency B cells. J Exp Med 172(6):1625–1631
183. Chun HY, Chung JW, Kim HA, Yun JM, Jeon JY, Ye YM et al (2007) Cytokine IL-6 and IL-10 as biomarkers in systemic lupus erythematosus. J Clin Immunol 27(5):461–466
184. Levy Y, Brouet JC (1994) Interleukin-10 prevents spontaneous death of germinal center B cells by induction of the bcl-2 protein. J Clin Invest 93 (1):424–428
185. Rousset F, Garcia E, Defrance T, Péronne C, Vezzio N, Hsu DH et al (1992) Interleukin 10 is a potent growth and differentiation factor for activated human B lymphocytes. Proc Natl Acad Sci 89 (5):1890
186. Park YB, Lee SK, Kim DS, Lee J, Lee CH, Song CH (1998) Elevated interleukin-10 levels correlated with disease activity in systemic lupus erythematosus. Clin Exp Rheumatol 16(3):283–288
187. Houssiau FA, Lefebvre C, Vanden Berghe M, Lambert M, Devogelaer JP, Renauld JC (1995) Serum interleukin 10 titers in systemic lupus erythematosus reflect disease activity. Lupus 4 (5):393–395
188. Beebe AM, Cua DJ, de Waal MR (2002) The role of interleukin-10 in autoimmune disease: systemic lupus erythematosus (SLE) and multiple sclerosis (MS). Cytokine Growth Factor Rev 13(4–5):403–412
189. Peng H, Wang W, Zhou M, Li R, Pan H-F, Ye D-Q (2013) Role of interleukin-10 and interleukin-10 receptor in systemic lupus erythematosus. Clin Rheumatol 32(9):1255–1266
190. Ishida H, Muchamuel T, Sakaguchi S, Andrade S, Menon S, Howard M (1994) Continuous administration of anti-interleukin 10 antibodies delays onset of autoimmunity in NZB/W F1 mice. J Exp Med 179(1):305–310
191. Llorente L, Richaud-Patin Y, García-Padilla C, Claret E, Jakez-Ocampo J, Cardiel MH et al (2000) Clinical and biologic effects of anti-interleukin-10 monoclonal antibody administration in systemic lupus erythematosus. Arthritis Rheum 43(8):1790–1800
192. Damjanov N (2020) Proof-of-concept study with BT063 in subjects with systemic lupus erythematosus (BT063 in SLE). Identifier NCT02554019: national library of medicine s[updated January 27. Available from: https://clinicaltrials.gov/ct2/show/NCT02554019 .
193. St Clair EW (1999) Interleukin 10 treatment for rheumatoid arthritis. Ann Rheum Dis 58(suppl 1): I99
194. Burmeister AR, Marriott I. (2018) The interleukin-10 family of cytokines and their role in the CNS. Front Cell Neurosci 12:458. https://doi.org/10.3389/fncel.2018.00458

Intracellular Targets in SLE

8

Alberta Hoi

Abstract

In SLE, the magnitude or persistence of the disease is determined by the perpetuation of an exaggerated innate and adaptive immune responses towards self-antigens. Soluble mediators acting on cellular surface receptors bring signals to a variety of immune cells, and then there is a complex network of intracellular signalling that leads to a variety of transcriptional changes. These pathways can be linked to the particular cytokines or chemokines but can also be activated by receptor independent responses. The kinases that initiate these pathways are generally classified according to their structure and primary functions. While there has been much interest to develop small molecule drug targeting these intracellular targets, the attrition rate of those made to clinical development has been quite high. In SLE, a number of emerging targets has been studied and developed. Some of these are closer to the therapeutic horizon than others and will be explored specifically in this chapter. Key T cell intracellular pathways such as the calcium-calcineurin-Nuclear Factor of an Activated T cells (NFAT) pathway, and Mitogen-Activated Protein Kinases (MAPK), and B cell receptor mediated pathways such as Brunton's tyrosine kinase (BTK), or spleen tyrosine kinase (SYK) pathway, and the Janus kinase (JAK), ubiquitin–proteasome system (UPS), and the Mechanistic Target of Rapamycin (mTOR) pathways that are present in broad ranging types of immune cells will be discussed.

A. Hoi (✉)
Rheumatology Research Group, Centre for Inflammatory Diseases, School of Clinical Sciences at Monash Health, Monash University, Clayton, VIC 3168, Australia
e-mail: alberta.hoi@monash.edu

A. Hoi
Department of Rheumatology, Monash Health, Clayton, VIC 3168, Australia

A. Hoi
Department of Rheumatology, Austin Health, Heidelberg, VIC 3084, Australia

A. Hoi (ed.), *Pathogenesis of Systemic Lupus Erythematosus*,
https://doi.org/10.1007/978-3-030-85161-3_8

Why is it Attractive to Consider Developing Therapeutics Against Intracellular Targets

In the last two decades, there has been growing interest and development of monoclonal antibodies, soluble receptors or fusion proteins against pathogenic proteins that mediate diseases in rheumatology. Collectively we refer to these drugs as biologicals. While these agents have shown to be impressively efficacious, there are limitations to their production and use in a more widespread way. The biologicals must be administered parenterally as otherwise these proteins will be denatured by the gastrointestinal tract. The manufacturing, transportation, and storage of these drugs are also resource intensive and therefore translated into increased cost.

Small molecules are much cheaper to manufacture in the long run. Understanding the intracellular signalling pathways and their interaction has been an ongoing research interest in rheumatic diseases including SLE. In this chapter we will review a number of key targets that have gained interests in the last few years based on our currently understanding of immunopathogenesis in SLE, and these may potentially develop into valid therapeutic options for patients in the near future.

Intracellular Signalling Pathways

In SLE, there is a complex network of intracellular signalling that leads to a variety of transcriptional changes, and these were studied initially in the context of various lymphocyte subsets but now recognised to also play an important part within the innate immune system and its interaction with the adaptive immune system.

While there has been much interest to develop small molecule drug targeting these intracellular targets, the attrition rate of those made to clinical development has been quite high. In SLE, the emerging targets came from our understanding of different receptor and non-receptor activated kinases. These pathways have functions that cross over between inflammation, cell survival and proliferation. Gene methylation is another critical step that determines the final expression of inflammatory response.

In this chapter we will focus on a few of these pathways, chosen because of emerging therapeutics that are being developed for the treatment of SLE that target these specific areas. This is by no means a comprehensive description of all the intracellular pathways that are of importance in lupus pathogenesis, but we have focussed on ones that have progressed furthest in the therapeutics in human SLE.

Inhibition of the Calcium-Calcineurin-NFAT Pathway

One of the oldest small molecule targets is calcineurin, which a calcium and calmodulin dependent serine/ threonine protein phosphatase. Calcineurin inhibitors (CNI), such as ciclosporin, have been in use since the early 1980s, as one of the most commonly used immunosuppressants, with voclosporin being the latest CNI currently under investigation for treatment of SLE [1].

Ciclosporin has been used as immunosuppressant in SLE patients who have inadequate response to corticosteroids or other conventional immunosuppressants. Germano et al. [2] It is still among a short list of agents that has a reasonable safety profile for pregnant lupus patients. Recently randomised controlled trials have demonstrated noninferiority of tacrolimus to mycophenolate mofetil (MMF) or cyclophosphamide (CYC) for induction therapy of lupus nephritis. Mok et al. [3] Furthermore, a multi-target regimen of using mycophenolate in combination of tacrolimus has been shown to be effective in controlling local and systemic inflammation in patients with severe lupus nephritis, both as induction and maintenance therapy [4, 5].

The more novel generation of CNI such as voclosporin is of great interest to clinicians. Volcosporin is structurally very similar to ciclosporin but the small structural change resulted in a more predictable dose response compared to its predecessor ciclosporin. It has similar effects on T cell activation, stabilised podocytes and protects against proteinuria. In phase 2 and 3 studies where it was used in combination with mycophenolate it can induce a higher rate of renal remission responses. Rovin et al. [6]. It is currently under priority review by FDA for approval.

The mechanism of action of CNI has been well studied, particularly its downstream effect of inhibiting IL-2 production and IL-2 receptor expression, leading to a reduction in T cell activation. The effects of IL-2 on T cell mediated immune responses and dysregulation of IL-2 in SLE were discussed in Chap. 6.

Calcineurin inhibitors can be classified according to the immunophilins that are targeted. These immunophilins are family of highly conserved intracellular protein that has enzymatic (peptidyloprolyl cis–trans isomerase, PPIase) activities [7]. Ciclosporin targets cyclophilins, whereas tacrolimus bonds to FK506 binding protein (FKBP); these drug-receptor complex specifically binds to and inhibits calcineurin in a competitive manner. Calcineurin is a crucial intracellular target to modulate T cell signalling. It is a serine/ threonine phosphatase that is controlled by calmodulin, a calcium binding messenger protein and can sense intracellular calcium influx (Fig. 8.1). Calcineurin has the subunit A and B; subunit A (CnA) is the catalytic unit that is regulated by the calcium sensing subunit B (CnB). Activated CnA dephosphorylates the cytoplasmic nuclear factor of an activated T cell (cNFAT) and allow its translocation into the nucleus [8].

In addition to directly targeting calcineurin, the nuclear factor of an activated T cell (NFAT) has been identified as a major signalling pathway that may be modulated in the treatment of autoimmune diseases. NFAT function is now appreciated not just limited to T cell activation, via its upregulation of IL-2, but a family of NFATs can be involved in lineage-specific transcription factors during T helper cell differentiation into Th1, Th2 and Th17 and Tfh lineage cells. NFAT can regulate a long list of immune modulating cytokines and signalling molecules, such as GM-CSF, Foxhead Protein 3 (FOXP3), immunoglobulin kappa, IFN γ, CD5, CD25, CD28, CD40, IL-3, IL-4, IL-5, IL-13, IL-8, IL-

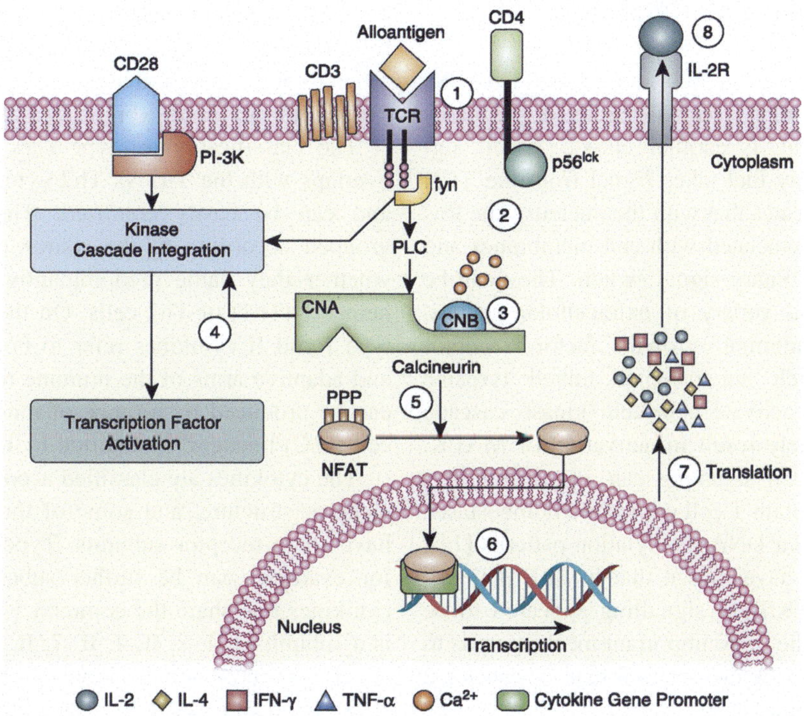

Fig. 8.1 Intracellular actions of calcineurin. Calcineurin consists of 2 subunits: CnA and CnB. Upon activation of the TCR by an alloantigen (1), an influx of intracellular calcium (2) serves to activate CnB (3). This allows CnB to unleash the phosphatase activity of CnA (4). The activated CnA of calcineurin can dephospharylate cytoplasmic NFAT (5) to allow translocation into the nucleus (6) where it uprgulate the expression of multiple cytokines and costimulatory molecules (7), including IL-2 which can serve as an autocrine factor on itself, to induce further activation and proliferation (8). PLC: phospholipase C; CnA: Catalytic subunit A; CnB: Catalytic subunit B; NFAT: Nuclear factor of activated T-cells; TCR: T cell receptor; Fyn: A Src family tyrosine-protein kinase. Courtesy of Azzi et al. Journal of Immunology [8]

31, Fas ligand, MIP-1α, SYK, COX-2, TNF α and BlyS. Lee et al. [9] The specific gene expression is dependent on the interaction of other transcription factors such as Activator Protein 1 (AP-1), and other interacting partner proteins such as T-bet, GATA-3, ROR γt, FOXP3, and Basic leucine zipper ATF-like transcription factor (BATF) [10, 11]. Additional kinases such as casein kinase (CK1), glycogen synthase kinase 3 (GSK3) and JUN N-terminal kinase (JNK) and p38 are important in maintaining NFAT in an inactive state [12, 13]. Manipulation of NFAT pathway, other than using direct calcineurin inhibitors, is under active investigation [9]. Drugs that inhibit calcineurin-NFAT interaction, NFAT migration or stability, or NFAT-DNA interaction are all potential emerging therapeutic agents.

The MAPK-ERK-AP1 Pathway

The Mitogen-Activated Protein Kinases (MAPK) or Extracellular signal-regulated kinases (ERK) pathway is another classic T cell signalling pathway that takes signal from the TCR and communicates that with the nucleus. The Ras proteins are associated with cell membranes and function as a binary signal switch. They can be activated by a variety of extracellular signals, such as epidermal growth factor receptor (EGFR), which are receptor linked tyrosine kinase. They activate a protein kinase cascade which ultimately result in activation of MAPK. Defective T cell signaling can also result in a failure to maintain T cell epigenetic homeostasis, and in particular DNA methylation patterns [14]. Some studies have shown that in SLE patients, level of MAPK/ERK signalling appeared to be reduced, but the more important mechanism is to reduction in DNA methylation that is associated with this change [15].

Another MAPK, c-Jun N-terminal kinase (JNK), seems to be promoted by inflammatory signals and has a more pro-apoptotic effect. The cross-talk between the two main MAPKs may determine the final outcome of cell survival [16].

Even though MAPK is ubiquitously expressed, its role in T cell functioning as one of the key pathways to regulate cell growth and proliferation in inflammatory conditions has been the subject of ongoing research. While there have been early interests in the development of MAPK inhibitor in SLE based on promising animal models, the development into therapeutics in human appears to be on hold in recent years [14, 17]. There have been greater interests in the development of specific DNA methyltransferase inhibitors and histone deacetylases (HDACs), that target the epigenetic programming and control gene expression [18, 19].

Janus Kinase Inhibition

The Janus Kinases (JAKs) are cytoplasmic tyrosine kinases that are associated with Type I and Type II cytokine receptors, and are responsible for mediating the intracellular signalling of over 50 cytokines, interleukins, interferons, growth factors and hormones (Fig. 8.2) [20].

The classification of type I vs II cytokines overlaps with the Th1 vs Th2 cytokines system and can be easily confused. The latter was grouped according to the source of cytokines, whether they came predominantly from the T helper 1 (Th1) or Th2 cells. On the other hand, type I and II cytokines refer to both the innate and adaptive arms of the immune responses and can be produced by a range of innate lymphoid cell (ILC) lineages, in addition to T helper cells.

The cytokines are classified according to their receptor structure, and some of these cytokines have share receptor subunits. Type I cytokines, for example, can be further subclassified into cytokines that share the common γ-chain, which is a subunit of IL-2, IL-4, IL-7, IL-9, IL-15 and IL-21. The common β chain cytokines include IL-3, IL-5 and GM-CSF. The gp130 receptor subunit tied in cytokines such as IL-6 IL-11, IL-27, and IL-31. The dimeric cytokine cytokines also belong to the type I cytokines, and include IL-12, IL-23 and IL-35. In contrast, type II cytokines include all the interferons (type I, II,

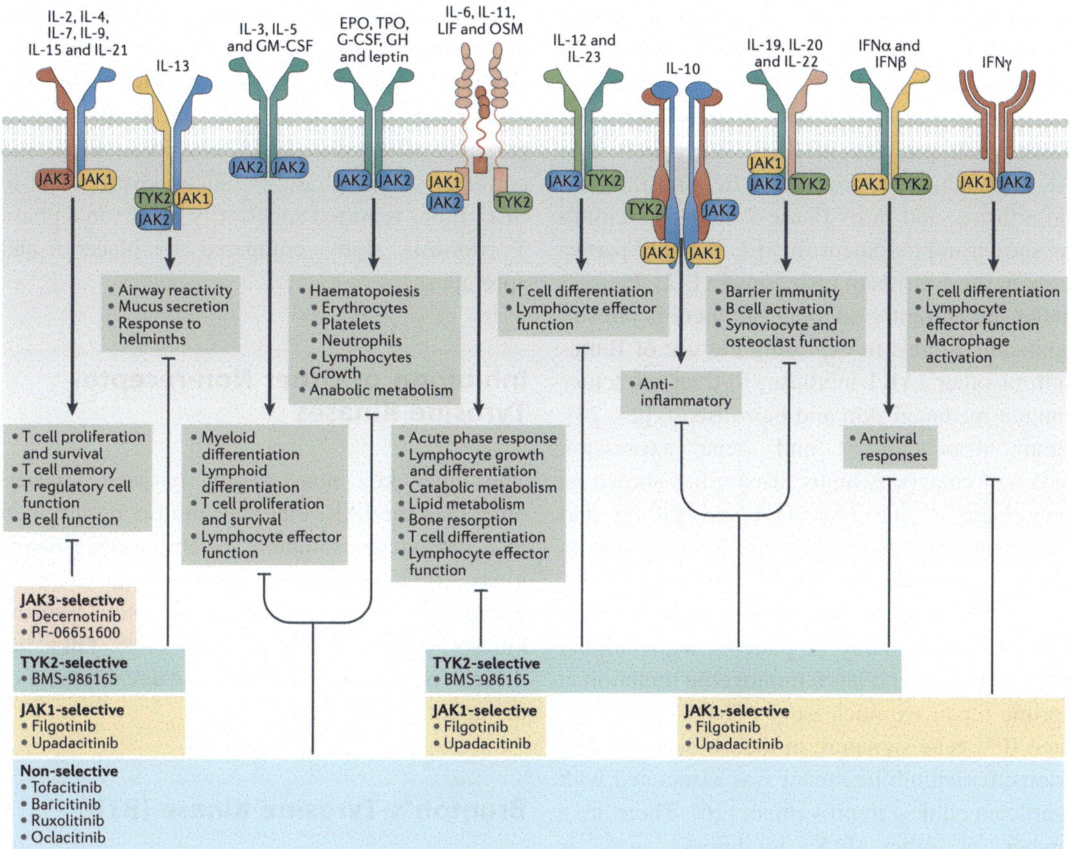

Fig. 8.2 Effects of different JAK targeting according to types of cytokine receptors. Cytokine receptors can be classified into Type 1 and II according to their structures, and they physically associate with Janus kinases and transduce intracellular signals. Different receptors associate with different JAKs and targeting a specific JAK can result in broad ranging downstream effects. Courtesy of Schwartz et al. Nature Reviews Drug Discovery [20]

and III) and IL-10 related cytokines (IL-10, IL-19, IL-20, IL_22, IL-24, and IL-26) [21].

Upon cytokine binding, the JAKs are activated. Even though JAKs does not have a transmembrane-spanning domain, they can phosphorylate the cytokine receptors, creating docking sites for signalling molecules such as the Signal Transducer and Activator of Transcription (STAT) where they induce transcription of target genes as the initiating step. The STAT proteins form homo or heterodimers and translocate into the nucleus where they induce transcription of target genes.

There are 4 isoforms of JAKs which physically associate with the receptors of a variety of type I and II cytokines, namely JAK1, JAK2, JAK3 and tyrosine kinase 2 (TYK2). These isoforms form different combinations, which mediate effects of different cytokine binding. Relevant to SLE, the emerging interest in targeting JAK is based on recognition that type I IFNs mediate their effects via signalling of the JAK-STAT pathway, and in particular, IFNα and IFNβ mediate their signalling via JAK1 and Tyk2 (Fig. 8.1). However, the effects of JAK inhibition are likely broad ranging, based on knowledge of the complexity and overlap of signalling pathways between a large range of cytokines.

While it may be an over-simplification to say that JAK1 inhibition blocks signalling of many important pro- inflammatory cytokines relevant to a number of inflammatory and autoimmune conditions, whereas JAK2 and JAK3 signalling pathways are more associated with growth factor signalling and host defence, recent observations from clinical trials using JAK1 selective agent has shown great promise in SLE. Baricitinib is a JAK1/2 inhibitor, approved for treating rheumatoid arthritis, and in its Phase 2 study baricitinib has shown improvement in SLE patients particularly in the arthritis or rash domain [22]. Phase 3 studies are currently ongoing but there is already significant interest in exploring the use of Baricitinib or other JAK1 inhibitor, to treat refractory immune-mediated skin and hair disease [23, 24]. Immunohistochemistry and gene expression studies of cutaneous lupus disease has shown an upregulated of the JAK/STAT signalling and inflammatory mediators that are associated with IFN gene upregulation [25]. While JAK1/2 inhibition affects more than IFN signalling, in a proof-of-concept study of patients with familial chilblain lupus, who has a monogenic mutation at 3 prime repair exonuclease 1 (TREX1) and elevated IFN gene signature in their blood and skin lesions, baricitinib treatment was associated with significant clinical improvement [26]. There are a number of other JAK inhibitors, such as upadacitinib, filgotinib currently under investigation for the treatment of SLE.

The 4 JAK isoforms are associated with different cytokine receptors, and in addition to JAK1, there is significant interest in exploring tyrosine kinase 2 (TYK2) which is the key component of the type I IFN signalling, as well as IL-6, and IL-12/23 signalling. TYK regulated pathways result in activation of Th17, Th1, B and myeloid cells, all critical in the pathogenesis of autoimmune diseases. Polymorphism of TYK2 have been well described in different SLE populations [27–29]. Furthermore, a deactivating coding variant has been described to be associated with protective effect from a number of common autoimmune diseases but no major signal on adverse effects of immunosuppression such as major infections [30, 31].

An orally active TYK inhibitor, Deucravacitinib (BMS-986165), acts by stabilising the pseudokinase domain of TYK2 in a conformational state and inhibiting receptor-mediated activation and activity of the catalytic domain. This novel allosteric mechanism avoids the binding to the active site on the kinase domain. Earlier in vitro and animal studies all support potential efficacy of BMS-986165 in SLE, and currently it is in phase 2 clinical development in SLE. It has reported superiority results in a phase 3 psoriasis study compared to placebo and apremilast.

Inhibition of Other Non-receptor Tyrosine Kinases

The other key non-receptor tyrosine kinases, other than the JAK, that play an important role in SLE pathogenesis include the Brunton's tyrosine kinase (BTK) and Spleen tyrosine kinase (SYK). There are many other non-receptor tyrosine kinases involved in significant cross talks but these two have led the way in the development of therapeutics in the area of SLE.

Brunton's Tyrosine Kinase (BTK)

The key B cell signalling molecule is called Brunton's tyrosine kinase (BTK). It is a non-receptor kinase, belonging to the TEC family, that is crucial for the signalling pathway downstream of the B cell antigen receptor (BCR). The downstream signalling upon B cell antigen receptor activation is mediated by BTK which is a crucial intracellular protein for B cell development. In Bruton's X-linked agammaglobulinemia in which BTK mutations are observed, we observed a complete absence of peripheral B cells and all immunoglobin classes [32]. BTK over-expression has been shown to increase the responsiveness of BCR stimulation with enhanced calcium influx.

BTK inhibition has gained interest for conditions that are thought to be largely driven by B cells. Ibrutinib and zanubrutinib, both oral BTK inhibitors, have been approved for the treatment of several B cell malignancies and [33, 34]. In some animal models of SLE, BTK inhibition has been shown clinical benefit [35].

In addition to its role on B cell activation, BTK is also expressed on a variety of other haematopoietic cells and can regulate signaling downstream to Fc receptors and Toll like receptors. BTK can be differentially regulated at different cell types and with other signalling pathways [36, 37]. Development of BTK inhibitors may have differential effects on these pathways depending on how it binds to the enzyme.

Spleen Tyrosine Kinase

The spleen tyrosine kinase (SYK) is a non-receptor tyrosine kinase that can mediate signalling by B cell receptors (BCR) and Fc receptors (FcR). Its function is not restricted to B cell signalling as it has been shown to be crucial in innate immunity and other non-immune functions. It is widely expressed in the hematopoietic lineage of the immune system, including B cells and T cells.

SYK activation leads to binding to the phosphorylated tyrosine residues in immunoreceptor tyrosine-based activation motifs (ITAMs) which in turn activates other receptor-proximal signalling molecules, such as the CARD9-BCL10-MALT1 pathway and the NLRP3 inflammasome [38]. Other classical pathway such as generation of membrane-associated phosphatidylinositol (3,4,5)-triphosphate (PIP3).

Over-expression of SYK is also seen in T cells of SLE patients. The normal engagement of the CD3/ T cell receptor complex is observed to be replaced by an alternative pathway of FcRγ and SYK in SLE. The FcRγ-SYK pathway is much more potent than the conventional CD3ζ and ζ-chain associated protein kinase (Zap70), in its activation of calcineurin. [39, 40] Inhibition of spleen tyrosine kinase (SYK) has been shown to downregulate T cell signalling.

SYK also mediates integrin signalling in neutrophil, macrophages and platelets. It is crucial as part of the innate immunity, in recognising fungal and other microbial pathogens. SYK is activated by C-type lectins and is required for the activation of NLR pyrin domain-containing 3 (NLRP3) inflammasome [38].

Fostamatinib, a Syk inhibitor was positioned to be a potential drug therapy in SLE, however its development was terminated due to poor results in the rheumatoid arthritis study.

Inhibition of the Ubiquitin/Proteasome System

In order to maintain normal protein homeostasis in the intracellular space, and thereby regulating a wide variety of cellular processes such as cell cycle progression differentiation, inflammatory response, antigen presentation and apoptosis, the proteolytic mechanism is tightly regulated to control protein degradation and turnover. The ubiquitin/ proteasome system (UPS) is a crucial pathway, acting as a recycler of misfolded or dysfunctional proteins.

The first step of the ubiquitin/ proteasome system to mark proteins destined by degradation by the attachment of ubiquitin. This process is activated through an ATP-dependent formation of a high energy thioester bond between the active site cysteine of the ubiquitin activating enzyme (E1) and the carboxyl terminus of ubiquitin. The activated ubiquitin is then transferred to the ubiquitin conjugating enzyme (E2), before it is moved to the protein substrates by the ubiquitin ligase (E3). Once the protein has been ubiquitinated by a monomer or polyubiquitinated chain, it is then recognised by the 26S proteasome which prepare the tagged substrate for protein degradation [41, 42].

Cereblon is a component of the E3 ubiquitin ligase complex called Cullin-4-RING ligase (CRL4) complex, which is formed with Damaged DNA Binding Protein 1 (DDB1), Cullin-4A (CUL4A) and RING box protein 1 (RBX1) [43]. There are now drugs that target cereblon, and activate the CRL4 complex, which in turn promotes the recruitment of substrates such as Ikaros (IKZF1) and Aiolos (IKZF3) to the E3 complex. These drugs are collectively referred to as Cereblon E3 ligase modulators (CELMoD), and include thalidomide and other related newer immunomodulatory agents such as lenalidomide, pomalidomide, and iberdomide. These novel CELMoDs are currently under investigation for the treatment of SLE or other haematopoietic malignancies such as multiple myeloma.

UPS dysfunction has been implicated in a number of neurodegenerative diseases as well as autoimmune diseases [44]. Tripartite motif-containing protein 21 (TRIM21) or Ro52 is an E3 ubiquitin ligase that has physiological and pathological roles in autoimmune diseases [45]. TRIM21 overexpression is associated with generation of the anti-Ro antibodies and increased apoptotic cell death in SLE patients.

Targeting Immunoproteasomes

The second major component of the UPS focuses on the function of proteasome in eliminating misfolded and dysfunctional proteins. Proteasome inhibitors have been developed and have been founded to be highly effective in haematopoietic malignancies such as multiple myeloma, where excessive proteins are produced, and the proteasome inhibitors stop to eliminate these dysfunctional proteins which result in rapid apoptosis of the malignant cells [46]. Proteasome inhibitors also have a number of other regulatory effects on other immune pathways, downregulating transcriptional regulators or activating protein kinases, but the exact mechanisms of action remain elusive. The success of proteosome inhibitors in the treatment of myeloma is limited due to some of its associated adverse effects particularly the gastrointestinal side effects and neurotoxicity, due to the ubiquitous nature of proteasomes [47].

Immunoproteasomes are inducible subset of proteasomes that are high expressed and upregulated by proinflammatory cytokines or oxidative stress in a range of immune cells. They are of particular interests as targets for the treatment of autoimmune diseases. The immunoproteasomes have three inducible β-subunits that substitute the β1c, β2c and β5c during de novo assembly. These inducible subunits are called low molecular mass polypeptide 2 (LMP2), multicatalytic endopeptidase complex-like 1 (MECL1), and LMP7 respectively [48].

The function of immunoproteasome has a shared functionality as other proteasomes in the generation of antigenic peptides for MHC Class I presentation and can regulate CD8 T cell response. Immunoproteasomes also play an important role in degrading the inhibitory IκB or p100 proteins in the NF-κB activation pathways, which result in the promotion of nuclear transloation of active NF-κB dimer for its transcriptional effects. In other animal models, inhibition of immunoproteasome has shown to reduce type I IFN activation and autoantibody production by plasma cells, suggesting a potential benefit in SLE [49].

Novel generation of proteasome inhibitors has greater selectivity for immunoproteasomes. ONX0914 is a potent inhibitor specific for the LMP7 subunit. Preclinical animal models have shown its therapeutic potential in a range of inflammatory disease models including murine lupus [49–51]. KZR616 is another novel immunoproteasome inhibitor with promising data in human SLE with and without nephritis, presented in ACR Convergence 2020. It has also been evaluated in other related immune diseases such as dermatomyositis, autoimmune haemolytic anaemia and thrombocytopenia.

Inhibition of Mechanistic Target of Rapamycin (mTOR) Pathway

Mechanistic target of rapamycin (mTOR) is another serine-threonine kinase, and play a significant part in cellular survival, proliferation and differentiation and cellular metabolism [52]. The first described drug targeting mTOR, is of course rapamycin itself, or it is also known by its other name sirolimus. Sirolimus is a macrolide that has been found to have ability to form a complex with the intracellular FK506-binding protein (FKBP12), and this complex can inhibit mTOR signalling. Tacrolimus, which also binds to FKBP12 but it blocks the calcium-calmodulin dependent calcineurin pathway instead.

Besides its known effect on cellular survival and proliferation, mTOR signalling is an interesting target in autoimmune diseases. Inhibition of mTOR has been shown to have important effects on restricting the expansion of T regulatory T cells, by inhibiting the activity and expression of FOXP3, and therefore has a specific role in the maintenance of immune tolerance at the level of regulatory T cells and dendritic cells [53]. The phosphoinositol-3-kinase (PI3K)/Akt pathway is an upstream activator of mTOR [54]. Network analysis of murine lupus nephritis model has showed that many of the genes upregulated in disease are linked to the mTOR pathway [53].

Inhibition of mTOR has shown efficacy in murine models [53] and proof of concept study of using sirolimus in human SLE with refractory disease have shown it is well tolerated and some clinical improvement [55]. The human study has revealed some insights into the significance of mTOR signalling on T cell differentiation, and specifically the expansion of naïve CD8 + T cells and depletion of CD8 + memory T cells were associated with mTOR inhibition using sirolimus [55]. It is believed that mTOR plays a central role in T cell lineage differentiation [56, 57]. Recent study has shown that inhibition of mTOR can also downregulate enhanced expression of stimulator of interferon gene (STING) and IFNα activity, demonstrating another mechanism by which mTOR inhibition can be beneficial in SLE [58].

Conclusions

This Chapter serves to provide an overview of some key intracellular signalling pathways that are associated with lupus pathogenesis. These biological pathways are complex and have significant crosstalks between them. As we identify more defects in the intracellular signalling pathways in SLE patients, there is a race to identify potentially effective small molecule drugs that can modulate these pathways and be utilised in the treatment of SLE. Attention is currently directed to a few of these key pathways, with emerging therapeutics that are being developed with some early promising results.

References

1. Li Y, Palmisano M, Sun D, Zhou S (2020) Pharmacokinetic disposition difference between cyclosporine and voclosporin drives their distinct efficacy and safety profiles in clinical studies. Clin Pharmacol 12:83–96
2. Germano V, Picchianti Diamanti A, Ferlito C, Podestà E, Salemi S, Migliore A et al (2011) Cyclosporine A in the long-term management of systemic lupus erythematosus. J Biol Regul Homeost Agents 25(3):397–403
3. Mok CC, Ho LY, Ying SKY, Leung MC, To CH, Ng WL (2020) Long-term outcome of a randomised controlled trial comparing tacrolimus with mycophenolate mofetil as induction therapy for active lupus nephritis. Ann Rheum Dis 79(8):1070–1076
4. Liu Z, Zhang H, Liu Z, Xing C, Fu P, Ni Z et al (2015) Multitarget therapy for induction treatment of lupus nephritis: a randomized trial. Ann Intern Med 162(1):18–26
5. Zhang H, Liu Z, Zhou M, Chen J, Xing C, Lin H et al (2017) Multitarget therapy for maintenance treatment of lupus nephritis. J Am Soc Nephrol 28(12):3671–3678
6. Rovin BH, Solomons N, Pendergraft WF, Dooley MA, Tumlin J, Romero-Diaz J et al (2019) A randomized, controlled double-blind study comparing the efficacy and safety of dose-ranging voclosporin with placebo in achieving remission in patients with active lupus nephritis. Kidney Int 95(1):219–231
7. Harikishore A, Yoon HS (2015) Immunophilins: structures, mechanisms and ligands. Curr Mol Pharmacol 9(1):37–47
8. Azzi JR, Sayegh MH, Mallat SG (2013) Calcineurin inhibitors: 40 years later, can't live without. J Immunol 191(12):5785–5791
9. Lee JU, Kim LK, Choi JM (2018) Revisiting the concept of targeting NFAT to control T Cell immunity and autoimmune diseases. Front Immunol 9:2747
10. Wu Y, Borde M, Heissmeyer V, Feuerer M, Lapan AD, Stroud JC et al (2006) FOXP3 controls regulatory T cell function through cooperation with NFAT. Cell 126(2):375–387
11. Macián F, López-Rodríguez C, Rao A (2001) Partners in transcription: NFAT and AP-1. Oncogene 20(19):2476–2489
12. Okamura H, Garcia-Rodriguez C, Martinson H, Qin J, Virshup DM, Rao A (2004) A conserved docking motif for CK1 binding controls the nuclear localization of NFAT1. Mol Cell Biol 24(10):4184–4195

13. Beals CR, Sheridan CM, Turck CW, Gardner P, Crabtree GR (1997) Nuclear export of NF-ATc enhanced by glycogen synthase kinase-3. Science 275(5308):1930–1934
14. Gorelik G, Richardson B (2010) Key role of ERK pathway signaling in lupus. Autoimmunity 43(1):17–22
15. Deng C, Kaplan MJ, Yang J, Ray D, Zhang Z, McCune WJ et al (2001) Decreased Ras-mitogen-activated protein kinase signaling may cause DNA hypomethylation in T lymphocytes from lupus patients. Arthritis Rheum 44(2):397–407
16. Bloch M, Kamminga J, Jayewardene A, Bailey M, Carberry A, Vincent T et al (2016) A screening strategy for HIV-associated neurocognitive disorders that accurately identifies patients requiring neurological review. Clin Infect Dis 63(5):687–693
17. Iwata Y, Wada T, Furuichi K, Sakai N, Matsushima K, Yokoyama H et al (2003) p38 Mitogen-activated protein kinase contributes to autoimmune renal injury in MRL-Fas lpr mice. J Am Soc Nephrol 14(1):57–67
18. Scharer CD, Blalock EL, Mi T, Barwick BG, Jenks SA, Deguchi T et al (2019) Epigenetic programming underpins B cell dysfunction in human SLE. Nat Immunol 20(8):1071–1082
19. Surace AEA, Hedrich CM (2019) The role of epigenetics in autoimmune/inflammatory disease. Front Immunol 10:1525
20. Schwartz DM, Kanno Y, Villarino A, Ward M, Gadina M, O'Shea JJ (2017) JAK inhibition as a therapeutic strategy for immune and inflammatory diseases. Nat Rev Drug Discov 17(1):78
21. Schwartz DM, Bonelli M, Gadina M, O'Shea JJ (2016) Type I/II cytokines, JAKs, and new strategies for treating autoimmune diseases. Nat Rev Rheumatol 12(1):25–36
22. Wallace DJ, Furie RA, Tanaka Y, Kalunian KC, Mosca M, Petri MA et al (2018) Baricitinib for systemic lupus erythematosus: a double-blind, randomised, placebo-controlled, phase 2 trial. Lancet 392(10143):222–231
23. Fornaro M, Coladonato L, Venerito V, Cacciapaglia F, Lopalco G, Iannone F (2019) Efficacy of baricitinib on refractory skin papulosquamous rash in a patient with systemic lupus erythematosus. Rheumatology (Oxford) 10:kez442. https://doi.org/10.1093/rheumatology/kez442. Epub ahead of print
24. de Oliveira AB, Alpalhão M, Filipe P, Maia-Silva J (2019) The role of janus kinase inhibitors in the treatment of alopecia areata: a systematic review. Dermatol Ther 32(5):e13053. https://doi.org/10.1111/dth.13053. Epub 2019 Aug 19. PMID: 31381252
25. Fetter T, Smith P, Guel T, Braegelmann C, Bieber T, Wenzel J (2020) Selective janus kinase 1 inhibition is a promising therapeutic approach for lupus erythematosus skin lesions. Front Immunol 11:344
26. Zimmermann N, Wolf C, Schwenke R, Lüth A, Schmidt F, Engel K et al (2019) Assessment of clinical response to janus kinase inhibition in patients with familial chilblain lupus and TREX1 mutation. JAMA Dermatol 155(3):342–346
27. Lee YH, Bae SC (2016) Association between TYK2 polymorphisms and susceptibility to autoimmune rheumatic diseases: a meta-analysis. Lupus 25 (12):1307–1314
28. Cunninghame Graham DS, Akil M, Vyse TJ (2007) Association of polymorphisms across the tyrosine kinase gene, TYK2 in UK SLE families. Rheumatology (Oxford) 46(6):927–930
29. Contreras-Cubas C, García-Ortiz H, Velázquez-Cruz R, Barajas-Olmos F, Baca P, Martínez-Hernández A et al (2019) Catalytically impaired TYK2 variants are protective against childhood- and adult-onset systemic lupus erythematosus in Mexicans. Sci Rep 9 (1):12165
30. Dendrou CA, Cortes A, Shipman L, Evans HG, Attfield KE, Jostins L et al (2016) Resolving TYK2 locus genotype-to-phenotype differences in autoimmunity. Sci Transl Med 8(363):363ra149. https://doi.org/10.1126/scitranslmed.aag1974. PMID: 27807284; PMCID: PMC5737835
31. Diogo D, Bastarache L, Liao KP, Graham RR, Fulton RS, Greenberg JD et al (2015) TYK2 protein-coding variants protect against rheumatoid arthritis and autoimmunity, with no evidence of major pleiotropic effects on non-autoimmune complex traits. PLoS One 10(4):e0122271. https://doi.org/10.1371/journal.pone.0122271. PMID: 25849893; PMCID: PMC4388675
32. Vetrie D, Vorechovský I, Sideras P, Holland J, Davies A, Flinter F et al (1993) The gene involved in X-linked agammaglobulinaemia is a member of the src family of protein-tyrosine kinases. Nature 361 (6409):226–233
33. Bond DA, Maddocks KJ (2020) Current role and emerging evidence for bruton tyrosine kinase inhibitors in the treatment of mantle cell lymphoma. Hematol Oncol Clin North Am 34(5):903–921
34. Tam CS, Opat S, D'Sa S, Jurczak W, Lee HP, Cull G et al (2020) A randomized phase 3 trial of zanubrutinib vs ibrutinib in symptomatic Waldenström macroglobulinemia: the ASPEN study. Blood 136 (18):2038–2050
35. Chalmers SA, Wen J, Doerner J, Stock A, Cuda CM, Makinde HM et al (2018) Highly selective inhibition of Bruton's tyrosine kinase attenuates skin and brain disease in murine lupus. Arthritis Res Ther 20(1):10
36. Bender AT, Pereira A, Fu K, Samy E, Wu Y, Liu-Bujalski L et al (2016) Btk inhibition treats TLR7/IFN driven murine lupus. Clin Immunol 164:65–77
37. Bender AT, Gardberg A, Pereira A, Johnson T, Wu Y, Grenningloh R et al (2017) Ability of Bruton's tyrosine kinase inhibitors to sequester Y551 and prevent phosphorylation determines potency for inhibition of Fc receptor but not B-cell receptor signaling. Mol Pharmacol 91(3):208–219
38. Mócsai A, Ruland J, Tybulewicz VL (2010) The SYK tyrosine kinase: a crucial player in diverse

biological functions. Nat Rev Immunol 10(6):387–402
39. Grammatikos AP, Ghosh D, Devlin A, Kyttaris VC, Tsokos GC (2013) Spleen tyrosine kinase (Syk) regulates systemic lupus erythematosus (SLE) T cell signaling. PLoS One 8(8):e74550. https://doi.org/10.1371/journal.pone.0074550
40. Krishnan S, Warke VG, Nambiar MP, Tsokos GC, Farber DL (2003) The FcR gamma subunit and Syk kinase replace the CD3 zeta-chain and ZAP-70 kinase in the TCR signaling complex of human effector CD4 T cells. J Immunol 170(8):4189–4195
41. Gong B, Radulovic M, Figueiredo-Pereira ME, Cardozo C (2016) The ubiquitin-proteasome system: potential therapeutic targets for Alzheimer's disease and spinal cord injury. Front Mol Neurosci 9:4
42. Bulatov E, Valiullina A, Sayarova R, Rizvanov A (2018) Promising new therapeutic targets for regulation of inflammation and immunity: RING-type E3 ubiquitin ligases. Immunol Lett 202:44–51
43. Chamberlain PP, Lopez-Girona A, Miller K, Carmel G, Pagarigan B, Chie-Leon B et al (2014) Structure of the human Cereblon-DDB1-lenalidomide complex reveals basis for responsiveness to thalidomide analogs. Nat Struct Mol Biol 21 (9):803–809
44. McNaught KS, Olanow CW, Halliwell B, Isacson O, Jenner P (2001) Failure of the ubiquitin-proteasome system in Parkinson's disease. Nat Rev Neurosci 2 (8):589–594
45. Yoshimi R, Ishigatsubo Y, Ozato K (2012) Autoantigen TRIM21/Ro52 as a possible target for treatment of systemic lupus erythematosus. Int J Rheumatol 2012:718237. https://doi.org/10.1155/2012/718237
46. Meister S, Schubert U, Neubert K, Herrmann K, Burger R, Gramatzki M et al (2007) Extensive immunoglobulin production sensitizes myeloma cells for proteasome inhibition. Cancer Res 67(4):1783–1792
47. Badros A, Goloubeva O, Dalal JS, Can I, Thompson J, Rapoport AP et al (2007) Neurotoxicity of bortezomib therapy in multiple myeloma: a single center experience and review of the literature. Cancer 110(5):1042–1049
48. Ferrington DA, Gregerson DS (2012) Immunoproteasomes: structure, function, and antigen presentation. Prog Mol Biol Transl Sci 109:75–112
49. Ichikawa HT, Conley T, Muchamuel T, Jiang J, Lee S, Owen T et al (2012) Beneficial effect of novel proteasome inhibitors in murine lupus via dual inhibition of type I interferon and autoantibody-secreting cells. Arthritis Rheum 64(2):493–503
50. Muchamuel T, Basler M, Aujay MA, Suzuki E, Kalim KW, Lauer C et al (2009) A selective inhibitor of the immunoproteasome subunit LMP7 blocks cytokine production and attenuates progression of experimental arthritis. Nat Med 15(7):781–787
51. Basler M, Dajee M, Moll C, Groettrup M, Kirk CJ (2010) Prevention of experimental colitis by a selective inhibitor of the immunoproteasome. J Immunol 185(1):634–641
52. Perl A (2016) Activation of mTOR (mechanistic target of rapamycin) in rheumatic diseases. Nat Rev Rheumatol 12(3):169–182
53. Reddy PS, Legault HM, Sypek JP, Collins MJ, Goad E, Goldman SJ et al (2008) Mapping similarities in mTOR pathway perturbations in mouse lupus nephritis models and human lupus nephritis. Arthritis Res Ther 10(6):R127
54. LoPiccolo J, Blumenthal GM, Bernstein WB, Dennis PA (2008) Targeting the PI3K/Akt/mTOR pathway: effective combinations and clinical considerations. Drug Resist Updat 11(1–2):32–50
55. Lai ZW, Kelly R, Winans T, Marchena I, Shadakshari A, Yu J et al (2018) Sirolimus in patients with clinically active systemic lupus erythematosus resistant to, or intolerant of, conventional medications: a single-arm, open-label, phase 1/2 trial. Lancet 391 (10126):1186–1196
56. Delgoffe GM, Pollizzi KN, Waickman AT, Heikamp E, Meyers DJ, Horton MR et al (2011) The kinase mTOR regulates the differentiation of helper T cells through the selective activation of signaling by mTORC1 and mTORC2. Nat Immunol 12(4):295–303
57. Pollizzi KN, Patel CH, Sun IH, Oh MH, Waickman AT, Wen J et al (2015) mTORC1 and mTORC2 selectively regulate CD8+T cell differentiation. J Clin Invest 125(5):2090–108. https://doi.org/10.1172/JCI77746. Epub 2015 Apr 20. PMID: 25893604; PMCID: PMC4463194
58. Murayama G, Chiba A, Kuga T, Makiyama A, Yamaji K, Tamura N et al (2020) Inhibition of mTOR suppresses IFNα production and the STING pathway in monocytes from systemic lupus erythematosus patients. Rheumatology (Oxford) 59 (10):2992–3002

Regulatory T Cells in SLE

9

Rachel Cheong and Joshua Ooi

Abstract

Autoimmune diseases such as type 1 diabetes and systemic lupus erythematosus are chronic inflammatory diseases caused by immune system dysfunction. When immune cells such as T cells and B cells recognize self-antigens, this can lead to cellular and tissue destruction. Multiple causes of autoimmune diseases have been suggested, such as genetic predisposition, molecular mimicry, and importantly, a loss in numbers and/or function of regulatory T cells. Regulatory T cells (or Tregs) are a subset of $CD4^+$ T cells essential for functional cellular immunity. They play an important role in preventing the induction of autoimmunity through various mechanisms, including, but not limited to, the secretion of immunosuppressive cytokines such as IL-10 and TGF-β. Their indispensability in protection against autoimmunity can be demonstrated in cases of immune dysregulation polyendocrinopathy enteropathy X-linked (IPEX) syndrome, where loss of Treg function results in severe autoimmunity and even death. Furthermore, a decrease in numbers of functional Tregs and/or loss of Treg function is often been implicated in the pathogenesis of autoimmune diseases. Consequently, there has been increased interest in the use of Tregs in immunotherapy (whether polyclonal or antigen-specific) to treat autoimmune diseases such as systemic lupus erythematosus. Recent studies have shown that antigen-specific Tregs are important for protection against autoimmune disease. In a paradigm shifting study, utilizing the model autoimmune disease, Goodpasture's, it was found that antigen-specific Tregs dominantly suppressed pro-inflammatory autoreactive conventional T cells and protected from disease. Therefore, there are now efforts to increase the numbers of antigen-specific Tregs to treat autoimmune diseases. One such way is to genetically engineer antigen-specific Tregs. Studies using animal models of disease and Jurkat cell lines have shown potential for the genetic engineering of antigen-specific Tregs for antigen-specific disease suppression. These Tregs can be conferred with antigen specificity through the use of lentiviral vectors containing specific T cell receptor (TCR) sequences. The use of antigen-specific Tregs as therapeutics in autoimmunity could also be applied to systemic lupus erythematosus (SLE), a chronic inflammatory autoimmune disease estimated to affect 1 in 1000 people. SLE is a highly heterogenous disease with a number of target autoantigens such as double-stranded DNA and the Smith

R. Cheong · J. Ooi (✉)
Regulatory T Cell Therapy Group, Department of Medicine At Monash Health, School of Clinical Sciences, Monash University, Clayton, VIC 3168, Australia
e-mail: joshua.ooi@monash.edu

R. Cheong
e-mail: rachel.cheong1@monash.edu

A. Hoi (ed.), *Pathogenesis of Systemic Lupus Erythematosus*,
https://doi.org/10.1007/978-3-030-85161-3_9

(Sm) antigen. Patients present with different disease manifestations, including, but not limited to, lupus nephritis, malar rash, and neuropsychiatric lupus. Amongst the autoantigens, the Sm antigen is most specific for SLE; and, autoreactivity to the Sm antigen is associated with worse prognosis, in particular a strong predisposition to develop lupus nephritis. Furthermore, there is a strong HLA association in anti-Sm + SLE patients, suggesting a dominant T cell directed response in these patients. Current treatments for SLE involve the use of corticosteroids and immunosuppressive agents. However, long-term intake of these medications can lead to severe side effects such as increased risks of infection and drug toxicity. Due to the importance of antigen-specific Tregs in protecting against autoimmune disease, there is now the potential to develop antigen-specific Tregs for use as treatment for SLE.

Regulatory T Cells

Regulatory T cells (or Tregs) are a subset of $CD4^+$ T cells most commonly characterized by their expression of the transcription factors forkhead box P3 (Foxp3) and Helios, high expression of the cell surface molecule CD25, and secretion of immunomodulatory cytokines such as IL-10 and TGF-β [1]. Tregs are vital for the continued and balanced function of the immune system, as seen in the case of immune dysregulation polyendocrinopathy enteropathy X-linked syndrome (IPEX), where a lack of functional Tregs results in widespread autoimmunity, often resulting in severe organ damage and even death [2]. This is due to the central role Tregs play in modulating the immune response of both T and B lymphocytes upon cognate antigen recognition and activation.

A decrease in the numbers of Tregs as well as a loss of their regulatory functions have been implicated in the pathogenesis of most autoimmune diseases, including type 1 diabetes, multiple sclerosis, rheumatoid arthritis, and SLE; further underscoring the importance of these cells in preventing unwanted autoreactive immune responses [3]. As such, there has been increasing interest in recent years in using both polyclonal and antigen specific Tregs to develop biological therapies to treat autoimmune diseases.

Types of Regulatory T Cells

There are different subsets of $CD4^+$ regulatory T cells involved in the adaptive immune response, namely, thymic Tregs (tTregs) and peripheral Tregs (pTregs), which develop as a part of central and peripheral tolerance mechanisms. Due to the large repertoire of TCRs generated during gene rearrangement in T cell precursors, tolerance mechanisms are essential for keeping autoimmunity in-check.

Thymic Tregs (also known as natural Tregs), are formed as a part of the central tolerance mechanism. They develop during the process of negative selection in the thymus, which involves the promiscuous presentation of tissue-restricted antigens (TRAs) to T cell precursors via autoimmune regulator (AIRE), a protein present on medullary thymic epithelial cells (mTECs). Cells with TCRs that bind to the TRAs with moderate affinity escape deletion, whereas cells that bind with high affinity are either deleted, or converted to tTregs [4].

TCR stimulation is important for the formation of tTreg cells [5, 6]. High affinity interactions between the TCR and MHC-peptide complex on thymic antigen-presenting cells (APCs) in combination with co-stimulation via CD28 and the presence of IL-2 can induce the expression of Foxp3, the master transcription factor for Tregs [7]. TCR stimulation is also involved in the demethylation of the conserved non-coding sequence region 2 (CNS2) in the *Foxp3* locus, which is important for stable Foxp3 expression in Tregs [8, 9]. However, Foxp3 expression alone is insufficient for the induction of a full and functional tTreg phenotype, with epigenetic mechanisms having been found to be essential in this process as well. Besides stabilizing Foxp3 expression, TCR stimulation is also able to induce CpG hypomethylation of genes associated with tTreg cell function, including *Foxp3*, *Il2ra* and *Ctla4* [8], with this Treg-specific hypomethylation considered to be a marker of Treg lineage commitment.

However, central tolerance mechanisms are unable to completely prevent the escape of autoreactive T cells into the periphery. As such, peripheral tolerance mechanisms are also essential for the avoidance of autoreactivity—one such mechanism is the formation of peripheral Tregs (pTregs). pTregs are formed when naïve $CD4^+$ T cells encounter and recognise their cognate antigen in the periphery, in combination with IL-2 and TGF-β stimulation [10].

Another mechanism of peripheral tolerance involves type 1 regulatory (Tr1) cells, a type of induced Treg, which are induced after exposure to high amounts of IL-10 alongside antigen recognition [11]. Tr1 cells can be characterized by their co-expression of the cell surface markers LAG-3 and CD49b [12], as well as IL-10 secretion [13].

Besides the $Foxp3^+$ thymic and peripheral Tregs and $Foxp3^-$ Tr1 cells, it has been suggested that there is another type of $CD4^+$ T cell with suppressive activity that is $CD52^{hi}$. In a study by Bandala-Sanchez et al. using a murine model of type 1 diabetes, the group found a small proportion of $CD4^+$ T cells with suppressor activity that were specific for the type 1 diabetes autoantigen GAD65. Interestingly, these suppressive cells did not express markers commonly associated with Tregs such as Foxp3, CTLA-4 and CD25. Instead, when the phenotypes of GAD65-specific cells with suppressive and non-suppressive abilities were compared, it was found that the suppressive cells were $CD52^{hi}$, while non suppressive cells were $CD52^{lo}$, suggesting CD52 contributed to the cell's suppressive ability [14]. The functional mechanisms of these various Treg subtypes will be discussed in further detail in the following section.

Functional Mechanisms and Roles of Regulatory T Cells

As seen from the section above, the differentiation of both thymic and peripheral Tregs form an important part of central and peripheral tolerance mechanisms in preventing unwanted immune responses. Functional Tregs are essential in preventing the induction of autoimmune responses through interactions with other immune cells as well as various mechanisms, including the secretion of immunosuppressive cytokines and cell-to-cell contact-dependent suppression.

Tregs are able to suppress other immune cells through contact-dependent mechanisms involving cell surface molecules such as CTLA-4 and LAG-3. Through the interactions of CTLA-4 on Tregs with CD80/86 on dendritic cells (DCs), Tregs are able to induce DC production of the enzyme indoleamine 2,3-dioxygenase (IDO). IDO then converts the amino acid tryptophan to kynurenine, a metabolite that suppresses the function of effector T cells [15]. LAG-3 on Tregs binds to MHC class II molecules, and in the case of immature DCs, LAG-3 binding induces inhibitory signals, resulting in reduced DC capacity to stimulate an immune response [16].

Another regulatory function of Tregs involves their secretion of immunosuppressive cytokines such as IL-10 and TGF-β. IL-10 can influence the downregulation of co-stimulatory molecules and MHC class II molecules on macrophage/monocyte APCs, thus reducing their functional ability to activate $CD4^+$ T cells. IL-10 signaling is also essential for Treg suppression of Th17 cell responses as a lack of IL-10 receptors on Tregs resulted in worse autoimmune murine colitis [17]. TGF-β signaling is another important mechanism for Th17 cell response suppression in Tregs, as *Tgfbr1*$^{-/-}$ Tregs in a murine colitis model had reduced Th17 suppression capacities [18].

The mechanism of suppression of $CD4^+CD52^{hi}$ T cells as mentioned in the previous section involves soluble CD52 [14]. Further studies by Bandala-Sanchez and colleagues revealed the binding of CD52 to the pro-inflammatory B box of damage associated molecular pattern HMGB-1 is necessary for CD52 to bind to the inhibitory receptor Siglec-10 [19]. Siglec-10, a type of sialic acid-binding immunoglobulin-like lectin, is found on the surface of immune cells including T cells and contains immunoreceptor tyrosine-based inhibitory motif (ITIM) cytoplasmic domains. The binding of CD52 to Siglec-10 results in the recruitment of the phosphatase SHP1 which dephosphorylates the TCR, thus leading to suppression of T cell activity.

Previously, it was thought that Tregs suppressed autoreactive B cell responses indirectly through helper T cells. However, Tregs have also been found to have direct suppressive activity on autoreactive B cells [20]. Using a mouse model of SLE, Iikuni et al. depleted endogenous T cells in mice with high serum anti-dsDNA antibody levels before administering syngeneic Tregs to the mice. IgG and anti-dsDNA antibody production was found to have significantly decreased in mice with depleted endogenous T cells, suggesting that Tregs are able to directly suppress B cell responses in the absence of helper T cells.

Tregs in Autoimmunity

As seen from the previous sections, the differentiation of Tregs form an essential part of both central and peripheral tolerance mechanisms. Tregs are also able to suppress cells from both the innate as well as adaptive arms of the immune system through the use of various cell surface receptors and secreted cytokines. Due to the importance of Tregs in helping to maintain immunological homeostasis, any loss of function or decrease in Treg numbers can result in the uncontrolled spread of autoreactive immune cells and autoimmunity.

Besides the reduction in numbers and functions of Tregs, some studies have suggested that Treg phenotype instability may also play a role in the pathogenesis of autoimmune diseases [21, 22]. It is possible that in the presence of pro-inflammatory cytokines such as IL-6 or in a lymphopenic environment, Tregs may lose their expression of Foxp3 and regulatory phenotype and convert to pro-inflammatory Th1 or Th17 cells [23]. In a murine model of cornea transplantation, Hua et al. found that after corneal transplantation in mice with dry eye disease (DED) and healthy controls, mice with DED had more cells that lost Foxp3 expression (exFoxp3 cells) in comparison to healthy controls. Furthermore, there was increased IFN-γ expression by the exFoxp3 cells in DED mice compared to healthy controls [21].

Similarly, in a study using autoimmune arthritis as a disease model, the transfer of $CD4^{+}Foxp3^{+}CD25^{lo}$ cells from donor mice to arthritic mice resulted in the loss of Foxp3 expression. When the percentage of IL-17 expressing donor and host $CD4^{+}Foxp3^{-}$ cells were compared, donor cells had a higher percentage of IL-17 expressing $Foxp3^{-}$ cells. Interestingly, when these exFoxp3 cells were phenotyped, it was found that they expressed Treg-associated markers such as CD25, CTLA-4 and CD39, suggesting they were not effector T cells transiently expressing Foxp3. These exFoxp3 cells were also found to express lower levels of *Ikzf2*, the gene encoding the transcription factor Helios [22]. Helios could play a role in the stability of the Treg phenotype and the expression of Foxp3 as $Foxp3^{+}Helios^{-}$ Tregs were more likely to lose Foxp3 expression in a lymphopenic environment as compared to $Foxp3^{+}Helios^{+}$ Tregs [24].

In the context of SLE, various studies have reported conflicting results on the numbers and functions of Tregs in patients. Some of these studies reported a reduction in Treg numbers and suppressive function [25–27], while others have reported no changes in the numbers of Tregs [28] in patients with SLE. These differing results in studies may be because SLE is a highly heterogenous autoimmune disease, where patients present with different disease manifestations varying in severity. Patient disease activity classification also varied between studies according to the SLE disease activity index

(SLEDAI) scores used to classify active or inactive disease states, which could also have affected the results.

However, studies using Tregs as treatment for SLE in both murine models and humans have shown the efficiency of Tregs in helping to reduce SLE disease activity [29, 30], suggesting altered Treg numbers and functions are indeed implicated in the pathogenesis of SLE.

Regulatory T Cell Therapies

Due to the important role of Tregs in preventing the onset of autoimmune diseases, their role in regulating immune responses, as well as the consequences of a reduction in Treg numbers and/or functionality, there has been interest in using Tregs as biological therapies for the treatment of autoimmune diseases as well as prevention of organ transplant rejection. Some of these potential therapies include the ex vivo expansion and transfer of polyclonal Tregs, the generation of chimeric antigen receptor (CAR) Tregs, and genetic engineering of antigen-specific Tregs [31].

Polyclonal Tregs

The transfer of polyclonal Tregs for use as immunotherapy has been used in models of autoimmune disease such as type 1 diabetes, as well as in organ transplants [32–34]. In a study by Bluestone et al., the transfer of autologous polyclonal Tregs expanded ex vivo into patients with type 1 diabetes showed the transferred Tregs retained their regulatory phenotype and persisted in circulation a year after transfer. There were also no adverse events recorded that were thought to be caused by the dose of administered Tregs, indicating the use of polyclonal Tregs is safe [33].

Polyclonal Tregs have also been used in clinical trials to prevent the rejection of organ transplants, such as liver transplants [34]. In a study by Todo et al., patients were started on immunosuppressive drugs such as calcineurin inhibitors and mycophenolate mofetil after liver transplantation. Ex vivo expanded autologous Tregs were then administered to the transplant recipients 13 days post transplantation. Six months post-transplantation, patients were gradually taken off the immunosuppressive drugs over the period of a year. Seven out of ten patients who received the Treg infusion were able to completely discontinue their intake of immunosuppressive drugs without transplant rejection for more than 12 months, thus suggesting polyclonal Treg therapy is indeed effective for the suppression of unwanted immune responses [34].

However, the use of polyclonal Tregs as immunotherapy has disadvantages, such as the possibility that these polyclonal Tregs could non-specifically suppress immune responses, including those against foreign antigens [35]. Although the safety of polyclonal Tregs has been shown in clinical trials, there was no improvement noted in clinical disease parameters even after high doses of cells (up to 2×10^9 cells per patient) were administered in the type 1 diabetes clinical trial [33]. Antigen-specific Tregs would be a more potent and efficacious treatment for autoimmune diseases, with various methods of developing antigen-specific Tregs proposed, including the use of CAR Tregs and genetically engineered antigen-specific Tregs.

Chimeric Antigen Receptor (CAR)-Tregs

Similar to the use of CAR-T cells as immunotherapy to treat various cancers including leukaemia, CAR-Tregs are a potential treatment for autoimmune diseases. CAR-T cells consist of an extracellular antigen-binding region (single chain variable fragment from an antibody), a transmembrane domain, as well as an intracellular domain that transmits cell activation signals. Upon antigen recognition and binding by the CAR extracellular region, cell activation signals are directly transmitted to the CAR T cell, thus eliminating the requirement for antigen processing and presentation as well as co-stimulation from an APC [36]. CAR-Tregs have also been used in a variety of disease models, including murine colitis, hemophilia A, and organ transplants [37–40].

Hemophilia A is a genetic disorder whereby coagulation factor Factor VIII (FVIII) genes are mutated, leading to a loss of function. It is often treated by administering high doses of FVIII to patients; however, patients often develop immune responses against FVIII. Yoon et al. created CAR-Tregs specific for human FVIII (known as ANS8 CAR-Tregs) and tested their function in vivo alongside FVIII-specific TCR-transduced Tregs (17,915 TCR-Tregs) [41] and non-specific Ob2F3 TCR-Tregs [38]. Mice were immunized with FVIII and administered ANS8 CAR, 17,915 TCR or non-specific Ob2F3 Tregs four hours later. The group found that both ANS8 CAR and 17,915 TCR Tregs were able to suppress the formation of anti-FVIII antibodies, while Ob2F3 Tregs were unable to suppress antibody formation. Interestingly however, when the mice were re-immunized with FVIII 8 weeks after the initial dose, the levels of anti-FVIII antibodies found in mice that received ANS8 CAR Tregs had increased such that they were similar to antibody levels found in mice that received non-specific Ob2F3 Tregs. Meanwhile, mice that received 17,915 TCR Tregs had a slight increase in anti-FVIII antibody levels; however, these levels remained significantly lower than those found in mice with ANS8 CAR and ObF23 Tregs. This suggests that CAR Tregs and TCR transduced Tregs act differently to suppress autoreactivity, and further studies should be performed to determine the most suitable type of antigen-specific Tregs to treat various autoimmune diseases [38].

TCR Transduced Tregs

Genetic engineering of antigen-specific Tregs through methods such as TCR lentiviral transduction is another potential way to generate Tregs for therapeutic functions. These engineered antigen-specific Tregs have been used in vitro and in vivo to suppress autoimmune cell proliferation [42] as well as treat murine models of experimental autoimmune encephalomyelitis (EAE) [43]. Lentiviral vectors are often used in the production of antigen-specific Tregs, where these vectors contain peptide-MHC specific TCR.

Hull et al. generated TCRs specific for two different type 1 diabetes autoantigens (IA2 and insulin), as well as an influenza-specific TCR that acted as a control [42] and tested them in vitro using the Jurkat cell line J76, as well as human Tregs, with B lymphoblastoid cell line (BLCL). PBMC co-cultures were used to determine TCR functionality. It was found that both TCR-transduced Jurkats and primary human Tregs had functional antigen-specific TCRs as measured by CD69 upregulation after stimulation with their respective cognate peptides.

The suppressive abilities of antigen-specific Tregs against polyclonal and specific antigens was also tested in vitro using proliferation dye labelled PBMCs stimulated with vaccines and/or specific peptide. It was found that all three TCR-Tregs were able to non-specifically suppress immune responses induced by vaccine stimulation up to a ratio of 1:50 (Tregs:PBMCs). Similarly, antigen-specific suppression against cognate peptide-induced responses was found to be effective, with Tregs found to be able to suppress responses even at a ratio of 1:1000 Tregs to PBMCs. This suggests that antigen-specific suppression by transduced Tregs is highly effective and also suggests that antigen-specific Tregs are also able to regulate immune responses non-specifically.

Antigen-specific TCR-Tregs specific for the autoantigen $MBP_{85\text{-}99}$ were also used in a murine model of EAE by Kim et al. [43]. In this study, TCRs specific for $MBP_{85\text{-}99}$ (known as Ob2F3) were lentivirally transduced onto both human T effector cells (Teffs) as well as Tregs. Besides the Ob2F3 TCR, a control TCR specific for coagulation factor FVIII (known as 17,195) was also transduced onto human Tregs. A Treg suppression assay was then performed by co-culturing Ob2F3 Teffs with either Ob2F3 Tregs or 17,195 Tregs, in the presence of $MBP_{85\text{-}99}$ peptide or anti-CD3ε antibodies at different ratios. Cell proliferation was measured using [^{3}H]-thymidine and it was found that Ob2F3 Tregs had significantly better suppression of the Teffs cultured with MBP peptide in comparison to the non-specific 17,195 Tregs.

The suppressive effects of Ob2F3 Tregs were also tested in vivo using a mouse model of EAE, where mice were first immunized with MOG_{35-55} before either Ob2F3 Tregs, 17,195 Tregs, or PBS were administered to the mice a week later. Any suppressive effects transferred Tregs had on disease was measured using disease severity scoring as well as immunofluorescence staining. As expected, mice that received Ob2F3 Tregs had a significantly lower disease score compared to mice receiving 17,195 Tregs and PBS. Higher numbers of Ob2F3 Tregs were also found in the brain and spinal cord in comparison to non-specific 17,195 Tregs, indicating antigen-specific Tregs migrate to the site of inflammation.

As seen from these two studies, there is potential for Tregs conferred with autoantigen-specific TCRs to effectively suppress activated $CD4^+$ T cell responses both in vitro and in vivo whether the effector T cells had identical or differing antigen specificities.

Limitations Associated with Treg Therapies

As with most experimental treatments, there are various limitations, advantages and disadvantages associated with the use of Tregs as a form of cell-based therapy. One such example would be the process of obtaining autologous Tregs from patients. As mentioned earlier, reduced numbers of Tregs have been associated with the pathogenesis of autoimmune disease [25], therefore, obtaining sufficient numbers of autologous patient Tregs for expansion may be more difficult. The isolation of human Tregs is also more complicated owing to the fact that Foxp3, the marker most often associated with Tregs, is an intracellular molecule and is transiently expressed in activated effector T cells in humans, unlike mice [44].

There are also no standardized protocols for the isolation and expansion of human Tregs for therapeutic use, where some groups may use magnetic-activated cell sorting (MACS) beads [45] while others may use fluorescence-activated cell sorting (FACS) [29, 33]. Depending on the methods used, the purity of the Treg population and/or numbers obtained could be affected, resulting in an impact on downstream processing.

Besides the limitations associated with the isolation of autologous Tregs from patients, there are also considerations to make when choosing the type of Treg therapy to pursue, whether polyclonal, CAR, or antigen-specific TCR transduced Tregs [31].

There are both advantages and disadvantages associated with the use of CAR-Tregs and TCR-Tregs as potential treatments for autoimmune diseases. An advantage of CAR-Tregs is their ability to recognize antigens without HLA restrictions; however, this means CAR Tregs are only able to recognize extracellular antigens. Conversely, one of the advantages of using TCR-Tregs is their ability to recognize intracellular antigens due to their requirement for antigen processing and presentation by APCs. TCR-Tregs are also able to recognize antigens present at a much lower concentration as compared to CAR-Tregs [31] and may be better suited at suppressing autoantibody-associated responses [38].

Regulatory T Cells in Systemic Lupus Erythematosus

Systemic lupus erythematosus (SLE), commonly referred to as lupus, is a chronic inflammatory autoimmune disease characterized by the presence of autoantibodies against parts of the cell membrane and nuclei. SLE can affect various organs such as the kidneys, nervous system, and skin, resulting in manifestations such as lupus nephritis, neuropsychiatric lupus and malar rash [46, 47]. The disease is more prevalent in females (especially those of child-bearing age) than males, and certain ethnicities including Asians, Hispanics, and those of African descent have been found to have a higher incidence and prevalence of SLE [47–49].

The aetiology for SLE is unclear, although various genetic and environmental risk factors have been identified. Deficiencies in genes such as C1q, a component of the complement system involved in clearing apoptotic cells and immune complexes, have been associated with the development of SLE [50]. The Epstein-Barr virus (EBV) has also been associated with an increased risk of lupus, potentially through molecular mimicry, as the viral antigen Epstein-Barr nuclear antigen 1 (EBNA-1) induces cross-reactive responses to SLE autoantigens [51].

HLA Association in SLE

As with most autoimmune diseases, there is an HLA association with the risk of developing SLE. HLA DR3 and HLA DR15 are associated with increased susceptibility to SLE, while HLA DR4 and DR11 confer resistance to the disease [52–55]. Lupus nephritis, a renal condition associated with more severe morbidity in lupus patients, was found to have an association with the class II HLA alleles DR3 and DR15 [56–58].

This association of particular HLA alleles with a risk of developing autoimmune disease could be due to differences in amino acids present in the peptide binding grooves of HLA molecules. These differences can result in altered peptide binding conformations during antigen presentation, leading to differing antigen recognition by T cells through T cell receptor (TCR) and peptide-HLA complex interactions. A recent study by Ooi et al., using Goodpasture's disease as a model autoimmune disease, showed that differences in the peptide-binding groove between the protective HLA-DR1 and the susceptible HLA-DR15 resulted in differing T cell phenotypes. Higher proportions of autoantigen-specific regulatory T cells were present in people with the protective HLA-DR1 allele, whereas higher proportions of autoantigen-specific pro-inflammatory conventional T cells were present in patients with the susceptibility-conferring DR15 allele [59]. In the context of HLA associations in SLE, a study by Kim et al. identified three amino acid positions in the peptide binding groove of HLA-DRB1 where amino acids differed between protective HLA alleles and alleles that confer susceptibility to SLE [60]. This suggests that there may also be differences in the conformations of autoantigens presented to CD4$^+$ T cells in SLE patients, resulting in a more pro-inflammatory immune response.

Autoantibodies and the Cognate Autoantigens Associated with SLE

SLE is an autoimmune disease best known for its characteristic presence of autoantibodies against various nuclear and cell membrane components, with anti-double stranded DNA (dsDNA) and anti-Smith (Sm) antibodies pathognomonic for SLE. Other autoantibodies present in lupus patients include anti-Ro60, anti-La, anti-phospholipid, anti-ribonucleoprotein (RNP) and anti-nuclear antibodies [61]. Due to disease heterogeneity, lupus patients will likely present with a combination of different autoantibody specificities, although there have been associations found between certain autoantibody specificities and clinical manifestations.

For example, the presence of autoantibodies against the Sm antigen has been associated with lupus nephritis [62, 63]. The immune response to the Sm antigen is positively associated with the HLA-DR3 and HLA-DR15 alleles [52, 53, 64–67]. The Sm antigen is a component of the small nuclear ribonucleoprotein complex and is made up of seven core proteins, namely B/B', D1, D2, D3, E, F, and G [61]. Studies on SmD using DR3 transgenic mice have found dominant T cell epitopes that are HLA DR3-restricted [65, 66], thus indicating the use of SmD in the context of HLA DR3 could be useful to study the role of antigen-specific CD4$^+$ T cells in SLE.

Pathogenesis of Systemic Lupus Erythematosus

SLE is a complex autoimmune disease involving numerous autoantigens, as well as a wide range of manifestations. As such, the pathogenesis of

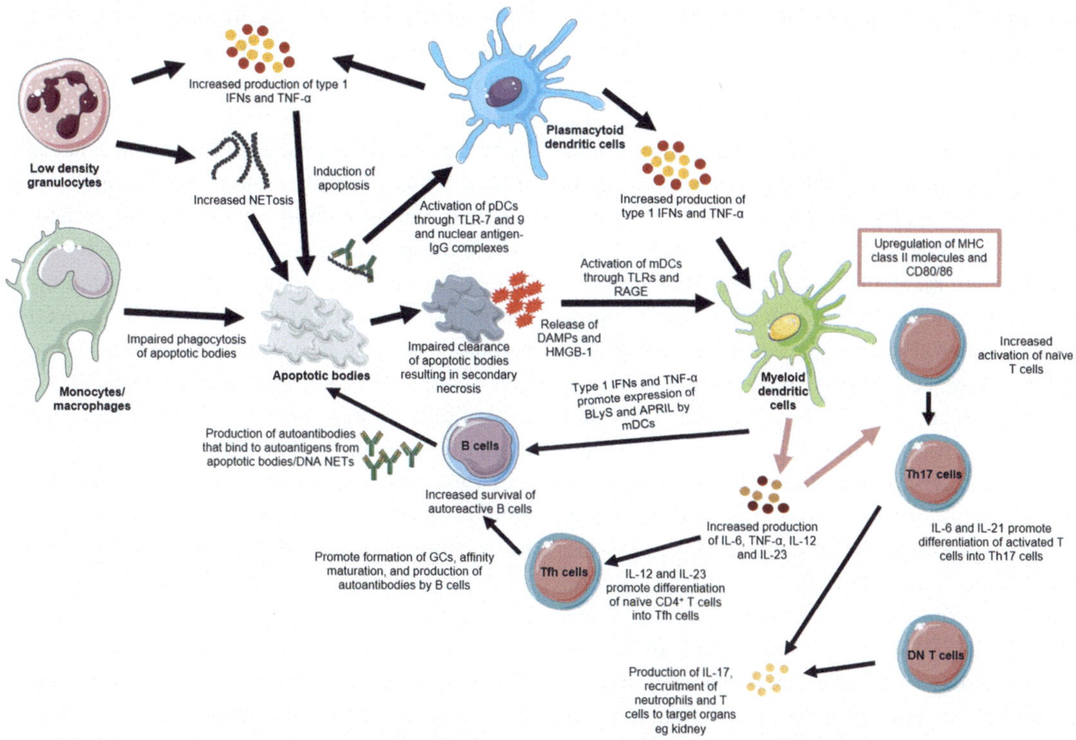

Fig. 9.1 **The pathogenesis of SLE**. Both the innate and adaptive arms of the immune system are implicated, including dendritic cells, B cells and T cells. *Source* (Diagram created using Servier Medical Art from smart.servier.com)

SLE is also complex, involving both the innate and adaptive arms of the immune system as seen below in Fig. 9.1.

Innate immune cells such as neutrophils, dendritic cells and macrophages have all been implicated in the pathogenesis of lupus [68, 69]. Studies focusing on neutrophils in the context of SLE have found the presence of a neutrophil subtype known as the low-density granulocyte (LDG) in the peripheral blood of lupus patients [69]. These LDGs present different nuclear morphology compared to neutrophils and monocytes, and were shown to produce significantly higher amounts of pro-inflammatory cytokines such as TNF-α and IFN-γ compared to neutrophils in healthy controls and lupus patients [70]. LDGs were also more likely to undergo NETosis, a form of neutrophil death where chromatin fibers containing antimicrobial peptides are released. Both the released chromatin fibers as well as apoptotic LDGs are a potential source of autoantigens that can be recognized by auto-reactive antibodies, thereby exacerbating lupus pathogenesis.

Macrophages and monocytes are well known for their phagocytic ability of foreign and apoptotic bodies. A study by Ren et al. on the role of macrophages in SLE pathogenesis showed that there was a decrease in the number of and capability of macrophages to phagocytose apoptotic bodies in vitro in polymorphonuclear neutrophil cultures isolated from patients as compared to those from healthy controls [71, 72]. This decreased macrophage ability to phagocytose apoptotic bodies could be due to a reduction in serum proteins such as the complement proteins C1q and C4 in patients with SLE [73]. This increase in apoptotic bodies present may have an impact on the activation of dendritic cells and, in turn, the adaptive arm of the immune system as seen in Fig. 9.1.

Both subsets of dendritic cells (myeloid dendritic cells (mDCs) and plasmacytoid dendritic cells (pDCs)) have also been associated with the pathogenesis of SLE. Due to a combination of increased apoptosis as well as impaired clearance of apoptotic bodies, secondary necrosis occurs, resulting in the release of damage associated molecular patterns (DAMPs) as well as the pro-inflammatory mediator high mobility group box protein-1 (HMGB-1). These DAMPs and HMGB-1 are then able to activate mDCs through toll-like receptors (TLRs) as well as the receptor for advanced glycation end products (RAGE), leading to the secretion of the pro-inflammatory cytokines IL-6 and TNF-α by the activated mDCs [74–76]. The role of pDCs in the pathogenesis of SLE involves their ability to produce large amounts of type 1 interferons (IFNs) upon activation through the nucleic acid-recognising TLRs TLR-7 and TLR-9 or through immune complexes comprising self-nuclear antigens and IgG [77, 78]. These secreted type 1 IFNs can induce increased cellular apoptosis and induce the activation and upregulation of major histocompatibility complex (MHC) II molecules and the co-stimulatory molecules CD80 and CD86, leading to more potent T cell activation abilities of the mDCs [79, 80]. Type 1 IFNs may also influence the production of the pro-inflammatory cytokine IL-17A and B cell survival factor (B lymphocyte stimulator, BLyS). A recent study by Lopez et al. showed that when peripheral blood mononuclear cells (PBMCs) were isolated from patients with SLE and healthy controls and stimulated with IFN-α in vitro, there was an increased amount of BLyS and IL-17A produced by the PBMCs from patients with SLE. In contrast, PBMCs isolated from healthy controls produced an increased amount of BLyS but did not produce increased amounts of IL-17A [81, 82]. These increased amounts of BLyS and IL-17A produced can further influence the survival of autoreactive B cells and the production of other pro-inflammatory cytokines such as IL-6 and TNF-α.

The pathogenesis of lupus also involves components of the adaptive immune system, B cells and T cells. One of the hallmarks of lupus is the presence of autoreactive antibodies against components of cell membranes and nuclei, showing the involvement of B cells in the pathogenesis of this disease. This activation and immunological function of B cells can be affected by the type 1 IFNs produced by pDCs, especially IFN-α, which is able to induce the secretion of B cell survival promoting cytokines BLyS and A proliferation-inducing ligand (APRIL) by dendritic cells and macrophages [83]. The binding of these pro-survival cytokines to their respective receptors such as BLyS receptor 3 (BR3), transmembrane activator-1 and calcium modulator and cyclophilin ligand-interactor (TACI) and B cell maturation antigen (BCMA) prevent apoptosis and thus promote the escape of autoreactive B cells from deletion [84, 85]. Accordingly, studies have shown increased serum BLyS levels in patients with SLE, thus implicating BLyS in the pathogenesis of SLE [86].

Autoantibodies produced by the autoreactive B cells can then bind to their cognate autoantigens from sources such as apoptosed LDGs or chromatin fibres released as a result of NETosis, and form immune complexes. Deposition of these immune complexes in target organs and recruitment of other immune cells through their Fc receptors can therefore lead to further tissue damage and inflammation.

T cells, particularly those of the $CD4^+$ T follicular helper (Tfh) and Th17 subsets, are also involved in the pathogenesis of SLE. This is due to their involvement in the activation of autoreactive B cells as well as their production of pro-inflammatory cytokines. Tfh cells have been implicated in the pathogenesis of various autoimmune diseases due to their role in promoting the formation of germinal centres, affinity maturation, and antibody production by B cells [87, 88]. The differentiation of naïve $CD4^+$ T cells into Tfh cells is induced by IL-12 and IL-23, cytokines produced by activated DCs and found in increased levels in the serum of patients with SLE [89–91]. The function of Tfh cells in SLE patients is also likely to be aberrantly affected by the lower amounts of IL-2 present, as IL-2 has been found to antagonize the differentiation of Tfh cells [92, 93].

Besides increases in IL-12 and IL-23, significantly higher amounts of IL-17 have also been found in the serum of patients with SLE as compared to healthy controls [94, 95]. In a study by Crispin et al., it was found that there was a higher proportion of double negative (DN) T cells ($TCR\alpha\beta^{+}CD3^{+}CD4^{-}CD8^{-}$) that produced significantly higher amounts of IL-17 in the peripheral blood of patients with SLE as compared to healthy controls. Furthermore, the study also found that $CD4^{+}$ T cells from SLE patients produced significantly higher amounts of IL-17 than $CD4^{+}$ T cells from healthy controls, thus implicating Th17 cells (IL-17 is the phenotypic cytokine of Th17 cells) [96]. IL-17, besides being a pro-inflammatory cytokine itself, can also initiate a positive feedback loop with other cytokines including IL-6 and TNF-α, leading to the exacerbation of inflammation [97].

Regulatory T Cells and Their Potential Sites of Suppression in SLE

Both the adaptive and innate arms of the immune system are implicated in the pathogenesis of SLE. However, the use of antigen-specific Tregs could target certain pathways to suppress autoreactivity and inflammation as seen below from Fig. 9.2.

Tregs could be able to target activated mDCs and pDCs through the secretion of IL-10 which can result in the downregulation of co-stimulatory molecules and MHC class II molecules. This would result in decreased ability of these DCs to stimulate autoreactive T cells and would also lead to a decrease in production of pro-inflammatory cytokines such as type 1 IFNs and TNF-α. Decreased stimulation of autoreactive T cells would also likely result in a decrease

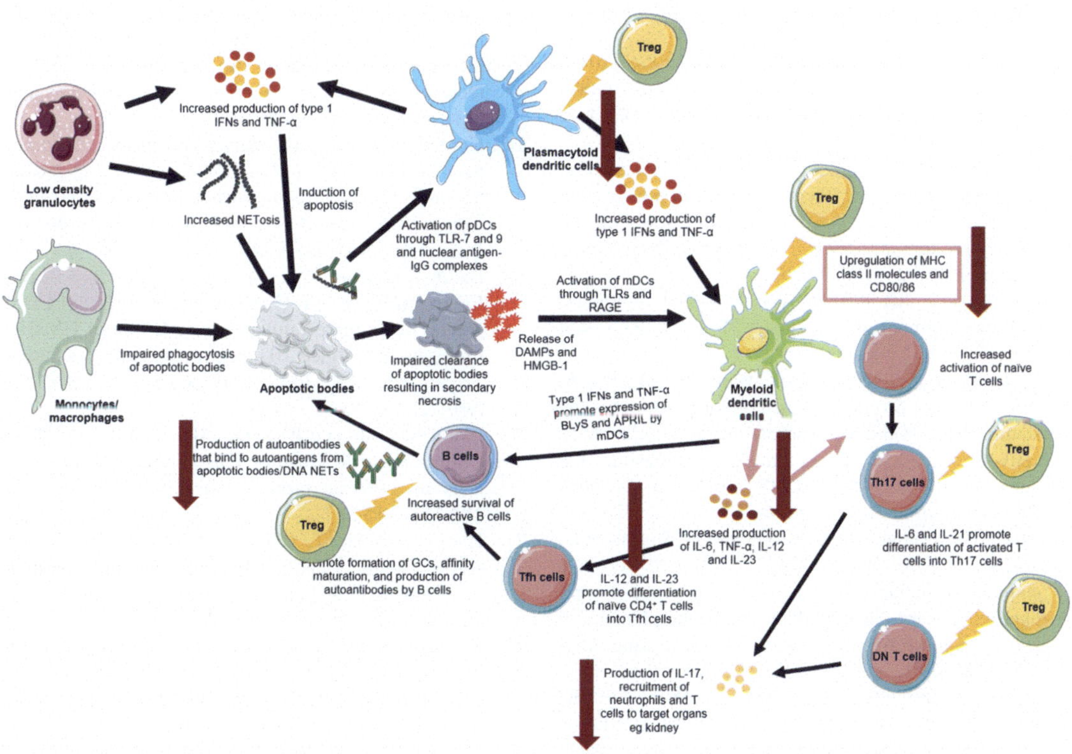

Fig. 9.2 Potential targets for antigen-specific Tregs in the pathogenesis of SLE. Multiple arms of the immune pathways that Tregs can target, for example activated mDCs and pDCs as well as direct suppression of autoreactive T and B cells. *Source* (Diagram created using Servier Medical Art from smart.servier.com)

in production of pro-inflammatory cytokines such as IL-17. Tregs may also directly suppress autoreactive B cells, leading to a decrease in production of autoantibodies that recognize autoantigens from apoptotic bodies and DNA NETs (20).

The development of antigen-specific Tregs is therefore potentially useful for the treatment of SLE. In particular, TCR-Tregs specific for the Sm antigen is a promising therapy for patients with intractable lupus nephritis.

References

1. Vignali DA, Collison LW, Workman CJ (2008) How regulatory T cells work. Nat Rev Immunol 8(7):523–532
2. van der Vliet HJJ, Nieuwenhuis EE (2007) IPEX as a result of mutations in Foxp3. Clin Dev Immunol 2007:89017
3. Dominguez Villar M, Hafler DA (2018) Regulatory T cells in autoimmune disease. Nat Immunol 19:665–673
4. Klein L, Kyewski B, Allen PM, Hogquist KA (2014) Positive and negative selection of the T cell repertoire: what thymocytes see (and don't see). Nat Rev Immunol 14:377
5. Morikawa H, Sakaguchi S (2014) Genetic and epigenetic basis of Treg cell development and function: from a FoxP3-centered view to an epigenome-defined view of natural Treg cells. Immunol Rev 259(1):192–205
6. Lee W, Lee GR (2018) Transcriptional regulation and development of regulatory T cells. Exp Mol Med. 50:e456
7. Wakamatsu E, Omori H, Kawano A, Ogawa S, Abe R (2018) Strong TCR stimulation promotes the stabilization of Foxp3 expression in regulatory T cells induced in vitro through increasing the demethylation of Foxp3 CNS2. Biochem Biophys Res Commun 503(4):2597–2602
8. Ohkura N, Hamaguchi M, Morikawa H, Sugimura K, Tanaka A, Ito Y et al (2012) T cell receptor stimulation-induced epigenetic changes and Foxp3 expression are independent and complementary events required for Treg cell development. Immunity 37(5):785–799
9. Toker A, Engelbert D, Garg G, Polansky JK, Floess S, Miyao T et al (2013) Active demethylation of the Foxp3 locus leads to the generation of stable regulatory T cells within the thymus. J Immunol 190 (7):3180
10. Kanamori M, Nakatsukasa H, Okada M, Lu Q, Yoshimura A (2016) Induced regulatory T cells: their development, stability, and applications. Trends Immunol 37(11):803–811
11. Levings MK, Gregori S, Tresoldi E, Cazzaniga S, Bonini C, Roncarolo MG (2005) Differentiation of Tr1 cells by immature dendritic cells requires IL-10 but not CD25+CD4+ Tr cells. Blood 105(3):1162–1169
12. Gagliani N, Magnani CF, Huber S, Gianolini ME, Pala M, Licona-Limon P, et al (2013) Coexpression of CD49b and LAG-3 identifies human and mouse T regulatory type 1 cells. Nat Med 19:739-46
13. Zeng H, Zhang R, Jin B, Chen L (2015) Type 1 regulatory T cells: a new mechanism of peripheral immune tolerance. Cell Mol Immunol 12:566–571
14. Bandala-Sanchez E, Zhang Y, Reinwald S, Dromey JA, Lee B-H, Qian J, et al (2013) T cell regulation mediated by interaction of soluble CD52 with the inhibitory receptor Siglec-10. Nat Immunol 14:741–748
15. Miyara M, Sakaguchi S (2007) Natural regulatory T cells: mechanisms of suppression. Trends Mol Med. 13(3):108–116
16. Liang B, Workman C, Lee J, Chew C, Dale BM, Colonna L et al (2008) Regulatory T cells inhibit dendritic cells by lymphocyte activation gene-3 engagement of MHC class II. J Immunol 180 (9):5916–5926
17. Chaudhry A, SamsteinRobert M, Treuting P, Liang Y, PilsMarina C, Heinrich J-M et al (2011) Interleukin-10 signaling in regulatory T cells is required for suppression of Th17 cell-mediated inflammation. Immunity 34(4):566–578
18. Konkel JE, Zhang D, Zanvit P, Chia C, Zangarle-Murray T, Jin W, et al (2017) Transforming growth factor-β signaling in regulatory T cells controls T helper-17 cells and tissue-specific immune responses. Immunity 46(4):660–674
19. Bandala-Sanchez EG, Bediaga N, Goddard-Borger ED, Ngui K, Naselli G, Stone NL et al (2018) CD52 glycan binds the proinflammatory B box of HMGB1 to engage the Siglec-10 receptor and suppress human T cell function. Proc Natl Acad Sci USA 115 (30):7783–7788
20. Iikuni N, Lourenco EV, Hahn BH, La Cava A (2009) Cutting edge: Regulatory T cells directly suppress B cells in systemic lupus erythematosus. J Immunol 183:1518–1522
21. Hua J, Inomata T, Chen Y, Foulsham W, Stevenson W, Shiang T et al (2018) Pathological conversion of regulatory T cells is associated with loss of allotolerance. Sci Rep 8(1):7059
22. Komatsu N, Okamoto K, Sawa S, Nakashima T, Oh hora M, Kodama T et al (2013) Pathogenic conversion of Foxp3+ T cells into TH17 cells in autoimmune arthritis. Nat Med 20:62–68
23. Kimura A, Kishimoto T (2010) IL-6: Regulator of Treg/Th17 balance. Eur J Immunol 40(7):1830–1835
24. Thornton AM, Lu J, Korty PE, Kim YC, Martens C, Sun PD et al (2019) Helios+ and Helios- Treg subpopulations are phenotypically and functionally

distinct and express dissimilar TCR repertoires. Eur J Immunol 49:398–412
25. Valencia X, Yarboro C, Illei G, Lipsky PE (2007) Deficient CD4+CD25high T regulatory cell function in patients with active systemic lupus erythematosus. J Immunol 178(4):2579–2588
26. Bonelli M, Savitskaya A, von Dalwigk K, Steiner CW, Aletaha D, Smolen JS et al (2008) Quantitative and qualitative deficiencies of regulatory T cells in patients with systemic lupus erythematosus (SLE). Int Immunol 20(7):861–868
27. Zhu Y, Huang Y, Ming B, Wu X, Chen Y, Dong L (2019) Regulatory T-cell levels in systemic lupus erythematosus patients: a meta-analysis. Lupus. 28 (4):445–454
28. Alvarado-Sánchez B, Hernández-Castro B, Portales-Pérez D, Baranda L, Layseca-Espinosa E, Abud-Mendoza C, et al (2006) Regulatory T cells in patients with systemic lupus erythematosus. J Autoimmun 27(2):110–118
29. Dall'Era M, Pauli ML, Remedios K, Taravati K, Sandova PM, Putnam AL et al (2019) Adoptive Treg cell therapy in a patient with systemic lupus erythematosus. Arthritis Rheumatol. 71(3):431–440
30. He J, Zhang X, Wei Y, Sun X, Chen Y, Deng J et al (2016) Low-dose interleukin-2 treatment selectively modulates CD4+ T cell subsets in patients with systemic lupus erythematosus. Nat Med 22:991–993
31. Ferreira LMR, Muller YD, Bluestone JA, Tang Q (2019) Next-generation regulatory T cell therapy. Nat Rev Drug Discov 18(10):749–769
32. Di Ianni M, Falzetti F, Carotti A, Terenzi A, Castellino F, Bonifacio E et al (2011) Tregs prevent GVHD and promote immune reconstitution in HLA-haploidentical transplantation. Blood 117(14):3921–3928
33. Bluestone JA, Buckner JH, Fitch M, Gitelman SE, Gupta S, Hellerstein MK, et al (2015) Type 1 diabetes immunotherapy using polyclonal regulatory T cells. Sci Transl Med 7(315):315ra189
34. Todo S, Yamashita K, Goto R, Zaitsu M, Nagatsu A, Oura T et al (2016) A pilot study of operational tolerance with a regulatory T-cell-based cell therapy in living donor liver transplantation. Hepatology 64 (2):632–643
35. Dawson NAJ, Levings MK (2017) Antigen-specific regulatory T cells: are police CARs the answer? Transl Res. 187:53–58
36. Guedan S, Calderon H, Posey AD Jr, Maus MV (2019) Engineering and design of chimeric antigen receptors. Mol Ther—Meth Clin D. 12:145–156
37. Blat D, Zigmond E, Alteber Z, Waks T, Eshhar Z (2014) Suppression of murine colitis and its associated cancer by carcinoembryonic antigen-specific regulatory T cells. Mol Ther 22(5):1018–1028
38. Yoon J, Schmidt A, Zhang A-H, Königs C, Kim YC, Scott DW (2017) FVIII-specific human chimeric antigen receptor T-regulatory cells suppress T- and B-cell responses to FVIII. Blood 129(2):238–245
39. MacDonald KG, Hoeppli RE, Huang Q, Gillies J, Luciani DS, Orban PC et al (2016) Alloantigen-specific regulatory T cells generated with a chimeric antigen receptor. J Clin Invest 126(4):1413–1424
40. Boardman DA, Philippeos C, Fruhwirth GO, Ibrahim MAA, Hannen RF, Cooper D et al (2017) Expression of a chimeric antigen receptor specific for donor HLA class I enhances the potency of human regulatory T cells in preventing human skin transplant rejection. Am J Transplant 17(4):931–943
41. Kim YC, Zhang AH, Su Y, Rieder SA, Rossi RJ, Ettinger RA et al (2015) Engineered antigen-specific human regulatory T cells: immunosuppression of FVIII-specific T- and B-cell responses. Blood 125:1107–1115
42. Hull CM, Nickolay LE, Estorninho M, Richardson MW, Riley JL, Peakman M et al (2017) Generation of human islet-specific regulatory T cells by TCR gene transfer. J Autoimmun 79:63–73
43. Kim YC, Zhang AH, Yoon J, Culp WE, Lees JR, Wucherpfennig KW et al (2018) Engineered MBP-specific human Tregs ameliorate MOG-induced EAE through IL-2-triggered inhibition of effector T cells. J Autoimmun 92:77–86
44. Wang J, Ioan-Facsinay A, van der Voort EIH, Huizinga TWJ, Toes REM (2007) Transient expression of FOXP3 in human activated nonregulatory CD4+ T cells. Eur J Immunol 37(1):129–138
45. Mathew JM, Voss H-J, LeFever A, Konieczna I, Stratton C, He J, et al (2018) A phase I clinical trial with ex vivo expanded recipient regulatory T cells in living donor kidney transplants. Sci Rep. 8(1):7428
46. Cojocaru M, Cojocaru IM, Silosi I, Vrabie CD (2011) Manifestations of systemic lupus erythematosus. Maedica. 6:330–336
47. Maidhof W, Hilas O (2012) Lupus: An overview of the disease and management options. P&T 37:240–249
48. Rees R, Doherty M, Grainge MJ, Lanyon P, Zhang W (2017) The worldwide incidence and prevalence of systemic lupus erythematosus: a systematic review of epidemiological studies. Rheumatology 56:1945–1961
49. Carter EE, Barr SG, Clarke AE (2016) The global burden of SLE: prevalence, health disparities and socioeconomic impact. Nat Rev Rheumatol 12:605–620
50. van Schaarenburg RA, Magro-Checa C, Bakker JA, Onno Teng YK, Bajema IM, Huizinga TW, et al (2016) C1q deficiency and neuropsychiatric systemic lupus erythematosus. Front Immunol 7:e00647
51. Poole BD, Scofield RH, Harley JB, James JA (2009) Epstein-Barr virus and molecular mimicry in systemic lupus erythematosus. Autoimmunity 39:63–70
52. Graham RR, Ortmann W, Rodine P, Espe K, Langefeld C, Lange E et al (2007) Specific combinations of HLA-DR2 and DR3 class II haplotypes contribute graded risk for disease susceptibility and autoantibodies in human SLE. Eur J Hum Genet 15:823

53. Jiang C, Deshmukh US, Gaskin F, Bagavant H, Hanson J, David CS et al (2010) Differential responses to Smith D autoantigen by mice with *HLA-DR* and *HLA-DQ* transgenes: dominant responses by *HLA-DR3* transgenic mice with diversification of autoantibodies to small nuclear ribonucleoprotein, double-stranded DNA, and nuclear antigens. J Immunol 184:1085–1091
54. Niu Z, Zhang P, Tong Y (2015) Value of HLA-DR genotype in systemic lupus erythematosus and lupus nephritis: a meta-analysis. Int J Rheum Dis. 18:17–28
55. Langefeld CD, Ainsworth HC, Graham DSC, Kelly JA, Comeau ME, Marion MC et al (2017) Transancestral mapping and genetic load in systemic lupus erythematosus. Nat Commun 8:16021
56. Marchini M, Antonioli R, Lleó A, Barili M, Caronni M, Origgi L et al (2003) HLA class II antigens associated with lupus nephritis in Italian SLE patients. Hum Immunol 64:462–468
57. Chung SA, Brown EE, Williams AH, Ramos PS, Berthier CC, Bhangale T et al (2014) Lupus nephritis susceptibility loci in women with systemic lupus erythematosus. J Am Soc Nephrol 25(12):2859–2870
58. de Holanda MI, Klumb E, Imada A, Lima LA, Alcantara I, Gregorio F et al (2018) The prevalence of HLA alleles in a lupus nephritis population. Transpl Immunol 47:37–43
59. Ooi JD, Petersen J, Tan YH, Huynh M, Willett ZJ, Ramarathinam SH et al (2017) Dominant protection from HLA-linked autoimmunity by antigen-specific regulatory T cells. Nature 545:243–248
60. Kim K, Bang S-Y, Lee H-S, Okada Y, Han B, Saw W-Y et al (2014) The HLA-DRβ1 amino acid positions 11–13-26 explain the majority of SLE-MHC associations. Nat Commun 5:5902
61. Riemekasten G, Hahn BH (2005) Key autoantigens in SLE. Rheumatology 44:975–982
62. Ishizaki J, Saito K, Nawata M, Mizuno Y, Tokunaga M, Sawamukai N et al (2014) Low complements and high titre of anti-Sm antibody as predictors of histopathologically proven silent lupus nephritis without abnormal urinalysis in patients with systemic lupus erythematosus. Rheumatology 54(3):405–412
63. Arroyo Ávila M, Santiago Casas Y, McGwin G, Cantor RS, Petri M, Ramsey Goldman R et al (2015) Clinical associations of anti-Smith antibodies in PROFILE: a multi-ethnic lupus cohort. Clin Rheumatol 34(7):1217–1223
64. Deshmukh US, Sim DL, Dai C, Kannapell CC, Gaskin F, Rajagopalan G et al (2011) HLA-DR3 restricted T cell epitope mimicry in induction of autoimmune response to lupus-associated antigen SmD. J Autoimmun 37:254–262
65. Chowdhary VR, Dai C, Tilahun AY, Hanson JA, Smart MK, Grande JP et al (2015) A central role for HLA-DR3 in anti-Smith antibody responses and glomerulonephritis in a transgenic mouse model of spontaneous lupus. J Immunol 195:4660–4667
66. Zhao Z, Ren J, Dai C, Kannapell CC, Wang H, Gaskin F et al (2019) Nature of T cell epitopes in lupus antigens and HLA-DR determines autoantibody initiation and diversification. Ann Rheum Dis 78:380–390
67. Xue K, Niu W-Q, Cui Y (2018) Association of HLA-DR3 and HLA-DR15 polymorphisms with risk of systemic lupus erythematosus. Chin Med J (Engl) 131(23):2844–2851
68. Moulton VR, Suarez-Fueyo A, Meidan E, Li H, Mizui M, Tsokos GC (2017) Pathogenesis of human systemic lupus erythematosus: a cellular perspective. Trends Mol Med 23:615–635
69. Kaplan MJ (2011) Neutrophils in the pathogenesis and manifestations of SLE. Nat Rev Rheumatol 7:691–699
70. Denny MF, Yalavarthi S, Zhao W, Thacker SG, Anderson M, Sandy AR et al (2010) A distinct subset of proinflammatory neutrophils isolated from patients with systemic lupus erythematosus induces vascular damage and synthesizes type I IFNs. J Immunol 184:3284–3297
71. Ren Y, Tang J, Mok MY, Chan AWK, Wu A, Lau CS (2003) Increased apoptotic neutrophils and macrophages and impaired macrophage phagocytic clearance of apoptotic neutrophils in systemic lupus erythematosus. Arthritis Rheum 48:2888–2897
72. Tsokos GC, Lo MS, Reis PC, Sullivan KE (2016) New insights into the immunopathogenesis of systemic lupus erythematosus. Nat Rev Rheumatol 12 (716–730)
73. Bijl M, Reefman E, Horst G, Limburg PC, Kallenberg CGM (2006) Reduced uptake of apoptotic cells by macrophages in systemic lupus erythematosus: correlates with decreased serum levels of complement. Ann Rheum Dis 65:57–63
74. Fransen JH, van der Vlag J, Ruben J, Adema GJ, Berden JH, Hilbrands LB (2010) The role of dendritic cells in the pathogenesis of systemic lupus erythematosus. Arthritis Res Ther. 12:207
75. Urbonaviciute V, Furnrohr BG, Meister S, Munoz L, Heyder P, De Marchis F et al (2008) Induction of inflammatory and immune responses by HMGB-1-nucleosome complexes: implications for the pathogenesis of SLE. J Exp Med 205:3007–3018
76. Blanco P, Palucka AK, Pascual V, Banchereau J (2008) Dendritic cells and cytokines in human inflammatory and autoimmune diseases. Cytokine Growth Factor Rev 19:41–52
77. Lövgren T, Eloranta M-L, Båve U, Alm GV, Rönnblom L (2004) Induction of interferon-α production in plasmacytoid dendritic cells by immune complexes containing nucleic acid released by necrotic or late apoptotic cells and lupus IgG. Arthritis Rheum 50(6):1861–1872
78. Sisirak V, Ganguly D, Lewis KL, Couillault C, Tanaka L, Bolland S et al (2014) Genetic evidence for the role of plasmacytoid dendritic cells in systemic lupus erythematosus. J Exp Med 211:1969–1976

79. Ding D, Mehta H, McCune WJ, Kaplan MJ (2006) Aberrant phenotype and function of myeloid dendritic cells in systemic lupus erythematosus. J Immunol 177:5878–5889
80. Simmons DP, Wearsch PA, Canaday DH, Meyerson HJ, Liu YC, Wang Y et al (2012) Type 1 IFN drives a distinctive dendritic cell maturation phenotype that allows continued class II MHC synthesis and antigen processing. J Immunol 188:3116–3126
81. López P, Scheel-Toellner D, Rodríguez-Carrio J, Caminal-Montero L, Gordon C, Suárez A (2014) Interferon-α-induced B-lymphocyte stimulator expression and mobilization in healthy and systemic lupus erthymatosus monocytes. Rheumatology 53:2249–2258.
82. López P, Rodríguez-Carrio J, Caminal-Montero L, Mozo L, Suárez A (2016) A pathogenic IFNα, BLyS and IL-17 axis in systemic lupus erythematosus patients. Nature 6:20651
83. Kiefer K, Oropallo MA, Cancro MP, Marshak-Rothstein A (2012) Role of type 1 interferons in the activation of autoreactive B cells. Immunol Cell Biol 90:498–504
84. Cancro MP, D'Cruz DP, Khamashta MA (2009) The role of B lymphocyte stimulator (BLyS) in systemic lupus erythematosus. J Clin Invest 119:1066–1073
85. Thien M, Phan TG, Gardam S, Amesbury M, Basten A, Mackay F et al (2004) Excess BAFF rescues self-reactive B cells from peripheral deletion and allows them to enter forbidden follicular and marginal zone niches. Immunity 20:785–798
86. Eilertsen GØ, Van Ghelue M, Strand H, Nossent JC (2011) Increased levels of BAFF in patients with systemic lupus erythematosus are associated with acute-phase reactants, independent of BAFF genetics: a case–control study. Rheumatology 50 (12):2197–2205
87. Tangye SG, Ma CS, Brink R, Deenick EK (2013) The good, the bad and the ugly—TFH cells in human health and disease. Nat Rev Immunol 13:412
88. Crotty S (2014) T follicular helper cell differentiation, function, and roles in disease. Immunity 41 (4):529–542
89. Schmitt N, Morita R, Bourdery L, Bentebibel SE, Zurawski SM, Banchereau J et al (2009) Human dendritic cells induce the differentiation of interleukin-21-producing T follicular helper-like cells through interleukin-12. Immunity 31(1):158–169
90. Koenig KF, Groeschl I, Pesickova SS, Tesar V, Eisenberger U, Trendelenburg M (2012) Serum cytokine profile in patients with active lupus nephritis. Cytokine 60(2):410–416
91. Dai H, He F, Tsokos GC, Kyttaris VC (2017) IL-23 limits the production of IL-2 and promotes autoimmunity in lupus. J Immunol 199(3):903–910
92. Ballesteros Tato A, León B, Graf Beth A, Moquin A, Adams Pamela S, Lund Frances E et al (2012) Interleukin-2 inhibits germinal center formation by limiting T follicular helper cell differentiation. Immunity 36(5):847–856
93. von Spee-Mayer C, Siegert E, Abdirama D, Rose A, Klaus A, Alexander T, et al (2016) Low-dose interleukin-2 selectively corrects regulatory T cell defects in patients with systemic lupus erythematosus. Ann Rheum Dis 75:1407–1415
94. Wong CK, Lit LCW, Tam LS, Li EKM, Wong PTY, Lam CWK (2008) Hyperproduction of IL-23 and IL-17 in patients with systemic lupus erythematosus: Implications for Th17-mediated inflammation in auto-immunity. Clin Immunol 127(3):385–393
95. Vincent FB, Northcott M, Hoi A, Mackay F, Morand EF (2013) Clinical associations of serum interleukin-17 in systemic lupus erythematosus. Arthritis Res Ther. 15(4):R97
96. Crispín JC, Oukka M, Bayliss G, Cohen RA, Van Beek CA, Stillman IE et al (2008) Expanded double negative T cells in patients with systemic lupus erythematosus produce IL-17 and infiltrate the kidneys. J Immunol 181(12):8761
97. Ogura H, Murakami M, Okuyama Y, Tsuruoka M, Kitabayashi C, Kanamoto M et al (2008) Interleukin-17 promotes autoimmunity by triggering a positive-feedback loop via interleukin-6 induction. Immunity 29(4):628–636

Balancing Strategies: GC and GILZ Axis

10

Champa Nataraja, Wendy Zhu, Wendy Dankers, and Sarah A. Jones

Abstract

Glucocorticoids (GC) are universally relied upon for treatment of SLE. Through binding of the glucocorticoid receptor (GR), they initiate a suite of mechanisms that substantially change the behaviour of cells. The GR binds directly to protein targets within the cytoplasm to interrupt their activity, and acts as a transcription factor to alter gene transcription on a scale that produces wholesale changes in the transcriptional output of cells. This results in a return to immunological quiescence that is a critical anchor of rheumatologic treatment, with no current parallels. Devastatingly, however, GC reliably bring about a collection of adverse effects that measurably and substantially contribute to poor outcomes in SLE. A safe alternative, with the same broad effects across cells and pathways of the immune system, is critically needed. Recent research has substantiated the concept that the glucocorticoid-induced leucine zipper (GILZ) protein may represent such an alternative. This chapter discusses the mechanisms utilized by the GR to bring about anti-inflammatory effects in lupus, and the harmful metabolic effects that are caused by GC. It also describes the protective functions of GILZ in comparison to GC, and preliminary evidence that GILZ may not evoke the metabolic pathways that make GC such a double-edged sword in the arsenal of lupus treatment.

C. Nataraja · W. Zhu · W. Dankers · S. A. Jones (✉)
Rheumatology Research Group, School of Clinical Sciences At Monash Health, Centre for Inflammatory Diseases, Monash University, Clayton, VIC 3168, Australia
e-mail: sarah.a.jones@monash.edu

C. Nataraja
e-mail: champa.nataraja@monash.edu

W. Zhu
e-mail: wzhu56@student.monash.edu

W. Dankers
e-mail: wendy.dankers@monash.edu

C. Nataraja
Department of Rheumatology, Monash Health, Clayton, VIC 3168, Australia

Management of SLE

The management of SLE is aimed at improving quality of life and preventing organ damage to prolong survival. However, due to the diversity in how this disease affects each individual patient, choosing the appropriate treatment can prove challenging to clinicians. Furthermore, the management options for SLE have scarcely changed since the 1950s, and most patients are prescribed non-specific immunosuppressants. These include antimalarials (such as hydroxychloroquine), cytotoxic and immunomodulatory

A. Hoi (ed.), *Pathogenesis of Systemic Lupus Erythematosus*,
https://doi.org/10.1007/978-3-030-85161-3_10

drugs (such as methotrexate, azathioprine, mycophenolate mofetil, and cyclophosphamide) and, in 80% of patients, glucocorticoids (GC) [1–3]. This is in contrast to the rising use of targeted therapies in other autoimmune conditions, such as rheumatoid arthritis [4], which has entirely changed the prognosis of these once debilitating diseases. Many attempts at developing an effective targeted therapy for SLE have failed, in part due to the heterogeneity of this disease.

The only approved novel targeted therapy for SLE is belimumab, an anti-B cell activating factor (BAFF) monoclonal antibody. Belimumab was originally seen as an exciting and potentially life-changing treatment option for SLE, however, in practice its clinical efficacy is modest at best, and it has mainly been investigated as an adjunct therapy in patients that meet select criteria [5]. However, with the successful completion of a phase III trial for the use of Belimumab in addition to standard therapy in lupus nephritis in mid-2020, there is promise for this therapy to expand its utility in SLE [6]. Another promising biologic is Anifrolumab, a monoclonal antibody which suppresses the effects of type I interferon through blocking interferon (IFN) alpha/beta receptor (IFNAR) [7]. In early 2020, successful results from the TULIP-2 Phase III clinical trial were published demonstrating that Anifrolumab significantly reduces disease activity in SLE patients as measured by the British Isles Lupus Assessment Group (BILAG)-based Composite Lupus Assessment (BICLA) [8]. Until Anifrolumab gains regulatory approval, the use of GC and immunosuppressants remains the standard of care, as they were more than half a century ago.

Glucocorticoids

Prednisone is the synthetic corticosteroid favoured for oral dosing in SLE, and its use is variable up to around 50 mg per day or 1 mg/kg per day. In life- or organ- threatening lupus or refractory major organ disease, initial pulse methylprednisolone given intravenously between 500—1000 mg over one to three days, followed by switching to high dose oral prednisolone, and subsequently tapering over weeks or months to achieve the minimum effective dose. Limiting GC exposure is a primary concern for treating physicians as the use of GC is accompanied by a litany of undesirable and predictable metabolic adverse effects that increase morbidity and mortality [9]. Additionally, multiple studies indicate that GC have a dose-dependent effect on disease damage accrual in SLE [1–3, 10]. There is significant unmet need for a drug that has the broad anti-inflammatory effects of GC, but without the debilitating metabolic adverse effects. To achieve this requires a deeper understanding of GC mechanisms of action.

GC Receptor: Mechanisms of Action

GC act chiefly through binding to glucocorticoid receptors (GR) in the cytoplasm [11]. The GR is present in virtually every cell in the human body, thus providing a wide range of cells in which GC can modulate gene expression (12). The GR mediates the effects of GC through the regulation of multiple genes via a number of mechanisms termed transrepression, transactivation, and cis-repression [11] (Fig. 10.1). Transactivation and cis-repression result in the induction or repression of target gene expression via the GR translocating to the nucleus and binding to target sequences termed GC response elements (GRE).

There are three methods via which GR affects gene transcription: (A) binding to GRE to induce gene expression or negative GRE to repress gene expression, (B) forming tethers with another transcription factor without touching DNA, and (C) composite binding in which GR binds a GRE while interacting with another transcription factor that concurrently binds an adjacent motif [13].

The gene encoding the GR (NR3C1) can produce two alternatively spliced isoforms, GRα and GRβ [14]. Present in the cytoplasm, GRα acts in a ligand-dependent fashion and can be considered the canonical GR form [15].

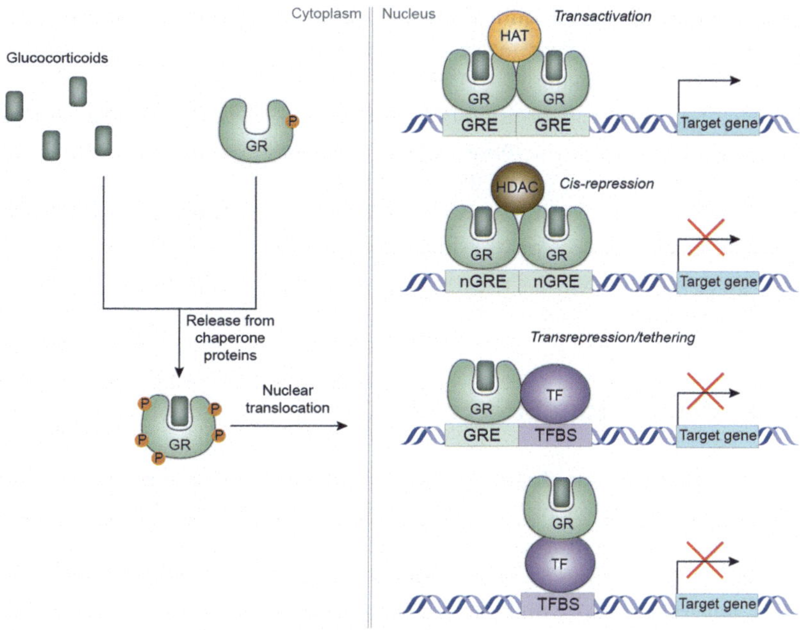

Fig. 10.1 Glucocorticoid binding to the glucocorticoid receptor (GR). Glucocorticoid binds to its receptor and chaperone proteins are released to allow for the nuclear translocation of GR. Within the nucleus, the GR can act in various ways to activate or inhibit gene expression

Conversely, GRβ is a probable negative regulator of GR functions that bears the DNA-binding capacity of the GRα isoform but not the GC binding capacity. It is present in the nucleus, is inactive on GC-sensitive genes and is correlated with GC resistance [16–18]. This effect may arise from its ability to compete with GRα for access to GREs, absorb transcriptional coregulators that act in concert with GRβ, or bind to GRα to form heterodimers that are inactive. GRβ can also directly modulate inflammatory gene expression independent of GRα, through recruitment of histone deacetylases and subsequent localised chromatin remodelling [18]. Other isoforms of the GR exist and have distinct properties, but their physiological roles are less clear [16].

The GR is dynamically regulated by post-translational modification. Free GR is phosphorylated and upon ligand binding, becomes hyperphosphorylated, causing its translocation to the nucleus [19]. Different phosphorylation sites affect the various downstream effects of GR activation, such as transcription factor suppression, nuclear translocation and genome binding, or GR protein degradation [19, 20]. GR function and stability is also regulated by acetylation, SUMOylation and ubiquitination [21–24].

GC-Mediated Gene Activation

Transactivation occurs via direct DNA binding of homodimers of GR to GRE in the promoter sequence of target genes and subsequent recruitment of histone acetyltransferases, which opens the chromatin and allows binding of the transcription machinery to initiate gene expression. Examples of genes targeted through this mechanism include glucocorticoid-induced leucine zipper (GILZ), Annexin-1, SLP1, IκBα, MKP1, the soluble decoy IL-1 receptor (IL-1R), and IL-10 [25–27].

GC-Mediated Gene Repression

Cis-repression occurs when homodimers of GR bind to negative GRE (nGRE), leading to the recruitment of histone deacetylases including HDAC2, which can close the chromatin at target loci to repress gene transcription [23]. This mechanism regulates several GC-sensitive proinflammatory modulators and cytokines including IL-1ß [11] and GM-CSF [28].

Transrepression, or tethering, involves direct or indirect binding of GR to suppress the action of a number of pro-inflammatory transcription factors, including nuclear factor kappa B (NF-κ B) and activator protein 1 (AP-1) [29]. These factors drive survival, proliferation, differentiation and expression of inflammatory effector genes such as cytokines. The direct inhibition of these pathways by ligand-activated GR has a broad blanketing effect on the immune system, extinguishing inflammatory processes at their molecular source.

Anti-Inflammatory Effects of GC

The inflammation-smothering, negative regulatory effects of GC apply to cells of both the innate and adaptive arms of the immune system. In parallel, GC also up-regulate multiple system-wide anti-inflammatory outcomes. The collective result of these effects of GC is the suppression of multiple components of the immune system that drive disease activity in lupus, returning the immune system to a state of quiescence (Fig. 10.2). Certain aspects of the immune system such as the Th17 cell axis and the type I IFN system are relatively resistant to GC treatment by virtue of having few interacting pathway components. In patients whose disease is driven by these features of the immune system, GC are poorly effective unless given at pulse doses.

Phagocyte Regulation by GC

In the innate immune system, GC block vasodilation, vascular permeability and chemokine expression to suppress the migration of immune cells to the site of inflammation [30]. GC induce pro-resolving factors to halt further immune cell infiltration of inflamed sites, and shut down TLR signalling and cytokine production by macrophages [31]. GC promote neutrophil apoptosis and enhance the phagocytic activity of monocytes with an anti-inflammatory phenotype while suppressing pro-inflammatory ones, leading to enhanced clearance of dead cells and toxins [31–33]. This is partly mediated through promoting the production of TGFβ and IL-10 and inhibiting the production of TNFα [32, 33]. Since apoptotic debris is a key source of auto-antigens and danger-associated molecular patterns (DAMPs), this prevents further activation of auto-reactive B

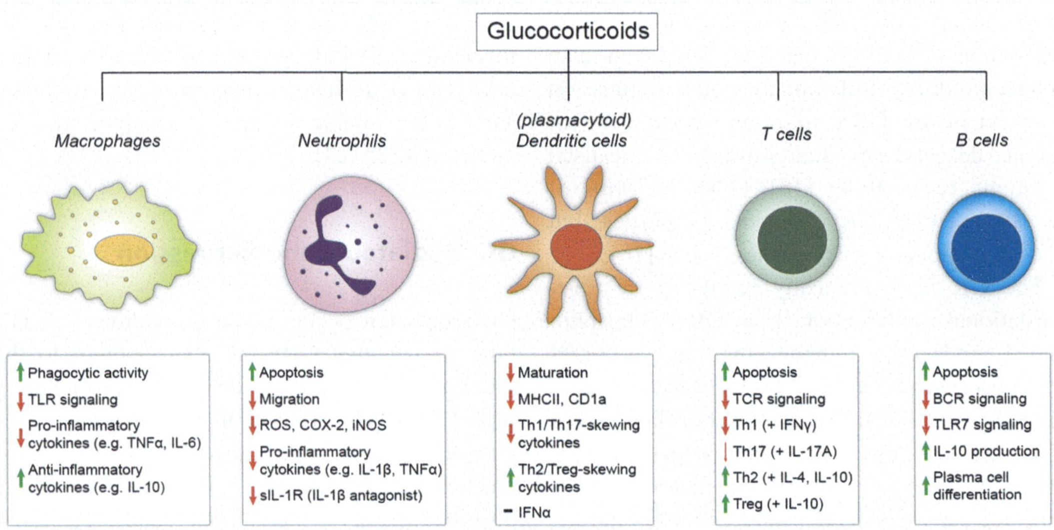

Fig. 10.2 Glucocorticoids have a multitude of anti-inflammatory effects across cells of the immune system. A number of mechanisms have been proposed in mediating the anti-inflammatory effect of glucocorticoid via a range of cell types

cells and ligation of DAMP-sensing receptors such as TLRs, which would otherwise lead to inflammatory cytokine expression and the amplification of disease activity.

Effects of GC on Lymphocytes

The outcomes of GC treatment on lymphocyte behaviour varies according to cell type and activation status, but collectively promote immunological homeostasis and a return to quiescence from an inflamed state [34]. In general, signalling through antigen receptors (such as the T cell receptor; TCR) is strongly suppressed by GC via multiple mechanisms of inhibition of a range of signalling mediators [35, 36]. GC also prevent T cell activation indirectly by inhibiting the maturation of antigen-presenting cells and downregulating their expression of MHC-II, CD1a, costimulatory molecules and pro-inflammatory cytokines [37]. Finally, apoptosis is induced in pro-inflammatory cells via GC-mediated inhibition of pro-survival factors and activation of caspases [38].

GC effects on specific lymphocyte types are more nuanced [34]. GC promote the differentiation of T helper 2 cells (Th2) while preventing the development of T helper 1 cells (Th1) [39]. Th1 cells produce more pro-inflammatory cytokines such as IFNγ and TNFα, while Th2 cells are known to produce cytokines that are less inflammatory in autoimmune conditions, such as IL-4, and the anti-inflammatory cytokine IL-10 [40]. Thus, the balance of these phenotypes is an important determinant of the inflammatory state and resultant autoimmunity that ensues.

GC may target Th17 to suppress development of autoimmunity in animal models of multiple sclerosis and arthritis, although their efficacy in preventing the activities of this highly inflammatory cell type appears to be more limited than for other lymphocyte types, meaning that Th17 cells are relatively resistant to GC [41, 42]. Conversely, GC have a clear role in increasing Treg formation through promoting the differentiation of tolerogenic dendritic cells (DC) which, in turn, induce the differentiation of Treg cells [43, 44].

While there is less research into the effect of GC on B cells, it has been demonstrated that GC are able to induce their apoptosis [42, 45]. GC also inhibit signalling through the B cell receptor (BCR) and TLR7 pathways in B cells and promote their transcription of IL-10 and the plasma cell differentiation factor BLIMP-1 [46]. While there is little knowledge on the mechanisms behind these effects, GC up-regulation of the mediator GILZ was a consistent feature in B cells [46], and further predictions may be made by extrapolating from studies of GC effects on T lymphocytes, which share similarities in activation and differentiation pathways.

Plasmacytoid DC, Key Sources of Type I IFN, Are Relatively Resistant to GC

A significant inflammatory pathway worthy of discussion here is the type I IFN system, a key driver of pathology in multiple autoimmune conditions, particularly SLE. In SLE, the expression of autoantibodies targeted against autoantigens, which together with complement components, form immune complexes that crowd the circulatory system. These are internalised by dendritic cells (DC), particularly plasmacytoid DC (pDC), triggering endosomal TLR7, TLR7/8 and TLR9 signalling. This activates downstream NFκB-dependent signalling pathways that result in upregulation of IRF7, the transcription factor that drives expression of type I interferons, particularly IFN. Plasmacytoid DC constitutively express interferon regulatory factor 7 (IRF7) [47], and are the major producers of type I IFN in SLE [48, 49].

As GC induce their anti-inflammatory effects in part by inhibiting NFκB activation, one would hypothesise that toll-like receptor (TLR)-induced interferon (IFN) production should be sensitive to GC. However, sustained stimulation of endosomal TLR7 and 9 in pDC results in a reduced ability of GC to inhibit NFκB, thus increasing the viability of TLR7 and 9 activated pDCs and resulting in persistently raised type I IFN expression [50]. Activation of TLR7 and 9 is an integral component of SLE pathogenesis, and this may explain why one-third of patients with SLE are resistant to GC [51]. Indeed, only high dose intravenous steroids can suppress interferon-stimulated genes (ISGs) in SLE patients [52] and clinical manifestations closely linked to IFN signature, including severe skin, renal, and haematological disease, are notoriously insensitive to GC.

Adverse Effects of GC

The ability of GC to target a wide range of genes and transcription factors makes it a potent immunosuppressant with broad actions. This mirrors SLE, which itself is characterised by heterogeneity and is why GC can effectively manage many of these disturbances. However, while GC have their place in SLE management, their long-term use comes with significant risk and morbidity. The following sections review the pathogenesis of GC-induced adverse effects on bone and glucose metabolism.

Normal Bone Remodelling

Normal bone remodelling is a dynamic process involving removal of foci of damaged bone and replacing them with new bone [53, 54]. During this process, osteoclast-induced bone resorption and osteoblast-induced bone formation are tightly regulated, such that there is no net loss in bone mass in normal, healthy individuals. A correct balance between these two cells is crucial to maintain skeletal integrity, but this balance is substantially swayed by GC towards bone loss, and osteoporosis commonly results from sustained GC treatment.

There are five phases involved in bone remodelling: activation, resorption, reversal, formation and termination [53, 54]. During the activation phase, detection of mechanical strain or microfractures induces expression of receptor activator of nuclear kappa-B ligand (RANKL), a member of tumour necrosis factor (TNF) superfamily, on quiescent osteoblasts. Interaction of RANKL with its receptor, RANK, expressed on osteoclast progenitors, triggers differentiation of osteoclasts. During the resorption phase, osteoclasts adhere to the bone surface and begin bone resorption by acidifying the bone matrix. Usually, osteoclasts then undergo apoptosis to avoid excess bone resorption. In the reversal phase, macrophage-like cells are involved in clearing debris prior to the recruitment of monocytes, osteocytes and pre-osteoblasts to initiate bone formation which will be completed in the formation phase. The final phase is the mineralization (calcification) phase; trabecular bone formation completes in 90 days and cortical bone formation completes in approximately 130 days [53, 54]. An imbalance in bone remodelling can eventually lead to bone disease. In the case of osteoporosis, there is increased bone resorption and/or decreased bone formation, resulting in net loss in bone mass.

Regulation of Osteoclast and Osteoblast Differentiation

Osteoclasts are multinucleated cells derived from the monocyte/macrophage lineage of haematopoietic cells. Their differentiation is induced by binding of RANKL to RANK and is inhibited by osteoprotegerin (OPG). A number of surrogate biomarkers exist which correlate with the metabolic products of bone resorption. However, the abundance of serum C-terminal telopeptide (CTX-1) is used for the measurement of osteoclastic activity and serum tartrate-resistant acid phosphatase-5b (TRACP-5b) level is used for the measurement of osteoclast activity and number [55, 56, 57]. Likewise, histochemical TRACP staining is the most common way to identify osteoclasts in tissue sections.

In contrast, osteoblasts are derived from pluripotent mesenchymal stem cells (MSC). Bone marrow derived MSC predominantly differentiate into osteoblasts and adipocytes, but can also give rise to chondrocytes, myocytes, and fibroblasts [58]. The relationship between the difference of osteoblasts and adipocytes is reciprocal; when the osteogenic pathway is blocked, the MSC enters the adipocyte pathway and vice versa [59]. Factors that promote expression of CCAAT/enhancer-binding protein δ (C/EBPδ) and peroxisome proliferator-activated receptor gamma 2 (PPARγ2) proteins inhibit osteoblast and induce adipocyte differentiation, and the opposite effect is true for those that suppress these proteins [60].

Osteoblasts secrete abundant proteins, predominantly osteocalcin and pro-collagen 1, a precursor form of type 1 collagen. Indices of osteoblastic activity are serum propeptide of type 1 N-terminal procollagen (P1NP) and osteocalcin, and an index of osteoblast number is alkaline phosphatase (ALP) level [55, 56, 57]. Therefore, for research purposes and sometimes clinically, CTX-1 and TRACP-5b are used as bone resorption markers and P1NP, osteocalcin and ALP are used as bone formation markers in osteoporosis.

Glucocorticoid-Induced Osteoporosis

GC-induced osteoporosis (GIOP) is the most common cause of secondary osteoporosis and the most common adverse effect of long-term GC therapy. The prevalence of reduced bone mineral density (BMD) among patients with SLE has been reported to be almost 50%, with a strong negative relationship between GC use and bone density [61]. GIOP is characterized by early and rapid bone loss which predominantly affects the cancellous bone and axial skeleton. This results in decreased trabecular bone volume and thickness, as well as decreased cortical thickness and increased marrow area. During the first 3 to 6 months of GC therapy, a rapid bone loss (up to 12%) occurs, which slows thereafter to approximately 3% annually [62]. An increased risk of fracture has been reported even with doses of prednisolone as low as 2.5 to 7.5 mg daily [63].

Mechanisms of GIOP are emerging. At pharmacological doses, GC induce adipocyte differentiation and suppress osteoblast differentiation by promoting binding of C/EBPδ to the PPARγ2 promoter, resulting in increased expression and transcription of PPARγ2 [64, 65]. Consequently, patients develop the typical Cushingoid truncal obesity as well as osteoporosis with long-term GC treatment. Dexamethasone can also inhibit osteoblast differentiation by suppressing *Bcl2*, an anti-apoptotic gene [66, 67]. Notably, GC promote bone resorption by increasing the expression of RANKL and decreasing the expression of osteoprotegerin, thus prolonging the survival of osteoclasts [68, 69, 70]. The net result is reduced BMD which, if left untreated, can develop into osteoporosis. Conversely, ceasing GC increases BMD and reduces fracture risk, although it is not known whether the level of risk returns to baseline when GC treatment is stopped [71].

Glucocorticoid-Induced Diabetes

Another widely known effect of GC treatment is the induction of new onset diabetes. GC-induced diabetes is defined as an abnormal increase in blood glucose in patients on GC treatment with or without prior history of diabetes mellitus. In a meta-analysis including 13 studies, the reported prevalence of GC-induced diabetes is 18.6% (ranged from 1.2% to 56%) and GC-induced hyperglycaemia is 32.2% (ranging from 20% to 45.2%) [72]. The incidence and prevalence of GC-induced diabetes varies with doses and duration of GC therapy as well as underlying co-morbidities [73, 72].

The pathophysiology of GC-induced diabetes is incompletely understood due to the effect of GC on multiple pathways of glucose metabolism. At the molecular level, GC-induced diabetes occurs by binding of ligand-bound GR to nGRE to inhibit insulin precursors and insulin receptor genes. Regulating expression of these genes leads to decreases in pancreatic beta cell number and function resulting in impaired insulin secretion, induction of insulin resistance by depletion or modification of insulin receptors, and altered fat distribution [73]. One study has demonstrated that GC-induced insulin resistance is due to the post-binding modulation of insulin receptor by GC [74]. Another study has shown that inhibition of insulin release is critical for the diabetogenic effect of GC [75].

In SLE patients, GC-induced diabetes has an estimated prevalence of 10–20% and it is reversible once GC therapy is ceased.[76–78]. Further to this, the diabetogenic effect of GC in SLE is dose and duration dependent [79]. Independent of GC therapy, the prevalence of insulin resistance is higher in SLE compared to other diseases due to underlying chronic inflammation [80].

Discovering a Safer Alternative to GC

The range, severity and penetrance of GC adverse effects are unacceptable. A steroid sparing therapy for SLE is long overdue. An ideal candidate would be one that has similarly broad, or even broader, immunosuppressive effects as GC, without utilising GR. Accumulating evidence suggests that the molecule GC-induced leucine zipper (GILZ) is a potential candidate.

GILZ

GILZ was first described in 1997, when it was discovered as a dexamethasone-induced gene that regulates apoptosis in T cells [81]. GILZ is a member of the transforming growth factor beta-stimulated clone 22 (TSC-22) family of proteins and is located on chromosome X. It is characterized by the presence of three domains: a TGFβ-stimulated clone (TSC) box, a central leucine zipper domain (dimerization domain), and a C-terminal domain [81]. There are five mouse *Gilz* isoforms, but the most studied isoform refers to 137 amino acid containing *Gilz-1*/ *Tsc22d3*-2. Human *GILZ* encodes 134 amino acids. The chicken GILZ protein was shown to have a short half-life of 2–3 h and to degrade quickly by ubiquitylation [82]. Within the *Gilz* sequence, several sites have been identified as post translational modification sites [82]. The GILZ gene contains six GREs in its promoter region and binding of GC-GR complex to these GRE sequences strongly enhances GILZ transcription [83].

GILZ mimics many of the anti-inflammatory effects of GC and bypasses the use of GR (Fig. 10.3). Like GCs, GILZ exerts anti-inflammatory actions by interacting with transcription factors, including NFκB and AP-1 [84, 85] In addition, GILZ also inhibits a key signaling pathway in the immune system, the MAPK/ERK pathway, by binding to the

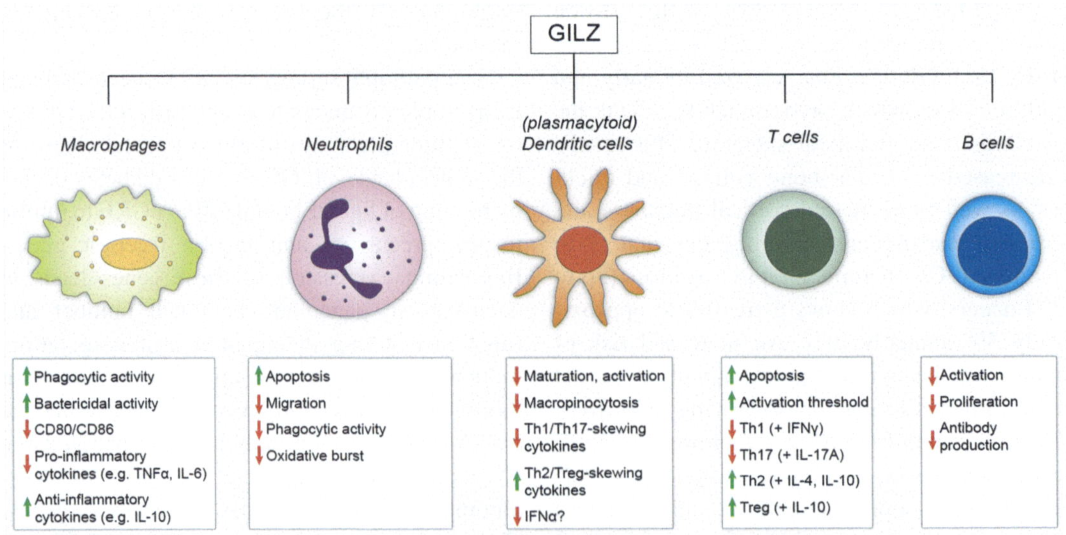

Fig. 10.3 **The broad cellular effects of GILZ**. Like glucocorticoids, GILZ has multiple anti-inflammatory effects in various cell types, without having the same metabolic profile as GC

upstream proteins Ras and Raf-1 [86, 87]. GILZ also regulates the mTORC2/AKT pathway by binding to mTORC2 [88] similarly to GC.

GILZ is one of the most GC-sensitive molecules in the genome [31, 81], and is ubiquitously expressed in mouse and human tissues [89, 90] [91]. It is rapidly and abundantly induced in various cell types in response to very low concentrations of GC, and has endogenous regulatory functions independent of the therapeutic action of GC [92, 93, 94, 95]. Recent discoveries show the anti-inflammatory role of GILZ in neuroinflammation [96], vascular and intestinal inflammation [97, 98], inflammation in liver [99], allergy [100], heart disease [101, 102], acute kidney injury [103] arthritis [92] and psoriasis [95]. In addition, active SLE in humans is associated with impaired GILZ expression (94) and loss of GILZ exacerbates inflammation which is exemplified by the development of a lupus-like phenotype in mice [94]. Moreover, GILZ is therapeutic in a model of autoimmune arthritis [104] and GILZ induction predicts patient response to GC [94, 105].

Effects of GILZ on Immune Cells

B Lymphocytes

GILZ is a non-redundant regulator of B cell activity which may have important clinical implications in SLE [94]. In patients with SLE, GILZ induction by GC in most B cell subsets is negatively correlated with SLE disease activity [94]. Deficiency of GILZ in B cells results in heightened states of activation and proliferation, and this is reversed by treatment with a cell permeable HHph-GILZ fusion protein [94]. Additionally, GILZ mRNA expression is negatively correlated with the SLE disease activity index (SLEDAI) as well as cytokines BAFF and IFNγ [106]. As discussed earlier, a lupus-like autoimmune phenotype is also seen in GILZ-deficient mice, as evidenced by development of antinuclear antibodies and immune complex mediated glomerulonephritis, indicating a crucial role of GILZ in regulating B cell activation and autoantibody production in SLE [94].

Outside of SLE, GILZ expression has also been implicated in the pathogenesis of other B cell disorders [107]. Lack of GILZ inhibits spontaneous B cell apoptosis, resulting in the development of B cell lymphocytosis [108]. B cell specific GILZ-deficient mice also produce more IFNγ through dysregulated AP-1 transcriptional activity, and this same murine model has an increased susceptibility to severe experimental inflammatory colitis. Conversely, treatment with a cell-permeable GILZ fusion protein, TAT-GILZ, inhibits B cell induced IFNγ production and protects the B cell specific GILZ-deficient mice against dinitrobenzene sulfonic acid (DNBS)-induced colitis [108].

T Lymphocytes

T cells are involved in the development of SLE by maturing the antibody response and contributing to organ damage. GILZ was first described as a regulator of T cell apoptosis [81]. Further to this, GILZ is an important stabilizer of the naïve T cell state, preventing activation until sufficient stimulation is experienced, and this effect is essential for the prevention of autoimmunity. In GILZ-transgenic (GILZ-Tg) mice which overexpress GILZ, T cells upregulate the expression of *Gata3* and *Stat6*, which are specific to the Th2 phenotype, while downregulating T-bet, a Th1 transcription factor. This effect of GILZ leads to the inhibition of delayed type hypersensitivity response (a Th1 response) and worsening of bleomycin-induced fibrosis (a Th2 response) in the GILZ-Tg mice [109]. Both murine and human cells demonstrate this skewing towards a Th2 phenotype and this is accompanied by a reduction of Th1 cytokines, such as IL-2 and IFNγ, and increased Th2 cytokines, such as IL-4 and IL-10 [109, 110]. Likewise, in the same GILZ-Tg mouse model, overexpression of GILZ inhibits NFκB nuclear translocation and attenuates the production of TNFα, IFNγ, and IL-1 in $CD4^+$ T cells from the intestinal lamina propria [98]. These mice are also less susceptible to Th1 mediated colitis and more susceptible to Th2 mediated colitis [98]. Moreover, GILZ can also suppress T cell activation via inhibition of NF-κB and AP-1, much like GCs, but also through the ERK cascade, which differentiates it from GC [11, 110].

GC have been demonstrated to enhance Th17 numbers in SLE patients and have also been found to induce Th17 cytokines in asthma patients [111, 112]. The differentiation of Th17 cells is dependent on the expression of transcription factors, such as RORγt, BATF, IRF4, and STAT3, which can bind directly to DNA [113]. In contract to GC, GILZ is able to negatively regulate these transcription factors [114]. Indeed, GILZ deficiency in mice leads to increased IL-17A and TNFα production, as well as increased IL-17 inducing cytokines IL-1ß, IL-23, and IL-6 [95, 104]. This is due to an expansion of Th17 cells and dendritic cells promoting the production of pro-Th17 cytokines [95]. Additionally, IL-6 suppresses GILZ expression which allows the upregulation of Th17-associated genes leading to Th17 proliferation, and supplementation with HHph-GILZ reverses this effect [95]. Consequently, in psoriatic skin lesions, which are pathologically driven by IL-17, GILZ is downregulated and loss of GILZ exacerbates imiquimod-induced psoriatic skin lesions in mice [95]. In contrast, another study has demonstrated that overexpression of GILZ causes worsening of skin inflammation due to heightened production of TGF [115].

GILZ is also a regulator of Treg proliferation, as are GCs. In GILZ deficient mice treated with dexamethasone, peripheral Treg numbers are reduced, most likely due to the inability to stimulate FoxP3. Accordingly, this impaired generation results in intestinal inflammation in these mice [116]. Conversely, overexpression of GILZ leads to increases the number of Tregs [116]. Similarly, in acute myocardial infarction and acute kidney injury, GILZ is markedly suppressed by inflammation, and treatment with exogenous GILZ results in decreased IL-17, increased Tregs and IL-10, and prevention of cell death. Therefore, GILZ demonstrates cardio- and renoprotective roles in T cells [102, 103]. This is highly relevant to SLE as patients with active disease have Treg depletion, Th17 expansion, and increased IL-17 production with a negative correlation between SLEDAI and IL-17 [117, 118]. Additionally, a negative correlation exists between GILZ mRNA and IL-17 level in PBMCs from SLE patients [106, 119]. Clearly, GILZ is an important regulator of T cell activation and differentiation. Further investigation of its effects on other T cell population such as cytotoxic T cells would be extremely informative.

Monocytes/Macrophages

GILZ is highly expressed in macrophages in the late stages of inflammation [120, 121]. GC and IL-10 induce GILZ expression in macrophages and the anti-inflammatory effect of GILZ on monocytes/macrophages has been described in several studies [120, 122]. Transfection of THP-1 cells (a human monocyte-like cell line) with GILZ decreases the production of the chemokines CCL3 and CCL5, through inhibition of NFκB p65 activation, as well as the expression of macrophage activation markers CD80 and CD86 upon treatment with lipopolysaccharide (LPS) [120]. Similarly, curcumin has anti-inflammatory actions which are, at least partly, mediated by upregulation of GILZ protein levels in macrophages [123]. Furthermore, GILZ downregulation sensitises Kupffer cells to secrete LPS-stimulated IL-6, CCL2, and TNFα, and GILZ overexpression reduces liver inflammation [99].

GILZ also appears to be beneficial in sepsis. In a mouse model of sepsis induced by caecal ligation and puncture (CLP), GILZ is downregulated. This same suppression of GILZ is also seen in patients with sepsis [124]. In parallel, overexpressing GILZ-Tg mice have reduced peritoneal bacterial counts secondary to increased phagocytic activity of macrophages [124]. Cell-specific GILZ overexpression in macrophages/monocytes also promotes their bactericidal functions and regulates systemic inflammation. Consequently, this leads to a favorable outcome in sepsis, which is in contrast to GC [125, 126]. GILZ therefore has protective anti-bacterial activity in addition to its anti-inflammatory effects on macrophage activation.

Neutrophils

While there is less research in this area, GILZ also appears to be a regulator of neutrophil function. in In GILZ-deficient mice with peritonitis, GC is unable to inhibit the migration of neutrophils and cannot upregulate annexin A1 expression (127). Likewise, in DNBS-induced colitis in mice, loss of GILZ leads to a significantly higher percentage of neutrophil infiltration in the colon. Neutrophils from these same mice are also more activated with increased phagocytic ability when exposed to *Candida albicans*. This regulatory function appears to be mediated by GILZ inhibiting the ERK cascade [128]. Finally, similarly to GC, GILZ is able to directly promote apoptosis of human neutrophils [129].

Dendritic Cells

GILZ is constitutively expressed in mouse and human tissues [130, 131]. This includes various leukocyte subsets with the highest levels of expression seen in macrophages [120], and the lowest in pDCs which are the primary producers of Type I IFN [132]. In relation to DCs, GILZ expression varies depending on subsets. Conventional DC (cDC)1 express higher GILZ compared to cDC2 and pDCs, and mouse bone-marrow derived DCs (BMDCs) express significantly lower GILZ [133]. Moreover, treatment with the dexamethasone upregulates GILZ expression in human DC in a dose-dependent manner [44]. GILZ expression in DC is also induced by IL-10, and TGFβ.

GILZ exerts inhibitory effects on DC maturation and activation, by inhibiting the expression of DC maturation markers such as CD80, CD83, and CD86 [134]. Accordingly, silencing GILZ with siRNA restores the expression of these markers [134]. This effect of GILZ is critical in SLE, as circulating apoptotic microparticles found in patients with SLE are known to activate DCs by increasing the expression of the costimulatory molecules CD40, CD80, CD83, and CD86. This change in surface marker expression is accompanied by increased production of IL-6, TNFα, and IFNα by pDCs and myeloid DCs [135]. Furthermore, GILZ-expressing DCs induce the development of antigen-specific Tregs [44], and GILZ overexpression in DCs promote Treg expansion [136]. GILZ is also an endogenous inhibitor of macropinocytosis in immature and recently activated DCs [133] where macropinocytosis, a receptor mediated endocytosis and phagocytosis, contributes to antigen cross presentation and T cell activation by $CD8a^+$ DCs in vivo. So far, there have been no published studies on the effect of GILZ on pDCs, the major producers of Type I IFN.

Effect of GILZ on Type I Interferon

Type I IFN are significant contributors to the pathogenesis of SLE, which is in turn frequently treated with GC. The effects of GC on IFN pathways, however, have been surprisingly under-studied.

GC regulate the transcriptional activity of factors that inhibit IFN gene expression (137) [138]. However, as previously mentioned, conventional doses of GC cannot suppress the IFN signature in SLE patients [52]. Early evidence suggests that, unlike GC, GILZ may be able to suppress Type I IFN production. Birds infected with Infectious Bursal Disease Virus (IBDV) develop an immunosuppressive disease named Infectious Bursal Disease (IBD). In HEK293T cells infected with this avian-IBDV, Type I IFN is suppressed through GILZ interacting with viral protein VP4 [139]. Consequently, knockdown of GILZ with siRNA abolishes VP4 suppression of Type I IFN, suggesting GILZ is key regulator of Type I IFN [139]. Further to this, IBDV VP4 can suppress IFNβ production by inhibiting K48-linked ubiquitylation of GILZ [82]. A major potential therapeutic avenue for SLE would be revealed if it were shown that, although GC scarcely inhibit IFN, GILZ does so. However, whether GILZ suppresses Type I IFN by pDC in SLE remains unexplored.

Differential Effects of GC and GILZ on Bone Metabolism

Given the widespread GC-independent anti-inflammatory effects of GILZ, it may represent the basis for a future approach to replace GC with a treatment that confers less metabolic adverse effects.

Effect of GILZ on Bone

In contrast to GC, GILZ exerts an opposite effect on MSC differentiation. In vitro studies on MSC have shown that GILZ binds directly to tandem repeats of C/EBPδ binding sites in the PPARγ2 promoter region, inhibiting PPARγ2 transcription which in turn inhibits adipogenesis [140]. Further studies in an overexpressing GILZ transgenic mouse showed that GILZ overexpression promoted osteoblastogenesis in MSC while inhibiting adipocyte differentiation. The opposite effect occurred where GILZ was knocked down [141]. Subsequent studies in an in vivo GILZ transgenic mouse model, where GILZ was overexpressed in bone cells, also demonstrated the osteogenic role of GILZ in bone formation [142], thus confirming the findings from the previous in vitro studies. Further to this, the effect of GILZ on osteoblasts and bone remodelling was studied in human foetal osteoblasts (hFOB) ex vivo [143]. GILZ inhibition in hFOB was associated with decreased osteoblast-related genes and increased osteoclastic activity, supporting the osteogenic role of GILZ in normal bone remodelling [143]. Thus, GILZ acts as a transcriptional repressor with effects that antagonise those of GC, shifting the balance away from adipocyte differentiation and towards favouring osteoblast differentiation.

In addition to this, the pro-inflammatory cytokines IL-6 and TNFα stimulate bone loss by inhibiting osteoblasts and stimulating osteoclast differentiation. GILZ antagonises the inhibitory effect of TNFα on MSC osteogenic differentiation by inhibiting MAPK/ERK activation and protein degradation [144]. Subsequent studies in a TNFα-GILZ double transgenic model where TNFa is overexpressed globally and GILZ overexpression is restricted to bone cells. Although GILZ overexpression in bone cells protected bones from TNFα induced bone loss, bone restricted GILZ was unable to inhibit TNFα induced arthritic inflammation due to the global overexpression of TNFα [145]. These results are indicative of the metabolically sparing action of GILZ on bones, in contradistinction to the effects of GC. However, this critical aspect of GILZ biology needs to be investigated by testing the effects of a GILZ-based therapy in a mouse model.

Effect of GILZ on Glucose Metabolism

There is limited evidence to show the effect of GILZ on glucose metabolism. GILZ deficiency in mice has been demonstrated to promote weight loss, particularly in fat mass, as they age [146]. These same mice express increased serum insulin with unchanged plasma glucose while on a normal diet. They also have upregulated glucagon secreting alpha cells in the pancreatic islet, with normal plasma glucagon, which may represent a compensatory mechanism for high plasma insulin [146]. However, when fed a high fat diet, loss of GILZ decreases hepatic steatosis, suggesting GILZ may have an adverse effect in this case [146].

Conclusions

To this day, the lifesaving immunosuppressive effects of GC has made them irreplaceable in the management of SLE. However, their continued use comes with significant morbidity which, combined with stagnant survival rates, represent a major unmet need for a therapeutic alternative. The development of a novel therapeutic which possesses similar anti-inflammatory effects to GC but without the adverse metabolic effects would be ground-breaking. The GC-induced protein, GILZ, is a potential candidate as its immunosuppressive actions are broad and possibly without the metabolic profile of GCs that may bear potential to fulfil this unmet need. Comprehensive analyses of the effects of GILZ on typical GC-mediated metabolic adverse effects such as osteoporosis and diabetes will be required as part of its validation as a safe target.

References

1. Apostolopoulos D, Morand EF. It hasn't gone away: the problem of glucocorticoid use in lupus remains. Rheumatology (Oxford). 2017;56((suppl_1)):i114-i22.
2. Apostolopoulos D, Kandane-Rathnayake R, Raghunath S, Hoi A, Nikpour M, Morand EF. Independent association of glucocorticoids with damage accrual in SLE. Lupus Sci Med. 2016;3(1):e000157.
3. Ruiz-Arruza I, Ugarte A, Cabezas-Rodriguez I, Medina JA, Moran MA, Ruiz-Irastorza G (2014) Glucocorticoids and irreversible damage in patients with systemic lupus erythematosus. Rheumatology 53(8):1470–1476
4. Singh JA, Furst DE, Bharat A, Curtis JR, Kavanaugh AF, Kremer JM et al (2012) 2012 update of the 2008 American College of Rheumatology recommendations for the use of disease-modifying antirheumatic drugs and biologic agents in the treatment of rheumatoid arthritis. Arthritis Care Res. 64(5):625–639
5. Tesar V, Hruskova Z (2017) Belimumab in the management of systemic lupus erythematosus—an update. Expert Opinion on Biological Therapy. 17 (7):901–908
6. Furie R, Rovin BH, Houssiau F, Malvar A, Teng YKO, Contreras G et al (2020) Two-Year, Randomized, Controlled Trial of Belimumab in Lupus Nephritis. N Engl J Med 383(12):1117–1128
7. Furie R, Khamashta M, Merrill JT, Werth VP, Kalunian K, Brohawn P et al (2017) Anifrolumab, an Anti–Interferon-α Receptor Monoclonal Antibody, in Moderate-to-Severe Systemic Lupus Erythematosus. Arthritis & Rheumatology. 69(2):376–386
8. Morand EF, Furie R, Tanaka Y, Bruce IN, Askanase AD, Richez C et al (2020) Trial of Anifrolumab in Active Systemic Lupus Erythematosus. N Engl J Med 382(3):211–221
9. Poetker DM, Reh DD (2010) A comprehensive review of the adverse effects of systemic corticosteroids. Otolaryngol Clin North Am 43(4):753–768
10. Apostolopoulos D, Kandane-Rathnayake R, Louthrenoo W, Luo SF, Wu Y-J, Lateef A et al (2020) Factors associated with damage accrual in patients with systemic lupus erythematosus with no clinical or serological disease activity: a multicentre cohort study. The Lancet Rheumatology. 2(1):e24–e30
11. Cheng Q, Morand E, Yang YH (2014) Development of novel treatment strategies for inflammatory diseases—similarities and divergence between glucocorticoids and GILZ. Frontiers in pharmacology. 5:169
12. Rhen T, Cidlowski JA (2005) Antiinflammatory Action of Glucocorticoids — New Mechanisms for Old Drugs. N Engl J Med 353(16):1711–1723
13. Oakley RH, Cidlowski JA (2013) The biology of the glucocorticoid receptor: new signaling mechanisms in health and disease. J Allergy Clin Immunol 132(5):1033–1044
14. Yudt MR, Cidlowski JA (2001) Molecular identification and characterization of a and b forms of the glucocorticoid receptor. Mol Endocrinol 15 (7):1093–1103
15. Gross KL, Cidlowski JA (2008) Tissue-specific glucocorticoid action: a family affair. Trends Endocrinol Metab 19(9):331–339
16. Oakley RH, Cidlowski JA (2011) Cellular processing of the glucocorticoid receptor gene and protein: new mechanisms for generating tissue-specific actions of glucocorticoids. The Journal of biological chemistry. 286(5):3177–3184
17. Kino T, Su YA, Chrousos GP (2009) Human glucocorticoid receptor isoform beta: recent understanding of its potential implications in physiology and pathophysiology. Cell Mol Life Sci 66 (21):3435–3448
18. Lewis-Tuffin LJ, Cidlowski JA (2006) The physiology of human glucocorticoid receptor beta (hGRbeta) and glucocorticoid resistance. Ann N Y Acad Sci 1069:1–9
19. Avenant C, Ronacher K, Stubsrud E, Louw A, Hapgood JP (2010) Role of ligand-dependent GR phosphorylation and half-life in determination of ligand-specific transcriptional activity. Mol Cell Endocrinol 327(1–2):72–88
20. Webster JC, Jewell CM, Bodwell JE, Munck A, Sar M, Cidlowski JA (1997) Mouse glucocorticoid receptor phosphorylation status influences multiple functions of the receptor protein. The Journal of biological chemistry. 272(14):9287–9293
21. Wallace AD, Cidlowski JA (2001) Proteasome-mediated glucocorticoid receptor degradation restricts transcriptional signaling by glucocorticoids. The Journal of biological chemistry. 276 (46):42714–42721
22. Druker J, Liberman AC, Antunica-Noguerol M, Gerez J, Paez-Pereda M, Rein T et al (2013) RSUME enhances glucocorticoid receptor SUMOylation and transcriptional activity. Mol Cell Biol 33(11):2116–2127

23. Ito K, Yamamura S, Essilfie-Quaye S, Cosio B, Ito M, Barnes PJ et al (2006) Histone deacetylase 2-mediated deacetylation of the glucocorticoid receptor enables NF-kappaB suppression. J Exp Med 203 (1):7–13
24. Charmandari E, Chrousos GP, Lambrou GI, Pavlaki A, Koide H, Ng SS, et al. Peripheral CLOCK regulates target-tissue glucocorticoid receptor transcriptional activity in a circadian fashion in man. PLoS One. 2011;6(9):e25612.
25. Barnes PJ (2011) Glucocorticosteroids: current and future directions. Br J Pharmacol 163(1):29–43
26. Pousset F, Cremona S, Dantzer R, Kelley KW, Parnet P (2001) Dexamethasone up-regulates type II IL-1 receptor in mouse primary activated astrocytes. J Neurochem 76(3):901–909
27. Barnes PJ (2010) Mechanisms and resistance in glucocorticoid control of inflammation. J Steroid Biochem Mol Biol 120(2–3):76–85
28. Bergmann MW, Staples KJ, Smith SJ, Barnes PJ, Newton R (2004) Glucocorticoid inhibition of granulocyte macrophage-colony-stimulating factor from T cells is independent of control by nuclear factor-kappaB and conserved lymphokine element 0. Am J Respir Cell Mol Biol 30(4):555–563
29. Clark AR (2007) Anti-inflammatory functions of glucocorticoid-induced genes. Mol Cell Endocrinol 275(1–2):79–97
30. Yang S, Zhang L (2004) Glucocorticoids and vascular reactivity. Curr Vasc Pharmacol 2(1):1–12
31. Vago JP, Tavares LP, Garcia CC, Lima KM, Perucci LO, Vieira EL et al (2015) The role and effects of glucocorticoid-induced leucine zipper in the context of inflammation resolution. Journal of immunology 194(10):4940–4950
32. Steer JH, Kroeger KM, Abraham LJ, Joyce DA (2000) Glucocorticoids suppress tumor necrosis factor-alpha expression by human monocytic THP-1 cells by suppressing transactivation through adjacent NF-kappa B and c-Jun-activating transcription factor-2 binding sites in the promoter. J Biol Chem 275(24):18432–18440
33. Ehrchen J, Steinmüller L, Barczyk K, Tenbrock K, Nacken W, Eisenacher M et al (2007) Glucocorticoids induce differentiation of a specifically activated, anti-inflammatory subtype of human monocytes. Blood 109(3):1265–1274
34. Taves MD, Ashwell JD. Glucocorticoids in T cell development, differentiation and function. Nature reviews Immunology. 2020.
35. Van Laethem F, Baus E, Smyth LA, Andris F, Bex F, Urbain J et al (2001) Glucocorticoids attenuate T cell receptor signaling. J Exp Med 193 (7):803–814
36. Baus E, Andris F, Dubois PM, Urbain J, Leo O (1996) Dexamethasone inhibits the early steps of antigen receptor signaling in activated T lymphocytes. Journal of immunology 156(12):4555–4561
37. Piemonti L, Monti P, Allavena P, Sironi M, Soldini L, Leone BE et al (1999) Glucocorticoids affect human dendritic cell differentiation and maturation. Journal of immunology 162(11):6473–6481
38. Thompson EB (1999) Mechanisms of T-cell Apoptosis Induced by Glucocorticoids. Trends Endocrinol Metab 10(9):353–358
39. Miyaura H, Iwata M (2002) Direct and Indirect Inhibition of Th1 Development by Progesterone and Glucocorticoids. J Immunol 168(3):1087–1094
40. Berger A (2000) Th1 and Th2 responses: what are they? BMJ 321(7258):424
41. Banuelos J, Shin S, Cao Y, Bochner BS, Morales-Nebreda L, Budinger GR et al (2016) BCL-2 protects human and mouse Th17 cells from glucocorticoid-induced apoptosis. Allergy 71 (5):640–650
42. Banuelos J, Cao Y, Shin SC, Lu NZ (2017) Immunopathology alters Th17 cell glucocorticoid sensitivity. Allergy 72(3):331–341
43. Peek EJ, Richards DF, Faith A, Lavender P, Lee TH, Corrigan CJ et al (2005) Interleukin-10–Secreting "Regulatory" T Cells Induced by Glucocorticoids and β2-Agonists. Am J Respir Cell Mol Biol 33(1):105–111
44. Hamdi H, Godot V, Maillot M-C, Prejean MV, Cohen N, Krzysiek R et al (2007) Induction of antigen-specific regulatory T lymphocytes by human dendritic cells expressing the glucocorticoid-induced leucine zipper. Blood 110(1):211–219
45. Gruver-Yates AL, Quinn MA, Cidlowski JA (2014) Analysis of Glucocorticoid Receptors and Their Apoptotic Response to Dexamethasone in Male Murine B Cells During Development. Endocrinology 155(2):463–474
46. Franco LM, Gadkari M, Howe KN, Sun J, Kardava L, Kumar P et al (2019) Immune regulation by glucocorticoids can be linked to cell type-dependent transcriptional responses. J Exp Med 216(2):384–406
47. Dai J, Megjugorac NJ, Amrute SB, Fitzgerald-Bocarsly P (2004) Regulation of IFN regulatory factor-7 and IFN-alpha production by enveloped virus and lipopolysaccharide in human plasmacytoid dendritic cells. J Immunol 173(3):1535–1548
48. Monaghan K, Jordan J, Sato T, Cesaroni M, Benson J, Ng M, et al. Depletion of plasmacytoid dendritic cells with JNJ-56022473 minimises induction of an interferon gene signature in response to TLR9 and SLE immune complex stimulation. Lupus Science & Medicine. 2017;4(Suppl 1):A42-A.
49. Zhan Y, Carrington EM, Ko HJ, Vikstrom IB, Oon S, Zhang JG et al (2015) Bcl-2 antagonists kill plasmacytoid dendritic cells from lupus-prone mice and dampen interferon-alpha production. Arthritis Rheumatol. 67(3):797–808

50. Guiducci C, Gong M, Xu Z, Gill M, Chaussabel D, Meeker T et al (2010) TLR recognition of self nucleic acids hampers glucocorticoid activity in lupus. Nature 465(7300):937–941
51. Stojan G, Petri M (2017) The risk benefit ratio of glucocorticoids in SLE: have things changed over the past 40 years? Current treatment options in rheumatology. 3(3):164–172
52. Bennett L, Palucka AK, Arce E, Cantrell V, Borvak J, Banchereau J et al (2003) Interferon and granulopoiesis signatures in systemic lupus erythematosus blood. J Exp Med 197(6):711–723
53. Rucci N (2008) Molecular biology of bone remodelling. Clin Cases Miner Bone Metab. 5(1):49–56
54. Katsimbri P. The biology of normal bone remodelling. Eur J Cancer Care (Engl). 2017;26(6).
55. Szulc P, Naylor K, Pickering ME, Hoyle N, Eastell R, Leary E (2018) Use of CTX-I and PINP as bone turnover markers: National Bone Health Alliance recommendations to standardize sample handling and patient preparation to reduce pre-analytical variability. Ann Biol Clin (Paris). 76(4):373–391
56. Vasikaran S, Eastell R, Bruyere O, Foldes AJ, Garnero P, Griesmacher A et al (2011) Markers of bone turnover for the prediction of fracture risk and monitoring of osteoporosis treatment: a need for international reference standards. Osteoporos Int 22 (2):391–420
57. Eastell R, Pigott T, Gossiel F, Naylor KE, Walsh JS, Peel NFA (2018) DIAGNOSIS OF ENDOCRINE DISEASE: Bone turnover markers: are they clinically useful? Eur J Endocrinol 178(1): R19–R31
58. Pittenger MF, Mackay AM, Beck SC, Jaiswal RK, Douglas R, Mosca JD et al (1999) Multilineage potential of adult human mesenchymal stem cells. Science 284(5411):143–147
59. Ahdjoudj S, Lasmoles F, Oyajobi BO, Lomri A, Delannoy P, Marie PJ (2001) Reciprocal control of osteoblast/chondroblast and osteoblast/adipocyte differentiation of multipotential clonal human marrow stromal F/STRO-1(+) cells. J Cell Biochem 81 (1):23–38
60. Rosen ED, Spiegelman BM (2001) PPARgamma : a nuclear regulator of metabolism, differentiation, and cell growth. J Biol Chem 276(41):37731–37734
61. Kipen Y, Buchbinder R, Forbes A, Strauss B, Littlejohn G, Morand E (1997) Prevalence of reduced bone mineral density in systemic lupus erythematosus and the role of steroids. J Rheumatol 24(10):1922–1929
62. LoCascio V, Bonucci E, Imbimbo B, Ballanti P, Adami S, Milani S et al (1990) Bone loss in response to long-term glucocorticoid therapy. Bone Miner 8(1):39–51
63. van Staa TP, Leufkens HG, Abenhaim L, Begaud B, Zhang B, Cooper C (2000) Use of oral corticosteroids in the United Kingdom. QJM 93 (2):105–111
64. Shi XM, Blair HC, Yang X, McDonald JM, Cao X (2000) Tandem repeat of C/EBP binding sites mediates PPARgamma2 gene transcription in glucocorticoid-induced adipocyte differentiation. J Cell Biochem 76(3):518–527
65. O'Brien CA, Jia D, Plotkin LI, Bellido T, Powers CC, Stewart SA et al (2004) Glucocorticoids act directly on osteoblasts and osteocytes to induce their apoptosis and reduce bone formation and strength. Endocrinology 145(4):1835–1841
66. Mocetti P, Silvestrini G, Ballanti P, Patacchioli FR, Di Grezia R, Angelucci L et al (2001) Bcl-2 and Bax expression in cartilage and bone cells after high-dose corticosterone treatment in rats. Tissue Cell 33(1):1–7
67. Silvestrini G, Ballanti P, Patacchioli FR, Mocetti P, Di Grezia R, Wedard BM et al (2000) Evaluation of apoptosis and the glucocorticoid receptor in the cartilage growth plate and metaphyseal bone cells of rats after high-dose treatment with corticosterone. Bone 26(1):33–42
68. Weinstein RS, Chen JR, Powers CC, Stewart SA, Landes RD, Bellido T et al (2002) Promotion of osteoclast survival and antagonism of bisphosphonate-induced osteoclast apoptosis by glucocorticoids. J Clin Invest 109(8):1041–1048
69. Sasaki N, Kusano E, Ando Y, Yano K, Tsuda E, Asano Y (2001) Glucocorticoid decreases circulating osteoprotegerin (OPG): possible mechanism for glucocorticoid induced osteoporosis. Nephrol Dial Transplant 16(3):479–482
70. Hofbauer LC, Gori F, Riggs BL, Lacey DL, Dunstan CR, Spelsberg TC et al (1999) Stimulation of osteoprotegerin ligand and inhibition of osteoprotegerin production by glucocorticoids in human osteoblastic lineage cells: potential paracrine mechanisms of glucocorticoid-induced osteoporosis. Endocrinology 140(10):4382–4389
71. De Vries F, Bracke M, Leufkens HG, Lammers JW, Cooper C, Van Staa TP (2007) Fracture risk with intermittent high-dose oral glucocorticoid therapy. Arthritis Rheum 56(1):208–214
72. Liu XX, Zhu XM, Miao Q, Ye HY, Zhang ZY, Li YM (2014) Hyperglycemia induced by glucocorticoids in nondiabetic patients: a meta-analysis. Ann Nutr Metab 65(4):324–332
73. Suh S, Park MK (2017) Glucocorticoid-Induced Diabetes Mellitus: An Important but Overlooked Problem. Endocrinol Metab (Seoul). 32(2):180–189
74. Caro JF, Amatruda JM (1982) Glucocorticoid-induced insulin resistance: the importance of post-binding events in the regulation of insulin binding, action, and degradation in freshly isolated and primary cultures of rat hepatocytes. J Clin Invest 69 (4):866–875
75. Delaunay F, Khan A, Cintra A, Davani B, Ling ZC, Andersson A et al (1997) Pancreatic beta cells are important targets for the diabetogenic effects of glucocorticoids. J Clin Invest 100(8):2094–2098

76. Angelopoulos TP, Tentolouris NK, Bertsias GK, Boumpas DT (2014) Steroid-induced diabetes in rheumatologic patients. Clin Exp Rheumatol 32 (1):126–130
77. Ha Y, Lee KH, Jung S, Lee SW, Lee SK, Park YB (2011) Glucocorticoid-induced diabetes mellitus in patients with systemic lupus erythematosus treated with high-dose glucocorticoid therapy. Lupus. 20 (10):1027–1034
78. Shaharir SS, Gafor AH, Said MS, Kong NC (2015) Steroid-induced diabetes mellitus in systemic lupus erythematosus patients: analysis from a Malaysian multi-ethnic lupus cohort. International journal of rheumatic diseases. 18(5):541–547
79. Raul Ariza-Andraca C, Barile-Fabris LA, Frati-Munari AC, Baltazar-Montufar P (1998) Risk factors for steroid diabetes in rheumatic patients. Arch Med Res 29(3):259–262
80. Chung CP, Oeser A, Solus JF, Gebretsadik T, Shintani A, Avalos I et al (2008) Inflammation-associated insulin resistance: differential effects in rheumatoid arthritis and systemic lupus erythematosus define potential mechanisms. Arthritis Rheum 58(7):2105–2112
81. D'Adamio F, Zollo O, Moraca R, Ayroldi E, Bruscoli S, Bartoli A et al (1997) A new dexamethasone-induced gene of the leucine zipper family protects T lymphocytes from TCR/CD3-activated cell death. Immunity 7(6):803–812
82. He Z, Chen X, Fu M, Tang J, Li X, Cao H et al (2018) Infectious bursal disease virus protein VP4 suppresses type I interferon expression via inhibiting K48-linked ubiquitylation of glucocorticoid-induced leucine zipper (GILZ). Immunobiology 223 (4–5):374–382
83. Ayroldi E, Riccardi C (2009) Glucocorticoid-induced leucine zipper (GILZ): a new important mediator of glucocorticoid action. FASEB journal : official publication of the Federation of American Societies for Experimental Biology. 23(11):3649–3658
84. Mittelstadt PR, Ashwell JD (2001) Inhibition of AP-1 by the glucocorticoid-inducible protein GILZ. The Journal of biological chemistry. 276 (31):29603–29610
85. Di Marco B, Massetti M, Bruscoli S, Macchiarulo A, Di Virgilio R, Velardi E et al (2007) Glucocorticoid-induced leucine zipper (GILZ)/NF-kappaB interaction: role of GILZ homo-dimerization and C-terminal domain. Nucleic Acids Res 35(2):517–528
86. Ayroldi E, Zollo O, Bastianelli A, Marchetti C, Agostini M, Di Virgilio R et al (2007) GILZ mediates the antiproliferative activity of glucocorticoids by negative regulation of Ras signaling. J Clin Invest 117(6):1605–1615
87. Ayroldi E, Zollo O, Macchiarulo A, Di Marco B, Marchetti C, Riccardi C (2002) Glucocorticoid-Induced Leucine Zipper Inhibits the Raf-Extracellular Signal-Regulated Kinase Pathway by Binding to Raf-1. Mol Cell Biol 22(22):7929–7941
88. Joha S, Nugues AL, Hetuin D, Berthon C, Dezitter X, Dauphin V et al (2012) GILZ inhibits the mTORC2/AKT pathway in BCR-ABL(+) cells. Oncogene 31(11):1419–1430
89. Ayyar VS, Almon RR, Jusko WJ, DuBois DC. Quantitative tissue-specific dynamics of in vivo GILZ mRNA expression and regulation by endogenous and exogenous glucocorticoids. Physiological Reports. 2015;3(6):e12382.
90. Cannarile L, Zollo O, D'Adamio F, Ayroldi E, Marchetti C, Tabilio A et al (2001) Cloning, chromosomal assignment and tissue distribution of human GILZ, a glucocorticoid hormone-induced gene. Cell Death And Differentiation. 8:201
91. Cari L, Ricci E, Gentili M, Petrillo MG, Ayroldi E, Ronchetti S et al (2015) A focused Real Time PCR strategy to determine GILZ expression in mouse tissues. Results Immunol. 5:37–42
92. Beaulieu E, Morand EF (2011) Role of GILZ in immune regulation, glucocorticoid actions and rheumatoid arthritis. Nat Rev Rheumatol 7 (6):340–348
93. Beaulieu E, Ngo D, Santos L, Yang YH, Smith M, Jorgensen C et al (2010) Glucocorticoid-induced leucine zipper is an endogenous antiinflammatory mediator in arthritis. Arthritis Rheum 62(9):2651–2661
94. Jones SA, Toh AE, Odobasic D, Oudin MA, Cheng Q, Lee JP et al (2016) Glucocorticoid-induced leucine zipper (GILZ) inhibits B cell activation in systemic lupus erythematosus. Ann Rheum Dis 75(4):739–747
95. Jones SA, Perera DN, Fan H, Russ BE, Harris J, Morand EF (2015) GILZ regulates Th17 responses and restrains IL-17-mediated skin inflammation. J Autoimmun 61:73–80
96. Mazzon E, Bruscoli S, Galuppo M, Biagioli M, Sorcini D, Bereshchenko O et al (2014) Glucocorticoid-induced leucine zipper (GILZ) controls inflammation and tissue damage after spinal cord injury. CNS Neurosci Ther 20 (11):973–981
97. Hahn RT, Hoppstadter J, Hirschfelder K, Hachenthal N, Diesel B, Kessler SM et al (2014) Down-regulation of the glucocorticoid-induced leucine zipper (GILZ) promotes vascular inflammation. Atherosclerosis 234(2):391–400
98. Cannarile L, Cuzzocrea S, Santucci L, Agostini M, Mazzon E, Esposito E et al (2009) Glucocorticoid-induced leucine zipper is protective in Th1-mediated models of colitis. Gastroenterology 136 (2):530–541
99. Robert O, Boujedidi H, Bigorgne A, Ferrere G, Voican CS, Vettorazzi S et al (2016) Decreased expression of the glucocorticoid receptor-GILZ pathway in Kupffer cells promotes liver inflammation in obese mice. J Hepatol 64(4):916–924

100. Kelly MM, King EM, Rider CF, Gwozd C, Holden NS, Eddleston J et al (2012) Corticosteroid-induced gene expression in allergen-challenged asthmatic subjects taking inhaled budesonide. Br J Pharmacol 165(6):1737–1747
101. Aguilar D, Strom J, Chen QM (2014) Glucocorticoid induced leucine zipper inhibits apoptosis of cardiomyocytes by doxorubicin. Toxicol Appl Pharmacol 276(1):55–62
102. Baban B, Yin L, Qin X, Liu JY, Shi X, Mozaffari MS (2017) The role of GILZ in modulation of adaptive immunity in a murine model of myocardial infarction. Exp Mol Pathol 102(3):408–414
103. Baban B, Marchetti C, Khodadadi H, Malik A, Emami G, Lin PC et al (2018) Glucocorticoid-Induced Leucine Zipper Promotes Neutrophil and T-Cell Polarization with Protective Effects in Acute Kidney Injury. J Pharmacol Exp Ther 367(3):483–493
104. Ngo D, Beaulieu E, Gu R, Leaney A, Santos L, Fan H et al (2013) Divergent effects of endogenous and exogenous glucocorticoid-induced leucine zipper in animal models of inflammation and arthritis. Arthritis Rheum 65(5):1203–1212
105. Quax RA, Koper JW, de Jong PH, van Heerebeek R, Weel AE, Huisman AM et al (2012) In vitro glucocorticoid sensitivity is associated with clinical glucocorticoid therapy outcome in rheumatoid arthritis. Arthritis Res Ther. 14(4):R195
106. Mohammadi S, Ebadpour MR, Sedighi S, Saeedi M, Memarian A (2017) Glucocorticoid-induced leucine zipper expression is associated with response to treatment and immunoregulation in systemic lupus erythematosus. Clin Rheumatol 36 (8):1765–1772
107. Bruscoli S, Biagioli M, Sorcini D, Frammartino T, Cimino M, Sportoletti P et al (2015) Lack of glucocorticoid-induced leucine zipper (GILZ) deregulates B-cell survival and results in B-cell lymphocytosis in mice. Blood 126(15):1790–1801
108. Bruscoli S, Sorcini D, Flamini S, Gagliardi A, Adamo F, Ronchetti S et al (2018) Glucocorticoid-Induced Leucine Zipper Inhibits Interferon-Gamma Production in B Cells and Suppresses Colitis in Mice. Front Immunol. 9:1720
109. Cannarile L, Fallarino F, Agostini M, Cuzzocrea S, Mazzon E, Vacca C et al (2006) Increased GILZ expression in transgenic mice up-regulates Th-2 lymphokines. Blood 107(3):1039–1047
110. Ayroldi E, Migliorati G, Bruscoli S, Marchetti C, Zollo O, Cannarile L et al (2001) Modulation of T-cell activation by the glucocorticoid-induced leucine zipper factor via inhibition of nuclear factor κB. Blood 98(3):743–753
111. Nanzer AM, Chambers ES, Ryanna K, Richards DF, Black C, Timms PM, et al. Enhanced production of IL-17A in patients with severe asthma is inhibited by 1α, 25-dihydroxyvitamin D3 in a glucocorticoid-independent fashion. Journal of Allergy and Clinical Immunology. 2013;132 (2):297–304. e3.
112. Suárez A, de Paz B, Prado C, Rodríguez-Carrio J, Gómez J, López P (2011) Glucocorticoids enhance Th17/Th1 imbalance and signal transducer and activator of transcription 3 expression in systemic lupus erythematosus patients. Rheumatology 50 (10):1794–1801
113. Ciofani M, Madar A, Galan C, Sellars M, Mace K, Pauli F et al (2012) A validated regulatory network for Th17 cell specification. Cell 151(2):289–303
114. Yosef N, Shalek AK, Gaublomme JT, Jin H, Lee Y, Awasthi A et al (2013) Dynamic regulatory network controlling TH17 cell differentiation. Nature 496:461
115. Carceller E, Ballegeer M, Deckers J, Riccardi C, Bruscoli S, Hochepied T et al (2016) Overexpression of Glucocorticoid-induced Leucine Zipper (GILZ) increases susceptibility to Imiquimod-induced psoriasis and involves cutaneous activation of TGF-beta1. Sci Rep 6:38825
116. Bereshchenko O, Coppo M, Bruscoli S, Biagioli M, Cimino M, Frammartino T et al (2014) GILZ promotes production of peripherally induced Treg cells and mediates the crosstalk between glucocorticoids and TGF-beta signaling. Cell reports 7 (2):464–475
117. Yang J, Chu Y, Yang X, Gao D, Zhu L, Yang X et al (2009) Th17 and natural Treg cell population dynamics in systemic lupus erythematosus. Arthritis Rheum 60(5):1472–1483
118. Tang Y, Tao H, Gong Y, Chen F, Li C, Yang X (2019) Changes of Serum IL-6, IL-17, and Complements in Systemic Lupus Erythematosus Patients. J Interferon Cytokine Res 39(7):410–415
119. Mohammadi S, Sedighi S, Memarian A (2019) IL-17 is Aberrantly Overexpressed Among Under-treatment Systemic Lupus Erythematosus Patients. Iran J Pathol. 14(3):236–242
120. Berrebi D, Bruscoli S, Cohen N, Foussat A, Migliorati G, Bouchet-Delbos L et al (2003) Synthesis of glucocorticoid-induced leucine zipper (GILZ) by macrophages: an anti-inflammatory and immunosuppressive mechanism shared by glucocorticoids and IL-10. Blood 101(2):729–738
121. Hoppstadter J, Kessler SM, Bruscoli S, Huwer H, Riccardi C, Kiemer AK (2015) Glucocorticoid-induced leucine zipper: a critical factor in macrophage endotoxin tolerance. J Immunol 194 (12):6057–6067
122. Godot V, Garcia G, Capel F, Arock M, Durand-Gasselin I, Asselin-Labat ML et al (2006) Dexamethasone and IL-10 stimulate glucocorticoid-induced leucine zipper synthesis by human mast cells. Allergy 61(7):886–890
123. Hoppstadter J, Hachenthal N, Valbuena-Perez JV, Lampe S, Astanina K, Kunze MM et al (2016) Induction of Glucocorticoid-induced Leucine Zipper (GILZ) Contributes to Anti-inflammatory

Effects of the Natural Product Curcumin in Macrophages. J Biol Chem 291(44):22949–22960
124. Ballegeer M, Vandewalle J, Eggermont M, Van Isterdael G, Dejager L, De Bus L et al (2019) Overexpression of Gilz Protects Mice Against Lethal Septic Peritonitis. Shock 52(2):208–214
125. Vandewalle J, Libert C (2020) GILZ in sepsis: "Poor is the pupil who does not surpass his master" . Eur J Immunol 50(4):490–493
126. Ellouze M, Vigouroux L, Tcherakian C, Woerther PL, Guguin A, Robert O et al (2020) Overexpression of GILZ in macrophages limits systemic inflammation while increasing bacterial clearance in sepsis in mice. Eur J Immunol 50 (4):589–602
127. Ricci E, Ronchetti S, Pericolini E, Gabrielli E, Cari L, Gentili M et al (2017) Role of the glucocorticoid-induced leucine zipper gene in dexamethasone-induced inhibition of mouse neutrophil migration via control of annexin A1 expression. FASEB J 31(7):3054–3065
128. Ricci E, Ronchetti S, Gabrielli E, Pericolini E, Gentili M, Roselletti E et al (2019) GILZ restrains neutrophil activation by inhibiting the MAPK pathway. J Leukoc Biol 105(1):187–194
129. Espinasse MA, Pepin A, Virault-Rocroy P, Szely N, Chollet-Martin S, Pallardy M et al (2016) Glucocorticoid-Induced Leucine Zipper Is Expressed in Human Neutrophils and Promotes Apoptosis through Mcl-1 Down-Regulation. J Innate Immun. 8(1):81–96
130. Ayyar VS, Almon RR, Jusko WJ, DuBois DC. Quantitative tissue-specific dynamics of in vivo GILZ mRNA expression and regulation by endogenous and exogenous glucocorticoids. Physiol Rep. 2015;3(6).
131. Cannarile L, Zollo O, D'Adamio F, Ayroldi E, Marchetti C, Tabilio A et al (2001) Cloning, chromosomal assignment and tissue distribution of human GILZ, a glucocorticoid hormone-induced gene. Cell Death Differ 8(2):201–203
132. Jones SA, Toh AE, Odobasic D, Oudin MV, Cheng Q, Lee JP, et al. Glucocorticoid-induced leucine zipper (GILZ) inhibits B cell activation in systemic lupus erythematosus. Ann Rheum Dis. 2015.
133. Calmette J, Bertrand M, Vetillard M, Ellouze M, Flint S, Nicolas V et al (2016) Glucocorticoid-Induced Leucine Zipper Protein Controls Macropinocytosis in Dendritic Cells. J Immunol 197(11):4247–4256
134. Cohen N, Mouly E, Hamdi H, Maillot MC, Pallardy M, Godot V et al (2006) GILZ expression in human dendritic cells redirects their maturation and prevents antigen-specific T lymphocyte response. Blood 107(5):2037–2044
135. Dieker J, Tel J, Pieterse E, Thielen A, Rother N, Bakker M et al (2016) Circulating Apoptotic Microparticles in Systemic Lupus Erythematosus Patients Drive the Activation of Dendritic Cell Subsets and Prime Neutrophils for NETosis. Arthritis Rheumatol. 68(2):462–472
136. Calmette J, Ellouze M, Tran T, Karaki S, Ronin E, Capel F et al (2014) Glucocorticoid-induced leucine zipper enhanced expression in dendritic cells is sufficient to drive regulatory T cells expansion in vivo. Journal of immunology 193(12):5863–5872
137. McCoy CE, Carpenter S, Palsson-McDermott EM, Gearing LJ, O'Neill LA (2008) Glucocorticoids inhibit IRF3 phosphorylation in response to Toll-like receptor-3 and -4 by targeting TBK1 activation. J Biol Chem 283(21):14277–14285
138. Levy DE, Marie I, Prakash A (2003) Ringing the interferon alarm: differential regulation of gene expression at the interface between innate and adaptive immunity. Curr Opin Immunol 15(1):52–58
139. Li Z, Wang Y, Li X, Li X, Cao H, Zheng SJ (2013) Critical roles of glucocorticoid-induced leucine zipper in infectious bursal disease virus (IBDV)-induced suppression of type I Interferon expression and enhancement of IBDV growth in host cells via interaction with VP4. J Virol 87(2):1221–1231
140. Shi X, Shi W, Li Q, Song B, Wan M, Bai S et al (2003) A glucocorticoid-induced leucine-zipper protein, GILZ, inhibits adipogenesis of mesenchymal cells. EMBO Rep 4(4):374–380
141. Zhang W, Yang N, Shi XM (2008) Regulation of mesenchymal stem cell osteogenic differentiation by glucocorticoid-induced leucine zipper (GILZ). J Biol Chem 283(8):4723–4729
142. Pan G, Cao J, Yang N, Ding K, Fan C, Xiong WC et al (2014) Role of glucocorticoid-induced leucine zipper (GILZ) in bone acquisition. J Biol Chem 289 (28):19373–19382
143. Lekva T, Bollerslev J, Kristo C, Olstad OK, Ueland T, Jemtland R (2010) The glucocorticoid-induced leucine zipper gene (GILZ) expression decreases after successful treatment of patients with endogenous Cushing's syndrome and may play a role in glucocorticoid-induced osteoporosis. J Clin Endocrinol Metab 95(1):246–255
144. He L, Yang N, Isales CM, Shi XM. Glucocorticoid-induced leucine zipper (GILZ) antagonizes TNF-alpha inhibition of mesenchymal stem cell osteogenic differentiation. PLoS One. 2012;7(3):e31717.
145. Yang N, Baban B, Isales CM, Shi XM. Role of glucocorticoid-induced leucine zipper (GILZ) in inflammatory bone loss. PLoS One. 2017;12(8): e0181133.
146. Suarez PE, Rodriguez EG, Soundararajan R, Merillat AM, Stehle JC, Rotman S et al (2012) The glucocorticoid-induced leucine zipper (gilz/Tsc22d3-2) gene locus plays a crucial role in male fertility. Mol Endocrinol 26(6):1000–1013

The manufacturer's authorised representative in the EU is Springer
Nature Customer Service Centre GmbH, Europaplatz 3, 69115 Heidelberg,
Germany. If you have any concerns regarding our products, please
contact ProductSafety@springernature.com

Printed and bound by CPI Group (UK) Ltd, Croydon, CR0 4YY
30/09/2026
02225506-0020